What Readers Are Saying About
iPhone SDK Development

I love this book's no-nonsense, straightforward approach to iPhone SDK development. Chris and Bill's approach is easy to follow, detailed, and, at the same time, not over-written or preachy. As someone just getting into iPhone development myself, this is definitely a must-have for my digital bookshelf.

> ▶ **Alex Lindsay**
> Founder, Pixel Corps

Dudney and Adamson's book streamlines the process of learning iPhone development by providing well-organized coverage of the topic that is both broad and deep in a way that any existing developer can understand. Read this book, and then come and join us at the next iPhoneDevCamp.

> ▶ **Raven Zachary**
> Founder, iPhoneDevCamp
> President, Small Society

If you are looking to build the next big iPhone application, this book is a great starting point. From the first chapters of Hello World to the advanced topics of dealing with audio, Dudney and Adamson will teach you what you need to know to build a great iPhone application.

> ▶ **Michael Sanford**
> Founder, FlipSide5, Inc.

Anyone who wants to learn and get better at iPhone development should read this book. Beginning developers will love Bill Dudney and Chris Adamson's easy-to-follow code examples. Experienced developers will find great in-depth examples that use the iPhone SDK. I strongly recommend it.

> ▶ **Henry Balanon**
> Founder and Lead Developer, Bickbot.inc
> Writer, GigaOm's TheAppleBlog.com

This is an excellent resource for the iPhone developer—the most thorough and comprehensive of its kind.

> **Marcus Crafter**
> iPhone Developer, redartisan.com

If you are looking for a solid foundation to learn iPhone development, I highly recommend this book. With extensive code examples and broad coverage of all things iPhone SDK, you'll hit the ground running.

> **John Muchow**
> Founder, iPhoneDeveloperTips.com
> CTO, 3 Sixty Software

iPhone SDK Development
Building iPhone Applications

iPhone SDK Development
Building iPhone Applications

Bill Dudney

Chris Adamson

The Pragmatic Bookshelf
Raleigh, North Carolina   Dallas, Texas

Many of the designations used by manufacturers and sellers to distinguish their products are claimed as trademarks. Where those designations appear in this book, and The Pragmatic Programmers, LLC was aware of a trademark claim, the designations have been printed in initial capital letters or in all capitals. The Pragmatic Starter Kit, The Pragmatic Programmer, Pragmatic Programming, Pragmatic Bookshelf and the linking *g* device are trademarks of The Pragmatic Programmers, LLC.

Every precaution was taken in the preparation of this book. However, the publisher assumes no responsibility for errors or omissions, or for damages that may result from the use of information (including program listings) contained herein.

Our Pragmatic courses, workshops, and other products can help you and your team create better software and have more fun. For more information, as well as the latest Pragmatic titles, please visit us at

> http://www.pragprog.com

Copyright © 2009 Bill Dudney and Chris Adamson.

All rights reserved.

No part of this publication may be reproduced, stored in a retrieval system, or transmitted, in any form, or by any means, electronic, mechanical, photocopying, recording, or otherwise, without the prior consent of the publisher.

Printed in the United States of America.

ISBN-10: 1-934356-25-5
ISBN-13: 978-1-934356-25-8
Printed on acid-free paper.
P2.0 printing, December 2009
Version: 2009-12-11

Contents

Foreword   **xiii**

1 Introduction   **1**
    1.1   In This Book 3
    1.2   Acknowledgments 5

2 Hello iPhone   **9**
    2.1   Gearing Up 9
    2.2   Create the Hello iPhone Project 11
    2.3   Creating a Simple Interface 13
    2.4   Rotating the Text—Adjusting the UI 16
    2.5   Rotating the Text—Adjusting the Code 17
    2.6   Code Navigation 18
    2.7   Running Your App on Your iPhone 19

3 iPhone Development Fundamentals   **23**
    3.1   The iPhone Software Architecture 24
    3.2   Coding in Objective-C 24
    3.3   Essential Cocoa Touch Classes 26
    3.4   Working with Xcode and Interface Builder 28
    3.5   Anatomy of Your iPhone Application 38
    3.6   Customizing Behavior with Delegation 40
    3.7   Managing Application Memory 43
    3.8   Accessing Variables as Properties 45
    3.9   Take-Away: Stuff to Remember 48

4 View Controllers   **51**
    4.1   Implementing a Button Action 51
    4.2   Building a Model 56
    4.3   Adding Outlets and Actions to the Controller 58
    4.4   Updating the UI 60
    4.5   Implementing the Controller 62
    4.6   Creating the New View Controller 64

	4.7	Building the UI	68
	4.8	Making the MovieEditorViewController	72
	4.9	The Editing View Controller in Interface Builder	73

5 Table Views   75
	5.1	Parts of a Table	75
	5.2	Setting Up Table-Based Navigation	77
	5.3	Modeling Table Data	78
	5.4	Table Cells	81
	5.5	Editing Tables	85
	5.6	Navigating with Tables	87
	5.7	Custom Table View Cells	94
	5.8	Sorting Table Data	98

6 Navigation   103
	6.1	Navigating Through Mail	103
	6.2	The Navigation Controller	104
	6.3	Navigation-Based Applications	105
	6.4	Pushing View Controllers	109
	6.5	Customizing the Navigation Bar	110
	6.6	Popping View Controllers	115

7 Tab Bar Controllers   117
	7.1	When to Use Tabs	117
	7.2	Creating a Tab Bar Controller	118
	7.3	View Controllers in Tab Controllers	120
	7.4	Many Controllers	125

8 File I/O   129
	8.1	Exploring Your Filesystem	130
	8.2	Creating Our Project	133
	8.3	Getting File Attributes	139
	8.4	Reading Data from Files	144
	8.5	Asynchronous File Reading	148
	8.6	Creating and Deleting Files and Directories	152
	8.7	Writing Data to Files	158
	8.8	Property Lists and NSCoding	161

9 Preferences 163
- 9.1 Displaying a Flippable Preference View in Your Application ... 163
- 9.2 Managing Preferences ... 166
- 9.3 Changing and Updating Preferences ... 169
- 9.4 Side Trip: Updating the Clock Label Every Second ... 172
- 9.5 Using the System Settings Application for Preferences ... 173
- 9.6 Loading Preferences Configured in the Settings Application ... 181

10 The SQLite Database 185
- 10.1 Creating Your Database ... 186
- 10.2 Creating the Sample Application ... 189
- 10.3 Putting Your Database on the Device ... 191
- 10.4 Using Your Database on the Device ... 194

11 Core Data 203
- 11.1 The Conference Application ... 204
- 11.2 The Core Data Stack ... 207
- 11.3 Building the Core Data Stack ... 210
- 11.4 Modeling ... 213
- 11.5 Track Table View ... 219
- 11.6 Fetching the Tracks ... 224
- 11.7 Change the Tracks ... 224
- 11.8 Navigation ... 228

12 Connecting to the Internet 235
- 12.1 Building a Browser in Ten Minutes with UIWebView ... 235
- 12.2 Reading Data from the Network ... 240
- 12.3 HTTP Authentication ... 246
- 12.4 Parsing XML from Web Services ... 253
- 12.5 Sending Mail from Your Application ... 264

13 Peer-to-Peer Networking 269
- 13.1 Using Ad Hoc Network Services with Bonjour ... 269
- 13.2 Bonjour Service Discovery ... 271
- 13.3 Game Kit Overview ... 277
- 13.4 Setting Up a Bluetooth-Networked Game ... 278
- 13.5 Setting Up a Peer Picker ... 280
- 13.6 Providing a Peer Picker Delegate ... 281
- 13.7 Network Game Logic ... 284
- 13.8 Communicating via the GKSession ... 286
- 13.9 Voice Chat ... 290

14 Video Playback — 293
- 14.1 Video Playback with MPMoviePlayerController — 293
- 14.2 Receiving Notifications from the Movie Player — 297
- 14.3 Supported Media Formats — 299

15 iPod Library Access — 303
- 15.1 Monitoring iPod Playback — 303
- 15.2 Controlling iPod Playback — 312
- 15.3 Using the iPod Library — 314
- 15.4 Browsing the iPod Library — 324

16 Playing and Recording Audio — 327
- 16.1 Creating an AVAudioRecorder — 327
- 16.2 Uncompressed Audio Formats — 331
- 16.3 Encoded Audio Formats — 335
- 16.4 Using the AVAudioRecorder — 339
- 16.5 Monitoring Recording Levels — 343
- 16.6 Playing Audio with the AVFramework — 348
- 16.7 Interacting with Audio Sessions — 353

17 Core Audio — 357
- 17.1 Using the Procedural-C APIs — 358
- 17.2 Playing System Sounds — 360
- 17.3 A Core Audio Overview — 366

18 Events, Multi-Touch, and Gestures — 373
- 18.1 Event Model — 373
- 18.2 Tracking Touches — 375
- 18.3 Tapping — 378
- 18.4 Multi-Touch Gestures — 379

19 Drawing in Custom Views — 385
- 19.1 Drawing Model — 385
- 19.2 Vector Drawing — 386
- 19.3 Paths — 387
- 19.4 Graphics Context — 392
- 19.5 Redisplaying a View — 394

20 Drawing Images and Photos — 397
- 20.1 Basic Image Drawing 398
- 20.2 Customizing the Image Display 400
- 20.3 Image Picker 402
- 20.4 Capturing Video 406

21 Core Animation — 409
- 21.1 Introduction to Core Animation 409
- 21.2 Animating UIView 410
- 21.3 Layers 416
- 21.4 OpenGL ES 419

22 Accelerometer — 423
- 22.1 Getting Device Orientation 424
- 22.2 Getting Shakes from the UIResponder Chain 425
- 22.3 Getting Raw Accelerometer Data 426
- 22.4 Filtering Accelerometer Data 432

23 Address Book — 441
- 23.1 Address Book UI 441
- 23.2 People Picker Delegate 442
- 23.3 Creating and Configuring the People Picker 444
- 23.4 Person Controller 445
- 23.5 Adding New Contacts 447

24 iPhone Location API — 451
- 24.1 Knowing Where 451
- 24.2 Location Updates 456
- 24.3 Compass 460

25 Map Kit — 463
- 25.1 Contact Mapper 463
- 25.2 Showing a Map 464
- 25.3 Map Annotations 467
- 25.4 Selecting an Annotation 477

26 Application Integration — 481
- 26.1 Launching Other Applications 481
- 26.2 Becoming Integration Ready 483

27 Debugging — 487
- 27.1 Understanding and Fixing Build Errors — 488
- 27.2 Understanding and Fixing Importing/Linking Errors — 490
- 27.3 Using iPhone SDK Documentation — 492
- 27.4 Understanding and Fixing Interface Builder Errors — 496
- 27.5 Debugging — 497
- 27.6 Finding Over-Released "Zombie" Objects — 505

28 Performance Tuning — 511
- 28.1 Investigating Performance with Shark — 512
- 28.2 Investigating Performance with Instruments — 517
- 28.3 Investigating Performance with the Clang Static Analyzer — 523

29 Before and After — 529
- 29.1 Starting Right — 530
- 29.2 Polish — 534
- 29.3 Other Features — 535
- 29.4 Beta Testing — 536
- 29.5 Getting into the Store — 537
- 29.6 Promoting Your Application — 538

A Bibliography — 541

Index — 545

Foreword

On January 9, 2007, at 9:42 a.m., the world changed forever. Something special was introduced to us, but not all of us knew just how special it would become in such a small amount of time.

The product we know as the iPhone is more than a gadget. It is a phenomenon. Many of us know news headlines, stock trends, and our day's schedule before we even get out of bed. And thanks to the iPhone SDK, there are more than 50,000 other things we can do, wherever and whenever we want. It truly is "your life in your pocket."

When I joined Apple in 2001, there was no App Store, no iPhone, and no iPod. Mac OS X was just a few weeks old. It is amazing to think how far things have come. The iPhone has created its own economy, where you no longer need patents or institutional investment to be wildly successful. All you need is an idea and the motivation to realize it.

As an iPhone developer, not only will you be participating in this phenomenon, but you will join a family of passionate, brilliant engineers, designers, and marketers who take more pride in their work than most of the people you've met. In many cases, their dedication dates back to the early releases of Mac OS X, and even the Classic Mac OS. They make conferences feel like reunions. They share knowledge with "competitors" because they know that better apps—whoever they're from—make a better platform. They love this technology. They live on it. They have built their careers on it. Working closely with them for more than eight years inspired me to leave my dream job and join their ranks.

This combination of technology and community has produced the historic times we now live in. Just one year after the App Store was born, it's hard to remember life before it, and yet there is surely so much more to come.

The platform is still growing, and we are all still learning. Your efforts will contribute to that story as it continues to unfold. This book is your first step in that special journey.

Welcome to our world. You're going to love it.

Matt Drance
July 11, 2009

Sent from my iPhone

Chapter 1
Introduction

The iPhone has changed everything.

Maybe by the time you read this you'll no longer notice that the latest cool new phone from every manufacturer looks and behaves more like the iPhone than like the models they used to sell before the iPhone was released. Meanwhile, at the time we write this, Apple's iPhone ads showcase the apps that you and other third-party developers are creating. The Apple marketing campaign is highlighting the iPhone as a platform for great software—software that you will create.

It's hard to overstate just how much the iPhone SDK and the App Store have changed the way we develop, release, and consume mobile software, even in the short time that they've been available.

Before the iPhone, writing software for the small device was rife with hazards. Many developers complained of *fragmentation*, the way that different devices would have different characteristics (screen sizes, color depths, input technologies, and so on) and different implementations of supposedly standard technologies, whose variant behaviors forced developers into an expensive and difficult game of "write once, test everywhere." And that's only in the cases where they had access to a market; in some cases, handset makers and mobile telephony carriers used the technology vendors' security practices as a means of enforcing business models, allowing only first-party applications on the device or crippling third-party apps by denying essential services, such as access to the network. It was enough to send developers fleeing back to the desktop, if not the server.

As an iPhone SDK developer, you don't need to worry about fragmentation. The iPhone and iPod touch are highly predictable, with only a few models thus far and highly consistent features and behavior between them. The SDK brings the tools and technologies of Mac OS X—refined by years of use creating world-class desktop applications—and makes its essential elements available for creating iPhone and iPod touch applications. While other mobile platforms throw an exception if you touch the networking stack, the iPhone gives you exceptional technologies such as the self-networking Bonjour. On other devices, playing a sound file is a hit-or-miss affair, while the iPhone offers extraordinarily deep support for media. And when you've finished writing your application, instead of a carrier demanding that you "partner" with them (and give them 99 cents of every dollar you make) and instead of being shut out entirely if you're not a big company touting a well-known intellectual property, the App Store lets you put your application before every iPhone and iPod touch user. And you set your own price with Apple taking a much smaller and quite reasonable cut.

The result has been revolutionary. At the rate the App Store is growing, there's no point quoting numbers because they'll be completely out-of-date by the time you read this book. Instead, it might be more helpful to notice the sea change in the industry that has resulted from the one-two punch of the SDK and the App Store. Carriers that had locked down their networks are now racing to open app stores of their own, terrified that every cool new app for the iPhone potentially lures away more of their customers. Some mobile developers are going iPhone-only. Given the realization that with iPhone users far more likely to download and pay for applications, it's a sucker's bet to work through the miseries of other mobile platforms—the aforementioned fragmentation and obstructionist carriers—in the hopes of reaching a user base that doesn't download third-party apps anyway. And in the enterprise, companies are writing their own custom iPhone applications and deploying them en masse in the field.

If you're reading this book, chances are you'd like to get in on this mobile revolution.

And we're glad you do, because we're eager to help see you through the journey that will take you from downloading the SDK to releasing your first application.

We begin by assuming only two things:

- You have a computer capable of running the current iPhone SDK. As of this writing, that means a Mac with an Intel CPU, running Mac OS X Leopard 10.5.4 or later.[1]

- You are familiar with a "curly brace" programming language (C, C++, C#, Java, Ruby, etc.) and with object-oriented programming in general. If you're primarily familiar with scripting languages (JavaScript, ActionScript, PHP), then you may find some of the iPhone's programming practices challenging, but we hope not insurmountable.

1.1 In This Book

We begin your journey with an introduction to the platform. We start with a quick success in the form of a Hello World iPhone application. Fresh on the heels of your first success, we dive into some of the fundamentals of the platform that you will need going forward. We cover the basics of the tools (Xcode, Interface Builder) and Objective-C, which is the language of Cocoa Touch development.

In the next four chapters, we look at view controllers and the views that are integral to most iPhone applications. You'll begin with a look at the most generic views and view controllers and move to the UI metaphors that are key to almost any iPhone application: tab bars, navigation, and table views.

We recommend that you read these chapters no matter what sort of iPhone application you are planning on building. You can then work through the rest of the book in order, or you can pick and choose according to your needs. We've grouped the remaining chapters into roughly six sections.

If you need to persist data to your device, you'll find four chapters that cover the various techniques. We start with filesystem access and options for saving and retrieving user preferences. Then we show two approaches to using the built-in database: directly, with the SQLite API,

---

[1]. SDK requirements, particularly the minimum OS version, are highly likely to change going forward. Check http://developer.apple.com/iphone for the current specs.

or via the object-relational mapping power of the Core Data framework added in iPhone OS 3.0.

Your application may require communication with servers or other devices, so the next two chapters take you out to the network. The first connects to websites, web services, and email on the public Internet, while the second uses the self-networking abilities of Bonjour and the new-in-3.0 Game Kit framework to connect iPhone OS devices to one another.

The next section consists of four chapters on interacting with media on your device. We start with the simple framework for viewing video on the iPhone and then look at 3.0's new framework for interacting with the user's iPod music library. For developers who want to exploit the platform's audio capabilities, we have a chapter on playing and recording audio with 3.0's new AVFoundation framework, closing the section with an overview of the lower-level Core Audio framework.

You are probably going to want to customize the way your users interact with your application. We return to presentation with a look at events, Multi-Touch, and gestures. Custom drawing via custom views is the topic of discussion in the next three chapters. We discuss drawing content in custom and unique ways with the sophisticated Core Graphics library, as well as manipulating and drawing images, both those provided by your application as well as those from the user's image library. This section closes with a chapter on Core Animation, the technology that underlies all the beautiful and natural-feeling animations that you see throughout the iPhone UI.

Next, we cover the more specialized technologies that not every iPhone application will use but that, when used properly, really make your application stand out. Some developers will be interested in using the built-in accelerometer to sense the device's orientation and motion, allowing the user to control applications in new and interesting ways. Developers of productivity and social networking applications will be able to use the Address Book framework to interact with a user's contact database. And because the iPhone and iPod touch are devices on the move, we have two chapters on location: one on the Core Location framework that helps determine the device's location and a second on the new-in-3.0 Map Kit framework that lets you present map data in your application.

Our final section offers four chapters on the final steps you'll need to take to complete your application. A handful of "application integration" APIs allow your application to launch other apps and to be launched by others. You'll learn to perfect your application with chapters on debugging your code and improving its performance with Xcode's various tools. A final chapter puts it all together by helping you hone your development process, get your work onto the App Store, and promote it to the app-shopping public.

1.2 Acknowledgments

From Chris Adamson

When I was part of an office-emptying mass layoff in 2001, I swore that I was done with mobile application development. A few years later, crawling to the finish of my second book, I said I'd never write another one. So, to be writing an introduction to a book on iPhone development, it would seem I've got some explaining to do.

It really all comes down to one moment in March 2008. I was at a Java conference, skeptically following the web coverage of the iPhone SDK announcement. Few people, myself included, thought we'd get a substantial level of functionality in the SDK and were shocked when what was unveiled was most of Mac OS X's Cocoa, with the UI rethought for a touch-based handheld. While I'd never been able to get out to the network with carrier-crippled Java phones, the SDK offered *freakin' BSD sockets* and self-networking with Bonjour as a bonus. Media, often presented at a toy level of functionality on other platforms, was present in the form of Core Audio, the same code used by professional audio applications on the Mac.

And this was all for a phone? I was hooked. I call it my "All Along the Watchtower" moment: I immediately knew that everything I'd been doing for years was instantly rendered irrelevant and that this is what I wanted—needed—to be doing going forward. I got the SDK over a very slow DSL connection that night, coded my first Hello World on the flight home, and was recruited for this book a few months later.

Of course, I have to thank our editor Daniel Steinberg for reaching out to me and convincing me that I was a good choice to coauthor the book. He's kept this enormous project on track through months of writing and rewriting, a shutdown when we feared Apple's nondisclosure agreement might keep us from ever releasing it, and the first slam-it-together beta release when the NDA dropped. Prags Andy and Dave also deserve kudos for a very productive (and yes, I'll say it, *agile*) system for book writing. Bill has been a responsive and productive coauthor, someone whose strengths are a well-suited balance to my weaknesses. Both of us are grateful to our many technical reviewers and the huge community that has developed on http://pragprog.com/, in the book's forum and errata page, giving us tons of useful feedback. And thanks as always to my family (Kelly, Keagan, and Quinn), who somehow withstood the collateral damage and stress of not only this book-writing marathon but also a cross-country move just a couple weeks before beta 1.

Obligatory end-of-book tune check: this time it was Bend Sinister, My Chemical Romance, Rilo Kiley, The Polyphonic Spree, Immaculate Machine, ...And You Will Know Us by the Trail of Dead, The Cribs, Kaki King, and the CBC Radio 3 stream and podcasts.[2]

From Bill Dudney

Writing a book is a big task. But of course it's not just the authors who work hard to produce the content. The editors (at least the ones who have corrected my prose) work just as hard if not harder to turn the technobabble and passive writing into acceptable English, so thanks to Daniel for helping me over the hump again.

The iPhone engineering team of course deserves a huge thanks for working nights and weekends to make something that is so much fun to build for.

The reviewers were also very generous with their time. Many helped, but a few stick in my mind as being especially helpful, so in no particular order let me say a hardy thanks to the following people: Tim Isted, Steven Troughton Smith, Patrick Burleson, Jason Hunter, Andy Simmons, Marcus Crafter, Tom Hauburger, David Hodge, David Smith, and Dee Wu. The book is much better for all your hard work!

I'd also like to thank the many folks who have been in the iPhone Studio for asking just the right question to help me see things from a different perspective. Those questions and the shift it caused helped to make this book. Thanks!

My family was also great during the long journey that has been this book. Thanks to you all for allowing me to stay up late nights and work

---

2. You can find up-to-date stats at http://www.last.fm/user/invalidname.

early mornings (and most of the time in between) to finish this up. I'd especially like to thank my oldest son, Andrew, who has been a great help in finding the things that I'm constantly forgetting. And finally, I'd like to thank the 2,000-year-old Jewish carpenter for making my life more than I ever thought it would or could be.

We are really excited about this platform and what you are going to build with your newfound knowledge. It is a great time to be an iPhone developer, and it's great to have you as part of the community!

Chapter 2
Hello iPhone

The iPhone is an amazing platform. It's filled with really cool technology that makes you want to create. The App Store is filled with applications that should both inspire and challenge you. The possibilities are practically endless.

Let's get started with a simple Hello World application. The rest of the book will take you through many of the technologies and APIs. You'll see everything from location to the accelerometer and from view controllers to Core Animation. In this chapter, we are going to introduce you to the tools and the basic development cycle for the iPhone.

Our first application is purposefully simple. We're going to build an application that displays the static text "Hello iPhone" on the screen. You'll first check that you have all the developer tools you need installed. You'll then use two of them to create and run your application in the iPhone Simulator. Because this is an iPhone app, you'll rotate the text when the device is turned sideways. Finally, you'll download this application and run it on your iPhone.

2.1 Gearing Up

If you are already developing applications for Mac OS X, then you are familiar with Xcode, Interface Builder, and the other development tools. Developing applications for the iPhone will be different because of the device you are targeting, but many of the tools, APIs, and skills you will use will be the same. You will be able to breeze through this first chapter.

> **Terminology**
>
> From here on out when we say iPhone, we mean any iPhone OS device. So, if you are pushing to an iPod touch, don't feel left out, because we mean you too.

If you are coming to the iPhone from another platform—welcome. You have undoubtedly used an integrated development environment (IDE) in the past. Xcode fills that role for iPhone developers. Xcode provides all the features you'd expect such as file management, compilation, debugging, and error reporting. Once you get to know Xcode, you will find it as familiar as your existing IDE. If you are completely new to development, we trust you will find it fun and rewarding.

You will use three main tools all the time in your iPhone development: Xcode, Interface Builder, and Instruments. You'll use Xcode to manage your project and to write, run, and debug your code. You'll use Interface Builder to create your interface and to wire it to your code. Instruments will help you identify areas that are hurting your performance. This book will not fully plumb the depths of any of these tools. We'll get you up and going with each of them and point out features required to achieve a goal for the app we are building. Any time you would like to learn more about a particular feature of Xcode, Apple's documentation is excellent and is built into Xcode.

Let's make sure that you have the development tools installed. If you've not already done so, download the iPhone SDK from http://developer.apple.com/iphone. You need to sign up as an iPhone developer, but membership is free. Membership and the SDK are free. It'll cost you $99 USD to join the iPhone Developer Program, but you'll need to do that to get a certificate that allows you to run your applications on the device.

After you download the package, install it by double-clicking the .dmg file and then the iPhone SDK package. Follow the on-screen instructions, and choose the default location to install the developer tools. When the install is complete, Xcode will be at /Developer/Applications/Xcode.app. If you choose another root directory to install to, the developer tools should adjust the path accordingly.

Launch Xcode by double-clicking it. You'll probably want to keep Xcode, Interface Builder, and Instruments in your Dock for easy access.

When you launch Xcode, you see the friendly welcome screen that is a great hub of information. Across the top you will notice several sections. Each one takes you to a different set of kickoff points for documentation or other helpful information. We use the Xcode News [RSS] section to keep informed about what new examples and documentation sets Apple has posted. The iPhone Dev Center is filled with links to the various sections of the Development Center online. Keep this welcome screen active at launch for at least a while as you get to know Xcode. Take the time to poke through each of the sections of the screen and get to know the content. It's the best way to learn Xcode.

2.2 Create the Hello iPhone Project

Now that you have Xcode running, create your first project. Select File > New Project or use ⌘-⇧-N to launch the New Project Wizard. Select iPhone OS > Application to see the list of iPhone project templates.

We will explore many of the templates in the examples throughout this book. For now, choose the View-based Application template, and click the Choose button.

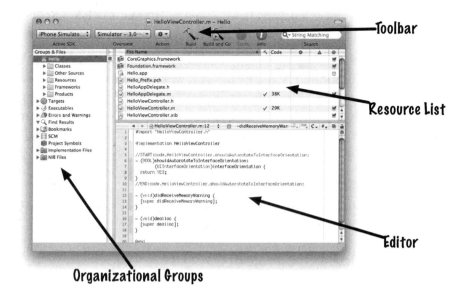

Figure 2.1: HELLO IPHONE PROJECT

When the open panel displays, choose a folder, and call the project Hello. Xcode uses the View-based Application template to make an application with a single view. Several other files are created for us, but we will look at many of them later, so don't worry about them now. You should have something that looks very similar to Figure 2.1.

On the left side of the window, you see a list of organizational groups. Each of these containers provides a bucket for you to put stuff into to help you keep organized. The groups are not implicitly tied to a filesystem structure, so there is a great deal of flexibility for you to organize your code and resources in any way you see fit. If you are just getting started, it is often easier to just learn the initial setup, but if you have something that works for you, feel free to reorganize to your heart's content.

Along the top of the window is the toolbar. This configurable space is where you can put the commands you most commonly use so that they are easy to invoke. We have the Overview pull-down menu that allows us to quickly switch between the iPhone Simulator and the device as the target to deploy our application on. If you don't see this Overview pull-down menu, it's probably a good idea for you to add it (some developers are seeing it by default; if it's already there, you don't need to do anything).

To customize your toolbar, select View > Customize Toolbar, or Ctrl+ click in the background of the toolbar and select Customize Toolbar. Select the Overview item, and drag it to the toolbar. As you drag it up, the cursor will change to a green circle with the + sign in it. Drop the Overview item on the left side of the toolbar.

On the right side below the toolbar is the file/resource list. This list displays files that are part of the group that is selected in the Groups & Files list on the left. Since the project is currently selected, all the files in the project are listed. As you click a file in this list, it is displayed just below in the editor pane. This list tracks the selection in the Groups & Files list; as you change the selected group, this list will change to reflect only the files that are part of that group.

The editor pane is where you will be doing most of your editing. It is where you access most of the code-centric features (completion, refactoring, and so on).

Now that we have our whirlwind tour of Xcode, let's run our new application to see what we get. Click the Build and Go button (or hit ⌘-↵), and Xcode will take care of the rest. iPhone applications, like Mac apps, are special directories called *app bundles*. During the build process, Xcode makes an app bundle with your compiled code, and your resources are copied to the appropriate places in the app bundle.

Now that Xcode has built your app and installed it into the simulator, the app will launch in the simulator. This first run will be a simple light gray background (a great flashlight app, if not quite as bright as a white background). Not much is here yet, which is what we expect given that we haven't customized our template yet. Next let's make some changes to the interface.

2.3 Creating a Simple Interface

Now it's time for you to meet Interface Builder. It's one of the major tools you will be using to build iPhone apps and the primary tool for user interface (UI) layout and configuration.

For now, the application isn't going to *do* anything, so we don't need to write any code. We'll use visual tools to create the visual interface. Interface Builder (IB), as the name implies, is responsible for enabling us to build the interfaces that our users interact with. IB understands the UIKit objects used to make interfaces and the Human Interface Guidelines (HIG) meant to help us make consistent UIs. It is a very powerful tool once you get your head wrapped around it. As with Xcode, this book can't possibly plumb the depths of IB, but through the experience you gain building the examples in this book, you will know enough to dig in and learn it yourself.

Let's build the interface for your first application. Open HelloViewController.xib by double-clicking it in Xcode (under the Resources group). We want to add a label to the interface to contain our "Hello iPhone" text. So, open the Library with ⌘-⇧-L (or Tools > Library).

The Library is where you will find all the interface and controller objects that you can instantiate into your nib file. In this case, we are looking for a label, so you can type label into the filter field at the bottom of the window.

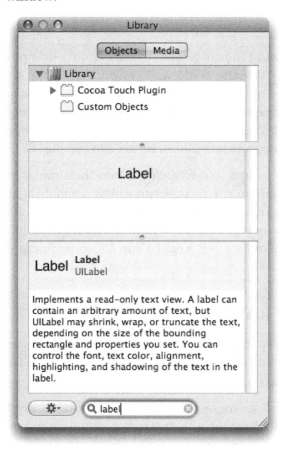

Drag the label out into the view, and move it until it snaps to the guides on the left side.

Select the label if it's not already selected. The little blue circles around the edge of the label allow you to resize the label. Make it stretch from the left guide to the right one so it takes the entire width of the view minus the border recommended by the guides. Also resize the height to 80 (you can drag with the mouse or just go to the Size inspector with ⌘-3). Now change the font by pressing ⌘-T and choosing a font size of 48. Double-click the label to select the text and edit it. Change it to Hello iPhone. You can also modify the color of the text and add a shadow and many other attributes in the Attributes inspector (⌘-1). Also, make sure to specify that you want the text centered in the Attributes inspector.

Here's what it looks like with the text centered, turned red, and with a shadow applied:

Now we are ready to run the application. Save your changes in Interface Builder. Switch back to Xcode, hit the Build and Go button, and check out your masterpiece. Once the simulator launches, you should see the text you set on the label show up in the center of the view on the gray

> **Joe Asks...**
> **What Is with XIB and NIB?**
>
> Interface Builder understands two file types, .xib and .nib. The .xib extension is simply the XML version of the .nib file. Since the .nib file type has been around since Mac OS X 10.0 (and even before in NeXTStep) and .xib is relatively new, people still refer to them as *nib files*. We will do the same throughout the book.

background. In the simulator, choose the menu item Hardware > Rotate Left or Hardware > Rotate Right. The simulator rotates the device left or right by 90 degrees, and the text rotates along with it. We'll make adjustments so that when the user rotates the device, the text remains oriented so that we can still easily read it.

2.4 Rotating the Text—Adjusting the UI

You have to perform two basic steps to properly respond when the device is rotated. You'll need to set some parameters on the UIView in Interface Builder, and you'll have to adjust a single line of code in the HelloViewController class using Xcode. In this section we'll make the UI adjustments, and then in the next section we'll make the changes to the code.

Select the label, and use the Size inspector (⌘-3), under the Autosizing section, to change the Struts and Springs configuration. Here, we want to keep the label centered when the device is rotated, so we turn off Struts for all four edges and turn on Springs for both directions.

The Struts and Springs configuration for views allows us to specify how the view should act when resized or when the view that contains it is resized. Struts runs along the outside edges of the view and allows us to specify that the distance between view and the edges of the view that contains it (the container view is referred to as the *superview*) remains constant. Springs allows us to specify that the view can grow or shrink in height or width. When the device rotates, the label's superview changes size and will therefore update the label according to the settings we put in place in the Size inspector. You also need to specify the text alignment on the text label to be centered.

Here are the Struts and Springs configuration with the correct setup:

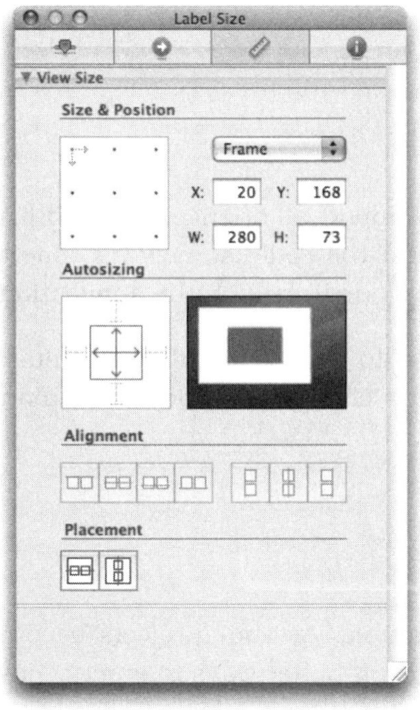

We are now finished with updating the UI. Let's go back to Xcode and make the necessary changes to the code so the app will properly respond to device rotations.

2.5 Rotating the Text—Adjusting the Code

In Xcode, select the Classes group, and then in the list of files on the top-right side select the HelloViewController.m file. Xcode will open the file in the editor pane below the list. Find the shouldAutorotateToInterfaceOrientation: method (make sure to uncomment it if it's commented out); this is where you will change code to make your application understand that the device has changed orientation.

The iPhone will ask your application whether it wants to rotate by sending shouldAutorotateToInterfaceOrientation: to it (well, technically this is not 100 percent accurate, but for now it's close enough; you will learn about this in much more detail in Chapter 4, *View Controllers*, on page 51). You respond by returning either YES or NO from this method. If you return YES, the iPhone will rotate your application to the orientation, and if you return NO, then your app will not rotate.

Here is the code for this method:

HelloiPhone/Hello/Classes/HelloViewController.m

```
- (BOOL)shouldAutorotateToInterfaceOrientation:
        (UIInterfaceOrientation)interfaceOrientation {
    return YES;
}
```

Since for this application you want to support all orientations, you just return YES. We have updated the UI and the code, so now it's time to run; click the Build and Go button in Xcode to launch the application.

Once it's launched, use ⌘-→ and ⌘-← to get the simulator to rotate. The label stays centered in the view and changes its orientation along with the device.

2.6 Code Navigation

We didn't have much code in this example. As your projects become larger and your classes grow, you will want better ways to quickly navigate your source code with Xcode. Once you have selected a file in the list, it is opened for you in the editor pane. The pane has several buttons along the top that give us a bunch of tools to navigate around in the code. Click the symbols pull-down menu to navigate immediately to any symbol.

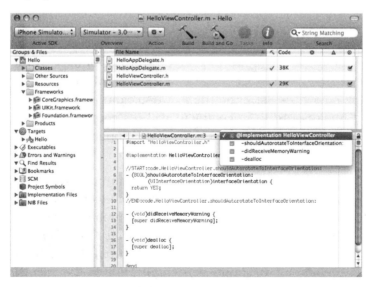

A simple click of the symbols list will display the symbols in the order they are found in the file. If it's easier to find what you are looking for from an alphabetically ordered list, you can hold down the option key and click the symbol list; then Xcode will display the symbols in alphabetical order.

2.7 Running Your App on Your iPhone

The simulator is a fine place to experiment, but there is no substitute for pushing the code to your device and watching it react in your hands.

Unfortunately, you won't be able to push to your phone until you have received your developer certificates from Apple. If you already have your certs installed, then the rest of this chapter takes you through the steps to get your app pushed onto your device.

If you don't have them installed and you are ready to get started, start with *iPhone Developer Program* [App08c] where you can apply for the program. Once accepted, you can go to *Obtaining Your iPhone Development Certificate* [App08f] to get more information about putting your certificates onto your devices and using them to sign your code from Xcode. If you don't have your certs, then please skip to the next chapter because the remainder of this chapter won't make much sense.

The change required to your project in Xcode is minimal to deploy an app to your phone. First you have to specify the proper application identifier in your project properties. Select the Hello target, hit ⌘-[i] (or right-click and choose Get Info), and then choose the Properties tab.

The identifier you use has to match one of the provisioning profiles you set up with your iPhone developer certificates. We created a specific provisioning profile for use with the examples in this book and suggest you do the same. If you don't want to make a specific profile, then feel free to change the identifier to one that matches an existing profile.

On the next page, we show the setup we used for this example.

Now you need to change from running in the simulator to running on the device. Select the Overview pull-down menu. Under the Active SDK list, choose Device – iPhone OS X (where X is the version installed on the device).

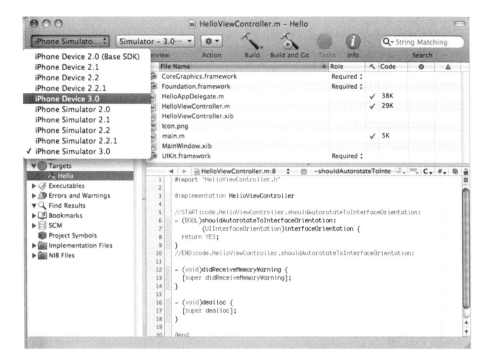

If you don't see this item, you probably have a problem in your developer certificates. Log onto the iPhone developer portal site, and review the direction in *Obtaining Your iPhone Development Certificate* [App08f] to make sure you have followed all the appropriate directions to get the certificates installed and working.

Once you have selected the device, click the Build and Go button again. This will take a few seconds to complete because Xcode has to rebuild all your code for running on the iPhone processor instead of for the processor on your Mac. Once it's done compiling, packaging, and signing your application, Xcode installs the application onto your device and starts it. Once it's started, try rotating the device to see how the "Hello iPhone" text follows when the device is rotated.

Congratulations! You have completed building and running your very own iPhone application. You have seen a bit of Xcode and IB and how the two tools are used to build the code and the interface for your applications.

Chapter 3
iPhone Development Fundamentals

Now that you've had a chance to get a simple application up and running with the iPhone SDK tools, let's take a step back and understand how we got here. In this chapter, we'll take a look at the organization of the iPhone's application stack, how the tools work in relationship to this architecture, and some of the concepts you'll need to master in pretty much any nontrivial iPhone application.

We barely touched the code in the previous chapter—we're going to dive more deeply into it now. iPhone applications are primarily written in *Objective-C*, a set of object-oriented extensions to the classic C language.

Our goal in this book is to make Objective-C programming for the iPhone understandable to developers with experience in any of the various C-based, curly-brace languages. If you've worked with C++, C#, or Java, many of Objective-C's concepts will be familiar to you, even if the syntax isn't. If your primary background is in scripting languages like Ruby, Python, or JavaScript, we expect that you'll be familiar with the concepts of object orientation, but some of the latent subtle C-isms, particularly involving pointers and memory management, may take some getting used to.

We will cover Objective-C's everyday essentials in this chapter, and if you want to dig deeper, you might want to look at Apple's *The Objective-C 2.0 Programming Language* [App09e], Bill's "Coding in Objective-C 2.0" screencasts at http://www.pragprog.com/screencasts, or a book on

Cocoa for the Mac, such as Daniel Steinberg's *Cocoa Programming: A Quick-Start Guide for Developers* [Ste09], also from the Pragmatic Programmers.

3.1 The iPhone Software Architecture

The iPhone's software stack is divided into a number of layers, with your application at the highest level of abstraction and core system services at the lowest level. From highest to lowest level, the stack can be summarized as follows:

- *Your application.*
- *Cocoa Touch*—A framework for developing touchscreen applications: UI elements, event dispatching, application life cycle, and so on. This also includes object wrappers around essential data types (strings, collections).
- *Media*—Graphics, animation, sound, video.
- *Core services*—Collections, strings, location awareness, SQLite database, Address Book, networking, and so on.
- *Core OS layer*—Unix services, standard I/O, threads, BSD sockets, power management, and so on.

You'll mostly want to work primarily with the GUI frameworks and the OO abstractions offered by the Cocoa Touch layer, and it's this layer that we primarily discuss in this book (with diversions deeper into the stack as necessary for specific topics). Most of the Cocoa Touch classes are designed to be called directly from your code; you can subclass these classes to add functionality, but you'll find that you need to do this far less frequently than in other languages.

3.2 Coding in Objective-C

The Cocoa Touch frameworks you'll be using are written in and called with Objective-C, which is a superset of the original C. Because of this, you can and will freely mix C and Objective-C syntax in your application code.

Classes in Objective-C are developed by creating a *header* file and an *implementation* file. These two files share the same name but different

> **Joe Asks...**
> **Do I Have to Learn C to Write iPhone Applications?**
>
> Well, we're not going to advise you to drop this book and read "K&R"* first, but if you're not already familiar with C, you will surely be picking up some of it along the way.
>
> Some iPhone APIs developed outside of Apple are used with what we'd describe as typical C coding practices. These include OpenGL, OpenAL, and SQLite. There are also Apple APIs called with plain C (Apple often uses the term *procedural C* to distinguish it from Objective-C) that use Apple-specific design patterns to craft an OO-like experience. These include Core Foundation, Core Audio, and Quartz/Core Graphics, which will be introduced in Section 17.1, *Using the Procedural-C APIs*, on page 358.
>
> Even within purely Objective-C Cocoa APIs, you'll use C syntax for things such as control flow, and you'll encounter C idioms, such as using pointers for all your Objective-C variables, providing their addresses (with the **&** operator) as a parameter to get return values from method parameters, variable-length argument lists, and so on.
>
> We learned C a long time ago, but we're sympathetic to how hard it is to go from modern languages back to C's functional style and memory management. We'll try to make sure to warn you of C's idiosyncrasies as we proceed.
>
> ---
> *. This is the short name for Brian Kernighan and Dennis Ritchie's much-cited *The C Programming Language* (KR98).

extensions; to create a class called Person, you would create both a Person.h file and a Person.m file. In the header file, you include the public parts of your class: the names and types of your instance variables and your method signatures, which describe the return type and parameters of a method. In the implementation file, you then implement these methods with code, declaring and using any needed local variables that are relevant only to the implementation and are not meant to be seen by other classes. Like C, you also need to indicate where the compiler can find other code you're using, by **import**ing the header file of classes you refer to in your headers or implementation.

In Objective-C, method calls actually represent *messages* dispatched by a small, lightweight runtime to the objects in memory. This is a subtle but important distinction, which allows for surprisingly dynamic behavior in what might initially seem like a fairly formal language. For example, if you call a method on an object reference that turns out to be **nil**, you don't crash or get an error (like Java's NullPointerException); the runtime realizes that sending a message to a nonexistent object would be meaningless and simply does nothing.

Typical Objective-C method calls are enclosed in square braces, beginning with the object and followed by parameter names and the values you're providing for those parameters. For example, if you had an NSString instance called myString, you would get its third character with a method invocation like this:

```
myChar = [myString characterAtIndex: 3];
```

The most significant difference from other languages, other than the square braces, is that the parameters are *always* named with the parameter-and-colon construct called a **keyword**. It does mean more typing, but it makes code a lot easier to read, compared to other C-like languages where you could have a half dozen parameters within a set of parentheses and no readily apparent idea what each one represents.

3.3 Essential Cocoa Touch Classes

Now that you know you'll be doing most of your coding in Objective-C, let's take a quick look at the classes provided in the iPhone SDK frameworks.

The Cocoa Touch application frameworks, first mentioned above in Section 3.1, *The iPhone Software Architecture*, on page 24, contain most of the classes you will use to develop your first applications. The term comes from *Cocoa*, the object-oriented frameworks developed for Mac OS X programming (and NextStep before that), along with GUI classes uniquely designed for use on a mobile, touchscreen device (hence the "Touch").

Cocoa's *Foundation* framework includes essential data classes, includes basic utilities, and establishes some core programming conventions that cannot be expressed by the Objective-C language alone, such as techniques for managing memory (which we'll visit in Section 3.7, *Man-*

aging Application Memory, on page 43). Nearly all Cocoa classes inherit from a root class, NSObject, defined in Foundation.

Perhaps the first and most important thing to discover in Foundation is its data management classes, which are used throughout Cocoa instead of the procedural C equivalents. For example, the traditional C string, the null-terminated **char** array, is almost *never* used in Cocoa. Instead, you use NSString, which represents not only the character data but also its encoding; with rich support for Unicode (and the UTF-8 and UTF-16 encodings), the NSString makes it easy to handle text in any of the dozens of character sets on the iPhone.

Cocoa also provides a deep set of *collection* classes, obviating the need for most uses of C arrays (or hand-rolled collections, such as linked lists and hashtables). Three classes are used for collecting Cocoa objects: NSArray for ordered collections of objects, NSSet for unordered collections, and NSDictionary for mapping key objects to value objects. These three collections are *immutable*—once initialized, they can't be changed. If you want to add, delete, or otherwise change their contents, use the mutable subclasses NSMutableArray, NSMutableSet, and NSMutableDictionary.

The collections can store only NSObjects. If you have C primitives, you can pass them around Cocoa with the wrapper classes NSData and NSMutableData, which wrap a byte buffer, and NSNumber, an object container for any of C's scalar (numeric) types, such as **int**, **float**, or **bool**.

Cocoa has a few more specific data classes, including NSURL for URLs (including file://-style URLs representing items on the local filesystem, though you often use NSString paths too), and timekeeping classes such as NSDate and NSTimeZone.

The "Touch" part of Cocoa Touch is largely represented by the UIKit framework, also imported by default in every iPhone application. This framework offers the drawing model, event handling, application life cycle, and other essentials for a touch-based application. You'll largely interact with it through the various user interface component classes it provides: UIButton, UITextView, UITableView, and so on.

Between the data types in Foundation and the UI components in UIKit, Cocoa Touch gives you a great foundation on which to start coding your application.[1]

3.4 Working with Xcode and Interface Builder

Now you might be thinking, "Hey, I barely wrote any code to create the application in the previous chapter, so why are we talking about code all of a sudden?" Fair enough. We've already introduced you to Xcode and Interface Builder as your primary tools, so let's take a look at the roles they play by looking at the contents of the Hello.xcodeproj project you created in Chapter 2, *Hello iPhone*, on page 9. With the project open in Xcode, you should have a window whose left side shows the Groups & Files list, which presents the contents of Hello as a set of folders. With Hello clicked, you'll see all the files listed in the right pane, but by clicking each folder individually, you can examine the parts of the project by type:

- *Classes*—Class files are the C and Objective-C source that you create to provide the functionality of your application. The template you used to create Hello set up two classes for you: HelloAppDelegate is responsible for handling life-cycle events (such as the application starting up or shutting down), and the HelloViewController manages what you see on-screen.

- *Other sources*—These are source files that are generated automatically and that you are largely not responsible for or interested in. HelloPrefix.pch is the *precompiled headers* file, created by Xcode to speed up processing of header files in your build. main.m is the implementation of the main() function that the system calls to launch your application. You could also put third-party library source here or procedural-C sources (.c files) that aren't "classes" per se.

- *Resources*—Resources are noncode files that are nevertheless needed by your application at runtime. Such files might include graphics or sound files, localization dictionaries, and the like. In

---

[1] Keep in mind that you can learn more about these classes at any time from Xcode's Help menu. See Section 27.3, *Using iPhone SDK Documentation*, on page 492 for an overview of Xcode's documentation viewers.

this project, the Info.plist[2] file includes basic settings for the application such as its icon and name. The resources also contain the nib files that contain the "freeze-dried" GUI components used to present the user interface. You edited the HelloViewController.xib file with Interface Builder in the previous chapter.

- *Frameworks*—These represent the frameworks that your application uses. By default, the Xcode template links in the frameworks Core Graphics and UIKit so you'll be able to call the various GUI classes, and Foundation for commonly used classes such as strings, collections, URLs, and the like. In later chapters, when we go beyond basic functionality, we'll add more frameworks to this folder.
- *Products*—This folder represents the files that will be created by the build process, in this case the Hello.app iPhone application.

Xcode is really the base for your application project. Although you'll use other applications to set up the GUI elements in the nib (Interface Builder), create images (any graphics application such as Photoshop or Pixelmator), and measure performance (Instruments), the job of managing all the pieces of the application and building it falls to Xcode. Typically, Xcode also provides your source editing, though you can configure external applications (perhaps BBEdit, TextMate, or even emacs) to do your source editing if you so desire.

Designing an Interactive Application

Our first application said "Hello iPhone," so let's return the favor by saying hi to the user. This will require prompting the user for their name.

Close out the Hello project by closing the project window if it's still open, and use ⌘ ⇧ N to create a new project. In the New Project window, you'll again choose a View-based Application, and when you're asked to name the project, choose a new name like HelloUser.

Just like in the example in the previous chapter, the template project provides us with two classes (HelloUserAppDelegate and HelloUserViewController), along with a few other classes that we'll explain in the next

---

2. Starting in iPhone SDK 3.0, Xcode prepends the project name to the filename, like Hello-Info.plist. We'll continue to use the term Info.plist as a generic reference to whatever file was created for your application's settings.

section. For now, it's enough to know that in a single-view application, we put stuff on-screen by working with the HelloUserViewController class and its associated HelloUserViewController.xib, which contains the freeze-dried GUI for this class. In the previous chapter, we edited only the .nib file, but this time, we're going to want to customize the class to provide some interactivity.

Declaring IBOutlets and IBActions

Our new application lets the user type his or her name into a text field. When they're done, we'll customize the message with the name. So, if you type Quinn, the label will change to "Hello Quinn." To do this, we will use a label for the message, a text field for the user's name, a button to accept the input, and a method that updates the label when the button is tapped.

To make this work, we need to create relationships between the logic of the application and the GUI objects in the nib. We'll use special keywords when we declare variables that refer to objects in the nib and when we declare methods to be called from events in the user interface, such as button taps. A reference from code to an object in the nib is called an *outlet* and is designated with the keyword **IBOutlet**. Similarly, a method that you want a nib's objects to be able to call is an *action* and is designated with the keyword **IBAction**.

Typically, you declare outlets and actions in code so that IB will be able to use them. Edit your HelloUserViewController.h header file so it looks like this:

HelloiPhone/HelloUser/Classes/HelloUserViewController.h

```
Line 1  #import <UIKit/UIKit.h>
     -
     -  @interface HelloUserViewController : UIViewController {
     -
     5          IBOutlet UILabel *helloLabel;
     -          IBOutlet UITextField *nameField;
     -
     -  }
     -
    10  - (IBAction) sayHello: (id) sender;
     -
     -  @end
```

This code declares two outlets and one action. Since the outlets are variables and therefore part of the class structure, they go in the curly-brace block of the @interface declaration. The first outlet, on line 5, declares a pointer to a UILabel called helloLabel, with the additional **IBOutlet** declaration signaling our desire to connect this reference to an object created in Interface Builder. With this connection, we'll be able to style the label in Interface Builder, and also alter it from code at runtime, in order to display the customized hello message. Similarly, on line 6, we declare a UITextField as an **IBOutlet**, so we'll be able to read its value from code. Finally, on line 10, we declare a method that will handle the button tap, and we declare it to be an action (note that we don't need an outlet to the button if we're not going to be manipulating it, such as changing its size or text, from code).

As you can see, method declarations in Objective-C have a unique syntax. They start with a single character (-) to declare an instance method, meaning that it is called on and applies to a single instance of the class, or + for a class method, which is independent of an instance. Following this is the return type in parentheses. (IBAction) is equivalent to (void) in that it returns no value, but using the **IBAction** keyword indicates our intent to connect the method to GUI events via Interface Builder. Finally, the arguments are provided as pairs of types and parameter names. In this case, the (id) type is a pointer to any object; the pattern (id) sender is used for all IBAction declarations so the called method can know what object (the sender) has called it.

Laying Out Your Interface in IB

Now you're ready to create the GUI in Interface Builder and connect its components to your code. First save the header file, if you haven't already.

Next, in Xcode's project window, double-click HelloUserViewController.xib to open it with IB. As in the previous chapter, IB will show a document window with three objects: two proxy objects (File's Owner and First Responder, which we'll explain later) and View. The view's preview window may already be open; if not, double-click View to open it. Your view starts as a blank area precisely the size of the iPhone screen. From the library, drag over three objects: a label, a text field, and a button.

Your preview should look something like this:

Obviously, this GUI could use a little bit of layout and polish. Let's customize the three components as follows:

- Select the label to show its handles, and drag the right and left edges until the dashed blue margin lines appear. This will let the label span all the way across the screen in case the user has a really long name. To center the text, bring up the Attributes inspector (⌘ 1), and where it says Alignment, click the "center text" icon (the middle one).

- Select the text field, and drag its edges to the margins too. In its Attributes inspector, you can use the placeholder to provide a grayed-out visual cue to tell the user what they should do with the text field, for example, Name.

- Finally, double-click the button, or edit its Title attribute, to set the button's text to Say Hello.

Here's the completed GUI. For simplicity, we won't worry about supporting rotation or resizing the components in this sample.

Now that we've set up our user interface, we're ready to connect it to our code. To do this, go back to the HelloUserViewController.xib document window and its three icons. We're going to be working with the File's Owner icon. This isn't a real object but, instead, a proxy object that represents an object assigned to be the "owner" of the nib when the nib is loaded.

Xcode has already set the class of File's Owner as HelloUserViewController (which you can examine with the Identity inspector, ⌘ 4), and this is key to getting our code connected to our interface. The object that owns this nib and its customized view is a HelloUserViewController, and we just created actions and outlets in that class by editing its header file. That means IB knows that the File's Owner will have those outlets and actions at runtime.

Because we've added those **IBOutlet**s and **IBAction**s, when you right-click (or Ctrl+click) the File's Owner to see its outlets, you'll see helloLabel and nameField.[3]

Now we're going to actually make the connections. In the gray heads-up-display (HUD) window's list of outlets, click the circle next to helloLabel, and drag to the label in the preview window. As you hover over the label, IB will show a box around it and open a window identifying your proposed connection as Label (Label), meaning that you're about

---

3. You can also examine an object's outlets and received actions in the Connections inspector (⌘ 2).

to connect to a label (specifically a UILabel), whose contents are currently the text "Label." Lift up on the mouse button to complete the drag and make the connection; the name of the connected component will appear back in the HUD window.

Now do the same thing with the nameField outlet, whose drag gesture is shown here:

You can connect the action with a similar gesture in the opposite direction, from the widget to the File's Owner. Right-click the button to show its list of events,[4] and drag a connection from the Touch Up Inside event to the File's Owner. When you release the mouse, a small HUD window will pop up and show the declared actions that you can connect to. Since you declared only one **IBAction**, it's your only choice here: sayHello:. End the drag by releasing the mouse button over this method name and it will flash to indicate that you have connected the action to this method.

This completes the wire-up of outlets and connections in Interface Builder. To keep things clean, you might want to set the label's title to an empty string so the application doesn't say "Label" when it launches. You're now done with IB and can save and quit.

---

4. Again, you could also see the list of connections with the Connections inspector. As a shortcut, Ctrl+dragging from a button to a receiver object connects the Touch Up Inside event without even bringing up the list of events.

Implementing the Action

Now you have a GUI, with connections from the IB-created widgets to instance variables in your custom class. If you run it right now, a tap on the Say Hello button will result in a call to sayHello:. Thing is, we haven't written that method yet, so our app will crash with an "unrecognized selector" error. Let's deal with that now.

Your method implementations go in the class's implementation file, which has a .m filename extension. Open HelloUserViewController.m from the project window in Xcode. The method can go anywhere between the lines containing the **@implementation** and **@end** keywords. Type in the following method implementation in the line following @implementation HelloUserViewController:

HelloiPhone/HelloUser/Classes/HelloUserViewController.m
```
Line 1  - (void) sayHello: (id) sender {
     2        NSString *userName = nameField.text;
     3        NSString *helloMessage =
     4              [[NSString alloc] initWithFormat: @"Hello %@", userName];
     5        helloLabel.text = helloMessage;
     6        [helloMessage release];
     7        nameField.text = @"";
     8  }
```

- On line 2, we refer to the instance variable nameField and then get the text that's been typed into it. Since UILabel provides text as a *property*, we get and set it with the dot operator. We'll have more to say about properties a little later in the chapter.

- Next, we create a new hello message by creating a new NSString on line 3. As is usually the case with Cocoa objects created in code, we first allocate memory for the object with alloc, and then we initialize its contents. To create the string substitution, we can use a format string and substitute in the user value. The leading @ lets us quickly allocate a static NSString for the format, which is just Hello %@. The %@ is a *format specifier* that allows us to insert a string representation of any Cocoa object.[5] Here, we insert the value of userName.

- With the format string prepared, we set it on the label in 5. We do this by getting the helloLabel instance variable and setting its text property.

---

5. Formatting strings and other format specifiers is described further in Section 27.5, *Logging to Standard Output*, on page 497.

- Finally, we have one little bit of housekeeping. We allocated memory for a string on line 3. As we no longer need this string, we are obligated to release it in order to free up the allocated memory. On line 6, we do this by sending the release message to the string. Technically, this doesn't free the memory; it says that we are no longer interested in the object, so if nobody else is claiming the string, it will be freed (it's not freed, by the way, since the UILabel also retained it when we set it as the label's text). We'll discuss the memory management system in detail later in this chapter.

Our application is now ready to run. Click Build and Go to run it in the simulator, and the resulting application will allow you to click in the text field, type your name, and tap Say Hello to display your name like this.

3.5 Anatomy of Your iPhone Application

So far, you've had to endure a bit of "type this here, do that there, never mind the little man behind the curtain"-type instruction, so let's step back and take a look at just how your application gets launched and how all these pieces interact. If this ends up being more detail than you need, you don't need to worry about it; we're covering it in order to demystify the nature of how an iPhone app works and why the project template is structured like it is.

As described earlier in Section 3.4, *Working with Xcode and Interface Builder*, on page 28, the main.m file has the implementation of the main() function that the system calls to launch your application. Open it up if you like, and you'll see it's just four lines:

HelloiPhone/HelloUser/main.m
```
Line 1  int main(int argc, char *argv[]) {
     2
     3      NSAutoreleasePool * pool = [[NSAutoreleasePool alloc] init];
     4      int retVal = UIApplicationMain(argc, argv, nil, nil);
     5      [pool release];
     6      return retVal;
     7  }
```

The signature on line 1 is a typical C main() function, designed to be called from the command line with a variable number of arguments. Line 3 sets up an *autorelease pool* for memory management (which will be explained later in Section 3.7, *Managing Application Memory*, on page 43), and line 5 releases it when the application ends.

On line 4, we call the UIApplicationMain function to start up the main event loop and start your application. The first two arguments pass the command-line arguments (if any), while the third and fourth indicate the application's main class and its *application delegate*, which is a class in your project that handles application life-cycle events. If these are **nil**, as they are here, then UIKit assumes that it needs to load the application from a nib file.

The Info.plist file provides the application's main nib bundle, typically MainWindow.xib. So, UIKit expects to find the app delegate in this nib. Double-click MainWindow.xib to open it in Interface Builder.

You should see something like this:

The nib contains an app delegate icon (HelloUser App Delegate), a view controller (HelloUser View Controller), a window (Window), in addition to the File's Owner and First Responder proxy objects that are always present.

Open the Connections inspector (⌘ 2) for the application delegate object, and you can start to see how the pieces fit together. Every iPhone application must have a single UIWindow object, and the application delegate has a connection to the nib's window object. It also has a connection to a view controller object, which is an instance of the HelloUserViewController class that we customized earlier in the chapter. And as you saw before, the view controller provides the logic associated with an on-screen view, the one that we customized in the HelloUserViewController.xib file. In fact, if you investigate the view controller with the Attributes inspector, you'll be able to see where it refers to that nib file, and if you double-click the view controller, the view's preview window says it's "Loaded from HelloUserViewController."

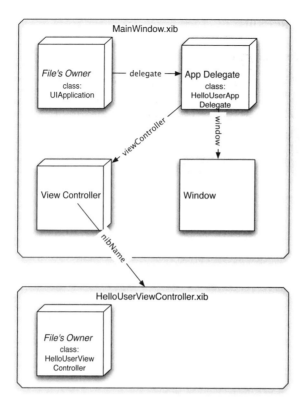

So, to summarize, the system starts by calling the main() function, which calls UIApplicationMain(), which uses Info.plist to look up the main nib, in which it finds an application delegate (connected to the app's one window) and a view controller, which loads from another nib file. Your application's first chances to provide custom logic come in the life-cycle methods associated with the app delegate (such as application-DidFinishLaunching:) and the view controller it loads (which gets callbacks like initWithCoder: when it's loaded from the nib and viewDidLoad when its view is loaded). But in our application, we don't need to do anything until the button is tapped. You'll probably never want or need to mess with main.m or MainWindow.xib, because these life-cycle callbacks offer more convenient points to insert start-up code, but now you know how all the stuff in your Xcode project works together to bring up your application.

3.6 Customizing Behavior with Delegation

The application delegate is an example of one of Cocoa's most significant design patterns: *delegation*.

The idea of delegation is that an object can have a single *delegate* object that it will call when certain events occur. From the delegate's point of view, it's sort of like a callback or notification scheme:[6] "Call me when this stuff happens." From the delegating object's point of view, it's more like handing off responsibility: "I don't know what, if anything, needs to happen when this event occurs, so you deal with it."

In this case, the application delegate gets callbacks when various events that affect the whole application occur: when it's launched from the home screen, when it's launched with a URL by another application (see Chapter 26, *Application Integration*, on page 481), when it's warned of a low-memory condition, and so on. This is an example of a formal delegate arrangement, one that is defined by an Objective-C *protocol*, UIApplicationDelegate. This protocol has its own documentation page, like a regular Cocoa class, but instead of describing how an existing class's methods work, it describes when the delegate methods will be called and what you as an implementor of those methods could or should do when they are. For a class to actually become an application delegate, it needs to declare in its header file that it implements the protocol and then implement all delegate methods not marked as "optional."[7]

There are a few delegate protocols that we could add to Hello User to make it behave more like a typical iPhone application. For example, you might have noticed when you ran the application that the keyboard didn't disappear when you tapped the Say Hello button, and the Return key did nothing. These are common tasks for iPhone applications, and the latter requires us to provide a delegate to the text field.

To dismiss the keyboard, you need to tell the text field to give up its role as the "first responder," meaning the component that initially receives the user input. It's easy enough to do this when the button is tapped; just add the following line to the bottom of sayHello:.

```
[nameField resignFirstResponder];
```

But what do we do when the Return key is tapped? That's not a button that comes from any of our GUI components—the keyboard is provided automatically by the text field—so there's no apparent way of wiring up

---

6. Cocoa also has a *notification* system that works with zero to many listeners, but its design and use are profoundly different from delegation, and the two are in no way interchangeable.
7. To Java programmers, Objective-C protocols are like Java **interface**s but with the potential for optional methods.

an event-handling method to it. However, if you look up the documentation[8] for UITextField, you'll see that it has a delegate property that is defined by the UITextFieldDelegate protocol, which is a defined group of related methods. Look up this protocol, and you'll see it has numerous methods that alert the delegate of events relating to the text field. One of them is textFieldShouldReturn, which looks like what we need in order to know when the user has tapped Return.

You indicate that an object implements some or all of the delegate methods by specifying the delegate name in the **@interface** header inside angle brackets following the name of the class that this class extends.[9] For our purposes, HelloUserViewController makes a good choice, since it already knows about the text field. So, declare that it implements the delegate protocol by editing the HelloUserViewController.h header file and adding the protocol, in angle braces, to the class declaration, like this:

HelloiPhone/HelloUser/Classes/HelloUserViewController.h

```
@interface HelloUserViewController : UIViewController <UITextFieldDelegate> {
```

Now you need to set the view controller as the text field's delegate. You can do this in code, but since we've defined the rest of our GUI in Interface Builder, it makes sense to set the delegate with IB too. Double-click HelloUserViewController.xib to open it in IB. By right-clicking the text field or using its Connections inspector, you'll see there's an empty connection called delegate. Drag from the circular connection point to the File's Owner (which represents the HelloUserViewController, because it's the object assigned to own this nib), and you'll declare the HelloUserViewController as the text field's delegate. Be sure to save before leaving IB.

With the view controller set as the text field delegate, it will get event-driven callbacks from the text field. Now you just need to provide implementations for whatever callbacks interest you.

Here is a simple method that dismisses the keyboard, which can go anywhere in between the **@implementation** and the **@end** in HelloUserViewController.m.

---

8. See Chapter 27, *Debugging*, on page 487 for a section on finding class documentation in Xcode.
9. In some simple cases, there is no formal delegate protocol, and the delegate just implements some or all of a list of methods listed in the documentation.

```
HelloiPhone/HelloUser/Classes/HelloUserViewController.m
- (BOOL)textFieldShouldReturn:(UITextField *)textField {
    [textField resignFirstResponder];
    return YES;
}
```

The signature of this method needs to exactly match the signature defined by the delegate protocol, so you may want to get in the habit of copy and pasting from the documentation or the header files if you choose to look them up.

As you can see, the method is passed a textField (which presumably is nameField, since our app has only one text field), to which we simply send the message resignFirstResponder. This tells the text field it's no longer going to receive text input (at least not until tapped again), which will make the text field remove the virtual keyboard. Then we return YES, because the protocol instructs us to do this if the button should implement its default behavior for the Return button tap.

Build and Go in Xcode, type in some text, and tap the Return button. You should see the keyboard disappear (if it doesn't, verify that you've connected the delegate to File's Owner in IB and that your method signature exactly matches the previous code).

It doesn't update the hello label, because the code to do that is in sayHello:, which is called only by the Say Hello button tap. In this common case, where you want the same code called by both a button tap and the Return key, you'd usually put the common functionality into a helper method that both event handlers call.

You might have noticed that one of the nice things about protocols is that you have to define only as much functionality as you're interested in. All the methods in UITextFieldDelegate are optional, so you can ignore any of the events that don't interest you. That's not always the case, since some delegate protocols declare required methods that the compiler will insist you implement. Still, it's a convenient pattern and an important one that you'll see again and again in Cocoa.

3.7 Managing Application Memory

Another defining trait of iPhone applications—one that's perhaps not as pleasant as delegation—is dealing with memory management. If you've been working in modern, garbage-collected languages (Java, C#, Ruby, and even Objective-C on the desktop, for example), you may give little thought to memory management, because the objects you no longer have references to get freed without any action on your part. Having to manually account for your memory use in iPhone applications is often one of the hardest things that developers have to deal with. Handled incorrectly, your application leaks memory and ultimately risks being terminated by the system.

Conceptually, though, memory management in iPhone apps isn't that hard to get your head around; it's the discipline of freeing up memory that takes some getting used to. Cocoa uses a reference-counting system that's actually fairly simple to understand. All objects are allocated with a *reference count* of 1. This reference count can be manipulated with the use of two instance methods: retain increments the reference count by one, while release decrements it. When the count goes to zero, the object is ready to be freed.

Along with this "plus one" description, another analogy for memory management is phrased in terms of interest. If you are interested in an object that you receive from a method call, you retain it. When you are no longer interested, you release it.

There is a *fundamental rule* for Cocoa memory management: you own any object you create by means of a method that uses the words alloc, new, or copy. And for any object you own, you must release it at some point.

The flip side of this rule is that you *don't* own any object you receive via any other means, such as one you get as a value from a method. You don't release these objects, because whatever code that created them is responsible for releasing them.

However, you might want to own such an object. In such a case, you can retain the object. By doing this, you become an owner and therefore must release it at some point.

One interesting twist is the idea of *autoreleased* objects. When you send the autorelease message to an object, you add it to a pool of objects that will all be sent a release message sometime in the future, typically when the event-dispatching call stack returns. Autoreleasing makes a lot of sense for methods that return objects that they themselves don't or can't manage. With autorelease, the caller gets a chance to retain the returned object, and if it doesn't, the object will eventually be released.

You might remember that we had a taste of memory management back in the sayHello: implementation in Section 3.4, *Designing an Interactive Application*, on page 29. We created an NSString with alloc and were therefore obligated to release it when we were done with it.

Some classes provide convenience constructors that can save you a step. Rather than the usual alloc and initXXX pair, you can sometimes use a class method that provides an autoreleased object. For example, NSString offers a class method stringWithFormat: that works just like the alloc and initWithFormat: calls, except that the returned string is autoreleased, meaning you don't (and shouldn't) explicitly release it yourself.

We do have a little bit of a leak in our application: the text field and the label. When objects are loaded from a nib, they are sent an autorelease message, which would be fine... except that if they are assigned to an instance variable, they are then sent a retain. So that means helloLabel and nameField each has a reference count of 1 that is never decremented, and therefore they're never freed. So when can we release them? Since they got the retain when the view controller was created, a good time would be when the view controller is being freed. You can do this by overriding the dealloc method, which is called for any object when its retain count goes to 0. You'll often want to release your instance variables in the dealloc method; here's what it should look like in HelloUserViewController.m:

HelloiPhone/HelloUser/Classes/HelloUserViewController.m
```
- (void)dealloc {
    [helloLabel release];
    [nameField release];
    [super dealloc];
}
```

3.8 Accessing Variables as Properties

One recent addition to Objective-C that you see throughout the iPhone APIs is the *property*. We dealt with properties earlier when retrieving the text from the text field and setting a different text property on the label.

Conceptually, properties are just instance variables with a naming convention for getter and setter methods. For example, the UILabel class has a text property that you could set via a setText method and retrieve via text, though you typically just use the dot operator to do assignments, so instead of [myLabel setText: myString], you write myLabel.text = myString, and instead of myString = [myLabel text],[10] you use myString = myLabel.text.

Actually, properties are deeper and more interesting than just a naming convention, since they allow you to declare the memory and thread-handling behaviors of the property and optionally have the compiler provide that behavior for you.

You define a property with a statement in the header file of the following form:

`@property (attributes) proptype propname;`

Then in the implementation file, you either provide implementations of getter and setter methods or let the compiler *synthesize* them for you:

`@synthesize propname;`

What's nice about properties is that you can use the attributes to provide information about your properties that you couldn't expose through instance variable or getter/setter methods alone. The most significant attributes address three traits:

- *Readabililty*: A property can be **readwrite** (the default) or **readonly**.

- *Memory management*: For setters, the default behavior is a simple **assign**, which is appropriate for primitive types. For properties that are objects, you can use **retain** to send a *retain* message to the object being set (ensuring that your object will be an owner of it), while also releaseing any previous value. The third option is to **copy** the object being set as the property, which also makes you an owner.

- *Thread handling*: By default, the synthesized getter and setter are thread-safe. Since most iPhone applications don't use multiple threads and just do their work as part of UIKit's main run loop, you can declare your properties as **nonatomic** for a thread-unsafe but higher-performance implementation.

Let's see how we declare properties for the helloLabel and nameField. In HelloUserViewController.h, after the sayHello: method declaration (really anywhere outside the **@interface** curly-brace block and before the **@end**), add declarations to create nonatomic, retaining property declarations for the two instance variables we've already declared.

---

10. Notice that by convention Cocoa getters don't actually use get in the method name.

> **Properties: Old and New**
>
> At the time of this writing, there's a difference in property support for the iPhone Simulator and on the actual device, a consequence of the fact that the device runs the "modern" Objective-C runtime and the simulator doesn't. The essential difference is that the modern runtime does not require you to explicitly back up a property with an instance variable; it's enough to just write the **@property** statement and let Xcode and the Objective-C runtime do the rest.
>
> However, if you do this, you'll find your code won't compile if your Active SDK is set to Simulator. For this book, we're making everything simulator-friendly by explicitly backing properties with instance variables (which we'll frequently abbreviate as *ivars*).
>
> If you want to distinguish a property from the instance variable backing it up, then you can use an = in the **@synthesize**. For example, to use the ivar myTheProp for the property theProp, you'd use @synthesize theProp = myTheProp; in your .m file. Also, note that you can declare **IBOutlet** on either the property or the ivar. We prefer the former.

HelloiPhone/HelloUser/Classes/HelloUserViewController.h

```
@property (nonatomic, retain) UILabel *helloLabel;
@property (nonatomic, retain) UITextField *nameField;
```

Now to get those properties created, we just need one line in HelloUserViewController.m, right after the **@implementation**.

HelloiPhone/HelloUser/Classes/HelloUserViewController.m

```
@synthesize helloLabel, nameField;
```

With this, our class (and others that call us) can refer to these properties with the dot operator, which is really convenient when you need to chain them together, such as when another class needs to call helloViewController.nameField.text. Moreover, our explicit use of **retain** makes it easier to keep track of memory-management issues. With just instance variables, it was easy to overlook the memory implications of helloLabel and nameField; now that we have property declarations that explicitly retain those objects, it's a good reminder that we need to release them in dealloc.

The UIKit classes use properties extensively, as you may have noticed in the use of a *text* property in both the UILabel and UITextField classes. And if you agree that the properties make the code a little more readable—in addition to expressing your intentions for readability, memory management, and threading—then we hope you'll use them for getters and setters, as we commonly do throughout the rest of book.

3.9 Take-Away: Stuff to Remember

We have covered an extraordinary amount of material in this chapter, having started with no interactivity at all and ending up with some sophisticated abilities for managing our class's members.

Rather than simply summarize, we'd like to wrap this chapter with a list of what we think are some of the most important things you should remember—stuff that will come up in more or less every iPhone application you write.

- Although other classes and nibs are loaded earlier as part of the application startup sequence, you usually want to look at your application delegate's applicationDidFinishLaunching: and any view controllers' initWithCoder: and viewDidLoad as places to add custom start-up behavior.

- Objective-C instance variables go in the header file, inside the **@interface** declaration, always with the type and, if they're objects, the pointer (*) character.

- Objective-C method declarations go in the header file, outside the **@interface** block, in the following form: - (returntype) methodName: (parameter1type) parameter1 parameter2Name: (parameter2type) parameter2... ;. Alternately, you can put a method declaration (or only its implementation) in the .m file, if you don't want other classes to be able to see it.

- Objects are created in code (usually with alloc and init) or with Interface Builder. Don't do both for the same object... you'll actually be creating two objects.

- To work with objects created in Interface Builder, declare instance variables as **IBOutlet**s and event-handling methods as returning **IBAction**. Then wire the connections in the nib file with IB. Be sure to save the file in IB before you build the project in Xcode.

- To implement a delegate protocol, add the protocol name in angle braces to your header file's **@interface** statement, after the name of the class you're subclassing. Then, in the implementation (.m) file, implement all required methods as well as any others you're interested in. Finally, make your object the delegate either by wiring a delegate outlet in IB or by setting a property in code (for example, textField.delegate = self;).[11]

- Declare properties with a **@property** declaration in the header file, and then use **@synthesize** in the implementation to have the compiler create the getters and setters for you.

- Add to the provided implementation of dealloc to release any instance variables your object may be retaining. That said, your code should never call dealloc directly; it will be called automatically when an object is being freed by the Objective-C runtime.

- Finally, remember the fundamental memory rule: you own any object you obtain by way of a method with alloc, new, or copy in its name, and you must eventually release (or autorelease) such objects. You do not own objects obtained through other means and must not release them, unless you choose to become an owner by calling retain.

---

11. We haven't mentioned **self** before, but you may have guessed it's how an object refers to itself, like **self** in Ruby or **this** in Java.

Chapter 4

View Controllers

Each screen in your iPhone application is managed by a *view controller* (VC) that is in charge of displaying the view and responding to just about every action the user can take on that view. Your view controller is the *C* in MVC (Model View Controller). It needs to communicate with both the view and model parts of your application. If you are unfamiliar with the MVC design pattern, you can get more information from the Apple documentation at *Cocoa Fundamentals Guide: Cocoa Design Patterns* [App06a].

We will start with a view controller that reacts to a tapped button by logging a message. In other words, we'll initially have no model and just use the view controller to interact with the view. Next we'll add to our view controller's responsibilities by building a Movie model class, creating an instance of it, and then displaying its data on the screen. In this phase, you'll see the view controller as the intermediary between the model and the view. In the third part of the example, we will add a second view controller and show you how control is passed between them.

4.1 Implementing a Button Action

Let's look in a little more detail at a simple view controller. Create a new project in Xcode using the View-based Application template. Remember, you get to New Project from the File menu or hit ⌘-⇧-N. Name the project Movie.[1] Once Xcode is done creating your new project, open

---
1. Although this initial example has little to do with movies, we will be using this example throughout the chapter, and the reason to call it Movie will be clear when we start building the second view controller.

the MovieViewController.xib file in Interface Builder by double-clicking it. This file should look familiar. Except for the name, it's the same file we started with in Section 2.2, *Create the Hello iPhone Project*, on page 11.

We are going to add a button to this initial view and have that button invoke an action method on our view controller. The first step is to add the button to the view. You will find buttons in the Library. To open the Library, choose Tools > Library from the menu, or press ⌘-⇧-L.

You should see something like this screenshot:

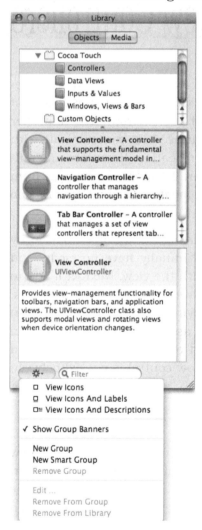

There are three ways to view the list of items. You can look at just the icons, which is the way most experienced UIKit developers prefer since it's smaller and they recognize the icons. You can choose the icons and a label, which is a little more descriptive. Or you can choose the icon and description. We recommend starting with the icon and description because the descriptions will really help you understand what each of the widgets is and when and how to use them. The previous screenshot shows the Library open and the pull-down selected.

Choose a Rounded Rect Button, and drag it into the view. Place it near the bottom, and set its title to Edit by double-clicking in the center of the button and then typing Edit. You should end up with something that looks more or less like this:

The button should do something when the user taps it. As you saw in Section 3.4, *Working with Xcode and Interface Builder*, on page 28, there are three steps to making this happen. In Xcode you will declare a method name in the MovieViewController header file. You will mark it with the IBAction return type. Then you'll return to Interface Builder and connect the button to this method. Finally, you'll head back to Xcode to implement the method.

Open the MovieViewController.h file in Xcode, and add this line to it just before the @end compiler directive:

ViewControllers/Movie01/Classes/MovieViewController.h

```
- (IBAction)edit;
```

Save your work in Xcode, and head back to IB to connect the button to this action. In IB select the button, and open the Connections inspector (⌘-2). Drag from the little circle to the right of Touch Up Inside to the File's Owner, and then mouse up and select the edit action from the pop-up window that appears. The button's Connection inspector should look like this:

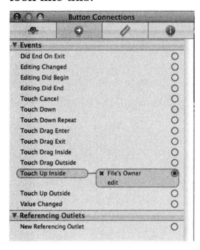

Congratulations, you just made a connection in Interface Builder. We will be doing a lot of connections like this in Interface Builder. You've just set a target/action pair (for more on target/action, see the sidebar on the next page). The target is the File's Owner, and the action is the edit method. Setting that target/action pair is what causes the button to invoke the edit method when it's clicked. We know it can still seem a bit like magic, but we will pull the curtain back a little bit at a time as we build more examples and make more connections.

We are done with IB for now, so save your work, quit IB, and switch back to Xcode. Now we need to implement the method in the MovieViewController.m file. Open the file in Xcode, and add some code between the @implementation and @end compiler directives that looks like this:

ViewControllers/Movie01/Classes/MovieViewController.m

```
- (IBAction)edit {
  NSLog(@"edit method invoked");
}
```

> **Target/Action Paradigm**
>
> IB makes extensive use of the target/action paradigm to make it easy for us to connect user interface elements to methods in our code. The *target* is the object that the control (that is, a button) will send a message to. The *action* is the message that will be sent. So, we say that the target is sent the action message when the control is activated.
>
> The UIControl defines this mechanism and allows us to connect several target/action combinations to handle the Multi-Touch nature of the iPhone interface using one of three method signatures to respond to events:
>
> - - (IBAction)action
> - - (IBAction)action:(id)sender
> - - (IBAction)action:(id)sender forEvent:(UIEvent *)event
>
> The first option is what we have chosen for our edit method because we don't need any information from the button for our code to function properly. For other controls (like a slider), you might want to get the object the user is interacting with (the sender argument) sent to you so you can use the current value in your code. Finally, if you need the event, for example, because you are tracking a Multi-Touch event, you can implement the third method style to get that extra object.
>
> Whichever approach you take, the target/action paradigm gives you a great deal of flexibility in how to get controls in the UI to invoke your code.

This code is invoked when the button is activated because of the connection we made in IB. It logs the message "edit method invoked" to the console every time the user releases the button.

Let's test where we are now; save your changes, and then click the Build and Go button in Xcode to run the app. When you tap the button, you should see the log message in the console (⌘-⇧-R). The view controller is responding to a user interacting with controls on the screen. When we get to more complex interactions such as adding and deleting table view items in Chapter 5, *Table Views*, on page 75, the basic interaction will be the same. The user will tap or drag some view on the screen, and then the view controller will have a method invoked by that action and will react.

4.2 Building a Model

We previously touched on the Model View Controller pattern, but we have not really built a model to speak of. A real model captures the essence of your application. In iTunes, classes like Song or Podcast and the actions they support like play form the model. In iPhoto, it's classes like Photo or Album. To build a serious application, you need a real model. The model does not have to be complex, but we do need something more than a string. We'll build a Movie class to play the part of our model. We will then use an instance of our Movie class to hold the data and modify the existing UI to display the data. The view controller will be the glue that is connected to both the model and the view.

Let's start by creating the Movie class. Select the Classes group in Xcode, and then right-click or ^-click the group and select Add > New File from the pop-up menu. In the dialog box that opens, choose Cocoa Touch Classes > Objective-C Class. In the "Subclass of" pull-down, make sure to select NSObject, and then click the Next button.

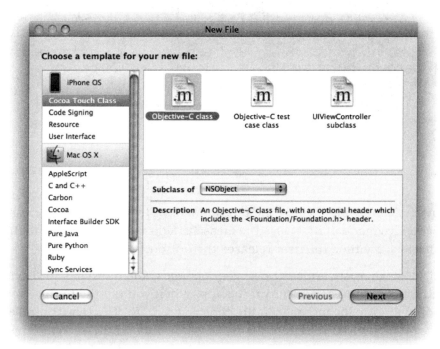

Name the class Movie, and make sure to check the two checkboxes for adding the .h file and adding the file to the target.

Now that we have our Model class defined, we will flesh it out by adding three properties: an NSString named title to hold the title of the movie, an NSNumber named boxOfficeGross to hold the box-office sales, and an

NSString named summary to hold the plot summary text. We'll declare three variables between the brackets in our header file that correspond to these properties and use @property for each so that the compiler will be able to build the get and set methods for us.

We also want to have a custom init method for our movie so we can initialize the three properties when we create a new movie. So, add a method called initWithTitle:boxOfficeGross:summary: to the class as well. Your header file should look like this:

ViewControllers/Movie02/Classes/Movie.h

```
@interface Movie : NSObject {
  NSString *title;
  NSNumber *boxOfficeGross;
  NSString *summary;
}

- (id)initWithTitle:(NSString *)newTitle
    boxOfficeGross:(NSNumber *)newBoxOfficeGross
           summary:(NSString *)newSummary;

@property(nonatomic, copy) NSString *title;
@property(nonatomic, copy) NSNumber *boxOfficeGross;
@property(nonatomic, copy) NSString *summary;

@end
```

Now that we have the interface done, complete the implementation by adding the @synthesize statements to create the get and set methods for the properties. We also need to implement the custom init method, and we'll need to clean up our memory.

ViewControllers/Movie02/Classes/Movie.m

```
Line 1  @implementation Movie
     -
     -  @synthesize title;
     -  @synthesize boxOfficeGross;
     5  @synthesize summary;
     -
     -  - (id)initWithTitle:(NSString *)newTitle
     -        boxOfficeGross:(NSNumber *)newBoxOfficeGross
     -               summary:(NSString *)newSummary {
    10      self = [super init];
     -      if(nil != self) {
     -        self.title = newTitle;
     -        self.boxOfficeGross = newBoxOfficeGross;
     -        self.summary = newSummary;
    15      }
     -      return self;
     -  }
```

```
 -  (void) dealloc {
20       self.title = nil;
 -       self.boxOfficeGross = nil;
 -       self.summary = nil;
 -       [super dealloc];
 -   }
25
 -   @end
```

This implementation has a couple of new concepts. First let's talk about the custom init method implementation. We have seen using custom init methods a couple of times in Chapter 3, *iPhone Development Fundamentals*, on page 23, but this is the first implementation we have built on our own. The code is not complex, but there are a couple of items we should talk about. In line 10, we've assigned [super init] to self. Assigning self to what the superclass returns from the init method gives the superclass the opportunity to return a different object. This somewhat obscure-sounding case shows up semiregularly in some frameworks (Core Data being the most common). So, it is important to include this line in your custom init methods. We then set all the properties to the values passed into our init method. And finally, we return self on the last line.

The dealloc starting on line 19 takes care of freeing up the memory used by this object.[2] Since the properties are marked as copy, each of them needs to be released in the dealloc method. We are accomplishing that by setting the properties to nil. As you learned in Section 3.8, *Accessing Variables as Properties*, on page 45, a property that copies or retains its object value will first release the old value before assigning the new value. So, in this case we set the property to nil, which first releases the existing object and then assigns the instance variable to nil.

Fantastic, you have just finished creating your first model class. We will use an instance of this class to manage the data for our application through the rest of this chapter. But first, save your work. We need to update our existing UI to display an instance of Movie.

4.3 Adding Outlets and Actions to the Controller

If you are new to developing with Cocoa, the process we are about to pursue will feel a little bumpy at first. A typical Cocoa Touch develop-

---

2. We had our first introduction to memory management in Section 3.7, *Managing Application Memory*, on page 43.

ment process requires jumping between IB and Xcode frequently. A big part of all the jumping back and forth is because of the nature of the view controller. It has one foot in the interface and one in the code. Since we edit the interface in IB and the code in Xcode, it is natural that we jump between the two tools. We promise the bumpy feeling won't last long. Soon you will build your intuition, and all the jumping back and forth will make sense (and you'll even come to like the fluid nature of it, promise).

Since we are already in Xcode, let's get started by adding the new properties we will need to our view controller. Open the MovieViewController.h file, and add four properties and the instance variables to back them. This is similar to what you just did in the Movie class. Here you'll add one property for the movie object that will be our model. Add three outlets, one each for the data in the movie object that we are going to display. Notice where the word IBOutlet appears for each property.

ViewControllers/Movie02/Classes/MovieViewController.h
```
Line 1  #import <UIKit/UIKit.h>
    -
    -   @class Movie;
    -
    5   @interface MovieViewController : UIViewController {
    -       Movie *movie;
    -       UILabel *titleLabel;
    -       UILabel *boxOfficeGrossLabel;
    -       UILabel *summaryLabel;
   10   }
    -
    -   @property(nonatomic, retain) Movie *movie;
    -   @property(nonatomic, retain) IBOutlet UILabel *titleLabel;
    -   @property(nonatomic, retain) IBOutlet UILabel *boxOfficeGrossLabel;
   15   @property(nonatomic, retain) IBOutlet UILabel *summaryLabel;
    -
    -   - (IBAction)edit;
    -   @end
```

The @class compiler directive on line 3 is a *forward declaration*, and it tells the Objective-C compiler that you know that it can't find a class named Movie and that it should not report errors or warnings about not being able to find it. Using these forward declarations is a common practice in header files so that we don't end up with include cycles (where one header includes another that also includes the first). We use forward declarations because the compiler provides poor error reporting on include cycles. We'd rather get into this habit than have to remember the bad warnings and what they mean.

Before we leave Xcode, let's add the @synthesize calls and the import to the implementation file. Open MovieViewController.m, and add the @synthesize statement for each of the properties. We always put these statements at the top of the @implementation block, so we usually do that first when building a new implementation.

ViewControllers/Movie02/Classes/MovieViewController.m
```
@synthesize titleLabel;
@synthesize boxOfficeGrossLabel;
@synthesize summaryLabel;
@synthesize movie;
```

Also, for every @class forward declaration you do in the header file, you almost always have to import the header file for that class in the implementation file. In this case, that means we have to import Movie.h:

ViewControllers/Movie02/Classes/MovieViewController.m
```
#import "MovieViewController.h"
▶ #import "Movie.h"
```

Let's turn our attention to the GUI side of our application, but don't forget to save everything before we head over to Interface Builder.

4.4 Updating the UI

Now that we have our header file, let's go back to Interface Builder and open MovieViewController.xib. Add three UILabels to our view, arrange them near the top, and stretch them across the view. Use the blue alignment lines to help you get them straight. After you place the labels where you want them, your view should look like the screenshot on the next page.

Once the labels are placed where you want them, connect the VC's outlets to them. We need these connections made so that our view controller knows about the labels and can place the data from the movie object into them at the appropriate time.

Connect the outlets to their respective labels. You make a connection to an outlet in one of two ways. First you can Ctrl+drag from the source of the connection (where the outlet is) to the destination object. When you lift up on the mouse, the list of valid outlets is displayed, and you choose the one you want to set by clicking it. That is all there is to it—once you click the outlet, the connection is made. Second, you can use the Connections inspector as you saw in Chapter 3, *iPhone Development Fundamentals*, on page 23.

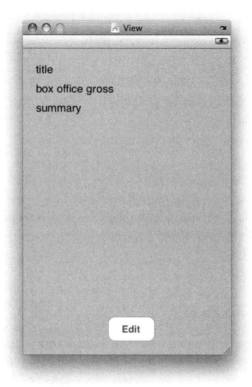

Once the connections are complete, select the File's Owner, and open the Connections inspector with ⌘-2. Your inspector should look similar to this:

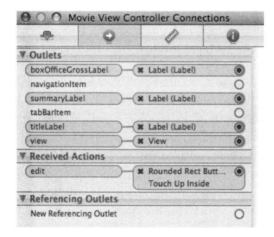

We're now ready to take advantage of all of this prep work to display the information for a specific movie.

4.5 Implementing the Controller

Now we're ready to update the implementation so that the movie's data is displayed on the screen. The first thing we will do is to create the movie object. Let's override the viewDidLoad method in the MovieViewController to create a new instance once the view that the MovieViewController controls is loaded. Here is the code:

ViewControllers/Movie02/Classes/MovieViewController.m

```
- (void)viewDidLoad {
    [super viewDidLoad];
    Movie *newMovie = [[[Movie alloc]
                        initWithTitle:@"Iron Man"
                        boxOfficeGross:[NSNumber numberWithFloat:650000000.00]
                        summary:@"Smart guy makes cool armor"] autorelease];
    self.movie = newMovie;
}
```

The code is straightforward. Allocate a new instance of Movie, set the property movie to the new Movie object and then release it. The interesting thing here is why choose viewDidLoad. To understand the reasoning behind that, we have to talk a bit about the way view controllers work. Let's start in the applicationDidFinishLaunching: method on the MovieAppDelegate class. This method is called by the UIApplication object when it's finished launching. The app delegate has two outlets that are connected in the MainWindow.xib nib file: window and viewController. The window property is connected to the main window of our application; the viewController property is connected to our movie view controller. Here is the code for applicationDidFinishLaunching::

ViewControllers/Movie02/Classes/MovieAppDelegate.m

```
- (void)applicationDidFinishLaunching:(UIApplication *)application {
    [window addSubview:viewController.view];
    [window makeKeyAndVisible];
}
```

The app delegate is asking the window to add the viewController's view as a subview. That is a simple enough looking line of code, isn't it? Well, behind the scenes, a bunch of very interesting stuff is going on. When a view controller is asked for its view, the first thing it does is check to see whether it already has one. If so, it just returns the already loaded view. If it does not have one, then it calls the loadView method. That process runs more or less like this:

- Is a nib filename set (typically set in IB but can be set via code)?
- If so, load the nib file passing the view controller in as File's Owner.

- After the nib file is loaded, assert that the view has been set. If not, throw an exception and print an error message.

If a nib filename is not set on your view controller, then you need to build the view manually in the loadView. Thankfully, we can set the nib filename via IB, and we rarely if ever need to manually code a loadView method.[3]

Once the loadView method finishes, the viewDidLoad method is called. This is a great place for us to do initialization (such as create our Movie object) because it's called only when the view is loaded. If because of memory warnings we release our view and set it to **nil**, we can also **nil** out the model object. Because we know that when the view is needed again, the whole process, including viewDidLoad, will be invoked again, and we will have a Movie object.

The next step in implementing our movie view controller is to implement the viewWillAppear: method. This method is called every time the view is about to become visible. Keep in mind that a view can become visible and then be removed from view several times during the normal flow of an application. For example, consider the Contacts app; the list of all your contacts is shown each time you finish looking at or editing an individual contact, and the viewWillAppear: method is called each time. This method is the ideal place to set the values we want to appear on the view when it does appear. Here is the code for our view controller:

ViewControllers/Movie02/Classes/MovieViewController.m
```
- (void)viewWillAppear:(BOOL)animated {
  [super viewWillAppear:animated];
  self.titleLabel.text = self.movie.title;
  NSNumberFormatter *formatter = [[NSNumberFormatter alloc] init];
  [formatter setNumberStyle:NSNumberFormatterCurrencyStyle];
  self.boxOfficeGrossLabel.text =
  [formatter stringFromNumber:self.movie.boxOfficeGross];
  [formatter release];
  self.summaryLabel.text = self.movie.summary;
}
```

Build and Go to run your application and test it. You should see the title of the movie that you choose appear in the text fields including a dollar sign in the box-office gross field.

---

3. If you want to dig into the MovieViewController's nib setting, open the MainWindow.xib file, and inspect (⌘-1) the MovieViewController. There is more detail in Section 3.5, *Anatomy of Your iPhone Application*, on page 38.

This code we added places the data for the Movie objects into the text fields. The one new class in this code is the NSNumberFormatter. We use formatters to convert back and forth between numbers and strings.

Congratulations! You have successfully built a full-fledged view controller and the model that it manages. Next you'll allow the user to edit the Movie object in a new view controller. You'll build another view controller in Xcode, set up its view in IB, and then wire the two view controllers together.

4.6 Creating the New View Controller

Let's now create a second view controller that we will use to manage the modal view. In Xcode, select the Classes group, Ctrl+click the group, and choose Add > New File from the pop-up menu. In the wizard that pops up, choose the "UIViewController subclass" item in the iPhone OS > Cocoa Touch Classes group. Don't select the "With XIB for user interface" checkbox, because we will be creating our interface shortly. Click Next, and name the class MovieEditorViewController.

We're imagining a UI with three text fields and one button, so we'll need to add the outlets and actions that we will need to interact properly with the UI to our header file. The text fields will allow the user to edit the title, box-office gross, and summary, respectively, and the button will signal the user is done with the edits. So, we need three ivar/property pairs, one for each of the text fields. We also need a single action method for the button to invoke. The last piece of this class is the instance of the Movie that our user will be editing, so we need to add an ivar/property pair for that too (don't forget to add the @class Movie; forward declaration). The interface for our class should look like this:

ViewControllers/Movie03/Classes/MovieEditorViewController.h
```
@class Movie;

@interface MovieEditorViewController : UIViewController <UITextFieldDelegate> {
  UITextField *titleField;
  UITextField *boxOfficeGrossField;
  UITextField *summaryField;
  Movie *movie;
}

@property(nonatomic, retain) IBOutlet UITextField *titleField;
@property(nonatomic, retain) IBOutlet UITextField *boxOfficeGrossField;
@property(nonatomic, retain) IBOutlet UITextField *summaryField;
@property(nonatomic, retain) Movie *movie;

- (IBAction)done;

@end
```

Our controller is going to need to respond to keyboard input, so we'll add a declaration at the top inside angle brackets for the UITextFieldDelegate protocol. That tells the compiler that this class intends to implement all required methods from that protocol. We first saw protocols in Chapter 3, *iPhone Development Fundamentals*, on page 23 when we discussed the UIApplicationDelegate protocol. The text field delegate protocol is very similar to the application delegate protocol in that methods are sent to the delegate during interesting points in the text field's life cycle. We will be digging in to the details shortly. Make sure to save the file so Interface Builder will know about the new outlets and actions we've added.

Don't forget to add the import of Movie.h to the top of the MovieEditorViewController.m file. And just inside the @implementation block, we need to add a @synthesize statement for each of the properties that were declared in the MovieEditorViewController.h file.

Now we need to implement the viewWillAppear: method to place the values from the Movie object into the text fields. We'll set the text property on the text fields.

ViewControllers/Movie03/Classes/MovieEditorViewController.m

```
- (void)viewWillAppear:(BOOL)animated {
  [super viewWillAppear:animated];
  self.titleField.text = self.movie.title;
  self.summaryField.text = self.movie.summary;
  NSNumberFormatter *formatter = [[NSNumberFormatter alloc] init];
  [formatter setNumberStyle:NSNumberFormatterCurrencyStyle];
  self.boxOfficeGrossField.text =
          [formatter stringFromNumber:self.movie.boxOfficeGross];
  [formatter release];
}
```

Notice the way that we convert the boxOfficeGross number to a string via an NSNumberFormatter. We are doing the same thing we did in the MovieViewController's viewWillAppear: implementation.

In the done method, we dismiss the current modal view controller with a call to the dismissModalViewControllerAnimated:

ViewControllers/Movie03/Classes/MovieEditorViewController.m

```
- (IBAction)done {
  [[self parentViewController] dismissModalViewControllerAnimated:YES];
}
```

Though we have not done so yet, when we build the UI, we will make our view controller the delegate of all the text fields so we can capture the input via the UITextFieldDelegate protocol. We are going to go over that implementation now, but the methods won't get called until we set the delegate for each text field in IB.

We need to implement two methods of the UITextFieldDelegate protocol. The textFieldShouldReturn: method is called by the text field when the user presses the Return button on the keyboard. This is the perfect spot for us to resign first responder status for the field (which in turn will cause the keyboard to animate out) by calling resignFirstResponder. It is typical to return YES from this method, but if you wanted to do some validation and keep the text field from resigning the first responder status, you can return NO. But keep in mind that the return value is simply advisory; if the user is taking a call, you won't be able to prevent the user from leaving the field, so if your application relies on the data being valid, this can't be the only place you check it.

ViewControllers/Movie03/Classes/MovieEditorViewController.m
```
- (BOOL)textFieldShouldReturn:(UITextField *)textField {
  [textField resignFirstResponder];
  return YES;
}
```

The textFieldDidEndEditing: method is called when the text field resigns its first responder status.[4] This method is a great spot to capture the data that the user typed into the field. We take the values from the fields and place those values into the movie object. Here is the code:

ViewControllers/Movie03/Classes/MovieEditorViewController.m
```
- (void)textFieldDidEndEditing:(UITextField *)textField {
  if(textField == self.titleField) {
    self.movie.title = self.titleField.text;
  } else if(textField == self.boxOfficeGrossField) {
    NSNumberFormatter *formatter = [[NSNumberFormatter alloc] init];
    [formatter setNumberStyle:NSNumberFormatterCurrencyStyle];
    self.movie.boxOfficeGross =
        [formatter numberFromString:self.boxOfficeGrossField.text];
    [formatter release];
  } else if(textField == self.summaryField) {
    self.movie.summary = self.summaryField.text;
  }
}
```

Since this one view controller is playing the part of delegate for each of the text fields, we need to distinguish which one was typed in so we know which property of the movie to update.[5]

Notice also that we are using a currency formatter to convert the data from the text field to a number for our movie's boxOfficeGross. This requires that we put a currency sign in the front of the number for it to parse properly. In a polished application, we'd want to test for the leading currency symbol and then choose the proper formatter for what

---

4. A text field will resign its first responder status because it was told to (as we did in the previous method) or when it is forced to by the system. The system forces text fields to resign first responder status when they are taken off-screen. Although in our example there is no way to send this view off the screen when the keyboard is up, later in Chapter 5, *Table Views*, on page 75 you will see how this can happen.

5. As your skill in using the Model View Controller pattern grows, you will likely find using one controller as the delegate of three text fields to be a little distasteful. We keep it all in one class here for simplicity's sake.

the user typed in. We'd also want to ensure that the value we get back from the formatter is not **nil**; if it was, we'd want to inform the user the number was some how invalid. For this app, though, in the interest of clarity, we are just going to leave it at the currency formatter.

4.7 Building the UI

Now we need to add a new nib file to the project that will contain the view for our new view controller. Still in Xcode, Ctrl+click the Resources group, choose Add > New File, click iPhone OS > User Interface, and then choose the View XIB item. Click Next, name the file MovieEditorViewController.xib, and then click Finish. Open MovieEditorViewController.xib in IB by double-clicking it.

We are going to modify the new nib so that it becomes the user interface for our MovieEditorViewController. The first step is to change the class of the File's Owner. Select this object, and open the Identity inspector with ⌘-4. Change the Class field to MovieEditorViewController. The inspector should update with all the outlets and actions that we added to the header and look like this:

Now that IB knows about our outlets and actions, we need to create the UI and make the connections. Open the view by double-clicking it. Add three text fields and one button. We also changed the background color just for fun:

Now we have a UI and the controller to manage it. All that remains is to make the connections. Select the File's Owner object, and connect each of the text field outlets to their respective text fields in the UI. Also connect the File's Owner's view outlet to the view.

We like the Ctrl+drag method, but you can also open the Connections inspector with ⌘-2 and drag from the buttons to the left of each of the outlets. You also need to connect the button to the done action method. Don't forget to connect each of the text field's delegates up to the File's Owner as well.

When we are finished with the connections, this is what our Connections inspector looks like:

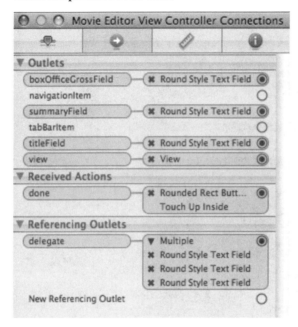

Configuring the Text Field

One of the coolest features of the iPhone is the Multi-Touch keyboard. With no physical keyboard, you get more screen real estate to use. Supporting different languages becomes far easier too. Any time you tap in a text region, the iPhone automatically brings up a keyboard. Technically what is happening behind the scenes is that the text field is becoming the first responder. The first responder is the object that is first in line to get certain types of events (keyboard taps being among the most important). All text-related controls in Cocoa Touch will bring up a keyboard whenever they become the first responder.

The Multi-Touch keyboard is a great feature for users but requires diligence on our part to use it correctly. The keyboard that the text field brings up when tapped is critical to good user experience. We want the correct keyboard to show up, and we want it to be properly configured so the capitalization works like the user wants/expects. Whenever you place a text entry on your UI, make sure to think through what the user is going to expect so you give them the best possible experience.

Here the text field is configured to capitalize words and take the defaults for correction, type, and appearance. The Return key is set to Done, so the user knows that when she is finished typing, she can simply touch it to finish editing.

When thinking through what keyboard to show your users, answer these questions about what the user will be expected to provide:

- Is the text going to have unusual words? Turn off Correction.
- Is the text a formal name? Capitalize each word.
- Is the text a phone number, email address, or URL? Use the correct keyboard type.
- What does the user expect to do when done—go to a URL, search on Google, or persist the data? Make sure to choose the correct Return key title.

Spend the time to make text entry right. Entry will feel more natural, and users will like your app all the more. If you don't and just accept the defaults, you could end up ruining the experience for your users.

Using the placeholder text is a great way to give the user a clue about what should be entered in the text field. This might be the First Name field or whatever, but putting what is expected into the placeholder helps the user keep context. We also specified that the keyboard should capitalize words because this text is going into a label and the user probably expects the words to be capitalized. Finally, the Return key is set to Done, so the user has a clear indication of what to hit to indicate they are finished.

This last step is optional. I like this app better if the text fields do not delete the text when editing starts. My suggestion is that you should try both and see which one you like better. This is another place where it's vitally important that we take the time to think through what our user is going to want and tweak the settings to match that. You can turn this feature on and off by switching the Clear When Editing Begins checkbox in the Attributes inspector for the text field.

4.8 Making the MovieEditorViewController

We are almost there. We have both view controllers and the user interface set up to display the movie as well as edit the movie. However, we have no way of getting to the second view where the user can edit the movie data. Here are the basic steps we need to follow:

1. Add an outlet to MovieViewController for the instance of MovieEditorViewController we are going to create.

2. Update MovieViewController's edit method to modally display the MovieEditorViewController.

3. Create an instance of MovieEditorViewController in MovieViewController's nib file, and make the connection from the outlet.

Adding the outlet is old hat now, so return to MovieViewController.h, and add an outlet named editingViewController to the MovieEditorViewController. Make sure to add the forward declaration:

ViewControllers/Movie03/Classes/MovieViewController.h

```
@class Movie;
▶ @class MovieEditorViewController;

@interface MovieViewController : UIViewController {
  Movie *movie;
  UILabel *titleLabel;
  UILabel *boxOfficeGrossLabel;
  UILabel *summaryLabel;
▶  MovieEditorViewController *editingViewController;
}

@property(nonatomic, retain) Movie *movie;
```

```
@property(nonatomic, retain) IBOutlet UILabel *titleLabel;
@property(nonatomic, retain) IBOutlet UILabel *boxOfficeGrossLabel;
@property(nonatomic, retain) IBOutlet UILabel *summaryLabel;
```
▶ `@property(nonatomic, retain) IBOutlet`
▶   `MovieEditorViewController *editingViewController;`

```
- (IBAction)edit;

@end
```

Next we need to update the edit method so it sets the editing view controller's movie instance and displays the new editing controller instead of just logging a message. The display is accomplished via the presentModalViewController:animated: method.[6] Also, don't forget to add the import statement to the implementation file for the MovieEditorViewController.h and the @synthesize statements for each property.

`ViewControllers/Movie03/Classes/MovieViewController.m`
```
- (IBAction)edit {
  self.editingViewController.movie = self.movie;
  [self presentModalViewController:self.editingViewController animated:YES];
}
```

4.9 The Editing View Controller in Interface Builder

Open the MovieViewController.xib file. We need to add a view controller to this nib file. From the Library (⌘-⇧-L), drag a view controller into the document window.

We need to change the class of this view controller to MovieEditorViewController. Select the new view controller, open the Identity inspector (⌘-4), and change Class to MovieEditorViewController. Next we need to set the nib filename for this view controller, so open the Attributes inspector with ⌘-1, and type MovieEditorViewController into the NIB Name field. Configuring our new view controller is complete.

All that remains is to make the connection from the File's Owner's editingViewController outlet to the new view controller. Once this connection is made, we have completed the application.

---

6. There is a bug in iPhone OS 2.2.1 (and previous) for modal views that causes the viewWillAppear: to be called before the view is loaded. To work around this bug, add [self.editingViewController view] in the edit method before the call to presentModalViewController:animated:.

The View Controller's View

Recall that when the view controller is asked for its view; if it's not there, the view controller checks to see whether its nib file is set. If so, then the view controller loads that nib file, passing self in as the File's Owner. And since we set the view property in our MovieEditorViewController.xib file, the view gets set. Here's the basic picture of what happens once the view is requested but not found:

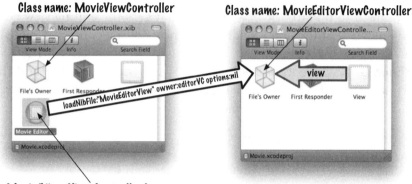

Now hit Build and Go to see your work in the simulator. Push the edit button, and the modal view should slide up from the bottom. Select one of the fields, type in a new value, and hit Done. You should be brought back to the movie view with the updated text showing in the appropriate label.

Fantastic! You have just built your very first view controller from scratch. As we go further and deeper into the rest of the UIKit, you will be building lots of view controllers. Some will be a bit more complex, but the underlying basics will always be the same. The view controller manages a view and helps that view display properly in its context on the screen. The view controller also interfaces very nicely with nib files, so we can easily chain together multiple parts of our interface. When you start building three- and four-level deep navigations into your app, the particulars will vary, but the basics will remain the same.

Chapter 5

Table Views

Table views are among the most common user interface idioms in iPhone OS. Many of the default applications use table views extensively, including Mail, Safari, Phone, iPod, iTunes, and more. In fact, there are fewer default apps that don't use tables than those that do. Because of tables' utility and convenience and the fact that your user will be thoroughly accustomed to tables, it is highly likely that your applications will want to present some of its interface with table views.

In this chapter, we'll explore how to present data in a table format by reusing the Movie class from Chapter 4, *View Controllers*, on page 51. In that chapter, we used view controllers to present and edit a single Movie object; here, we'll use a table view to show many Movie objects and navigate to a second view controller that will let us edit an existing object or create a new one.

5.1 Parts of a Table

On the iPhone, a table view is a one-dimensional, top-to-bottom list of items, optionally split into multiple sections. The sections actually make the list a two-dimensional data structure. Each section has a variable number of items, so a given item in a table is identified by its section and its *row* within that section.

In Figure 5.1, on the next page, we can see Interface Builder's presentation of table views, with dummy data that provides U.S. state names for section headers and provides cities for row titles. There are two visual styles for tables: a "plain" style that allows cells to stretch horizontally to the table's bounds and a "grouped" style that uses corner-rounding and indentation to group the rows of each section. The grouped table

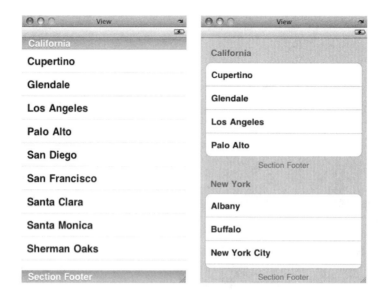

Figure 5.1: UITableViews in Interface Builder, with "Plain" style (left) and "Grouped" style (right)

in the figure shows two sections: one with four rows for the first section ("California") and three rows for the second ("New York").

An iPhone table consists of three things: a view, a data source, and a delegate. You start with a UITableView class that presents the table on-screen and handles user interaction, like a tap to select a row or a swipe to delete an item. The UITableView depends on one or more other objects to provide its functionality:

- A *table view data source* is an object that manages the relationship between the visual UITableView and its contents. Methods in the UITableViewDataSource protocol provide the number of sections and rows in the table, provide titles for headers and footers, and generate the views for each cell. The data source also has methods to handle the insertion, deletion, or reordering of table rows. Most of these features are optional: the only required methods are tableView:numberOfRowsInSection: and tableView:cellForRowAtIndexPath:.

- A *table view delegate* allows the host application a greater level of control over the table's visual appearance and behavior. An object implementing the UITableViewDelegate protocol is notified of various user actions like the beginning and end of row selection or editing. Other methods allow the delegate to provide customized views for headers and footers and to specify nondefault cell heights. The idea of Cocoa delegates was introduced in Section 3.6, *Customizing Behavior with Delegation*, on page 40, and in the previous chapter's Section 4.6, *Creating the New View Controller*, on page 64, we used a UITextFieldDelegate to customize keyboard behavior on a text field.

To use a table in your application, you will create a UITableView, typically in Interface Builder, and connect it to objects that implement the data source and delegate protocols. Often, the view controller that manages the table view will also serve as both the data source and the delegate. By making the view controller the data source for the table, you'll be required to implement tableView:numberOfRowsInSection: and tableView:cellForRowAtIndexPath: from UITableViewDataSource, and at a minimum you will usually also implement UITableViewDelegate's tableView:didSelectRowAtIndexPath: to handle the user tapping one of the table rows.

5.2 Setting Up Table-Based Navigation

The UITableView is used as a navigation metaphor throughout the iPhone OS. In Mail, you use tables to select an account, then a mailbox within that account, and then a message within that mailbox. For each of these steps, the data is presented as a table, and selecting a row *navigates* to a new view, drilling down either to another table or to a view that shows the contents of one message. To standardize this kind of behavior across applications, Xcode provides a Navigation-based Application project template that uses a table for its first view. We'll use this template to learn about tables and navigation in this chapter.

In Xcode, select File > New Project, and choose the Navigation-based Application template. Make sure the checkbox labeled "Use Core Data for storage" is not selected.

Name the project MovieTable, and Xcode will set you up with a project containing two classes (MovieTableAppDelegate and RootViewController), as well as two nibs (MainWindow.xib and RootViewController.xib).

This is a somewhat more advanced project template than you've seen before, and it helps to understand how the pieces go together. Open

Figure 5.2: NAVIGATION OBJECTS IN THE MAINWINDOW.XIB FILE

MainWindow.xib with Interface Builder, and switch the view mode from icons to list (the middle button of the view mode control). By expanding the tree structure, you'll see the arrangement shown in Figure 5.2. The nib has a *navigation controller*, an object that we'll use in code to navigate forward and backward between views. The navigation controller has two children: a *navigation bar*, which you'll recognize as the bar at the top of navigation screens (where it usually hosts buttons like Back or Edit), and the RootViewController, which has a *navigation item* object.

That's all well and good for navigating, but where's the table? If you inspect the RootViewController, you will see that it gets its view from RootViewController.xib. Open that nib and look: its default contents are a single UITableView object. The take-away for now is that a navigation application has this UINavigationController class that's responsible for navigation, which is a parent of the RootViewController, which is the view controller for the first view the user sees, which is a UITableView.

5.3 Modeling Table Data

Because the RootViewController owns the table view, let's take a look at the class. In its implementation file, RootViewController.m, you'll see default implementations for some of the table data source and table delegate methods, three of which are uncommented: numberOfSectionsInTableView:, tableView:numberOfRowsInSection:, and tableView:cellForRow-

AtIndexPath. If you look at RootViewController.xib with Interface Builder, you'll find that the table's dataSource and delegate outlets are connected to File's Owner. The net result is that this table is wired up and ready to run; the table expects the RootViewController to serve as its delegate and data source, and the class provides the minimum implementation of those protocols to run the application. The protocols are not explicitly declared in RootViewController.h, because the controller subclasses UITableViewController, whose declaration includes the two protocols. Keep in mind that if you ever use some other view controller with a table, you'll have to add <UITableViewDataSource, UITableViewDelegate> to the **@interface** in the header file to declare that you implement these protocols.

The default implementation provides for a table that has one section with zero rows. It also provides logic for creating cell views in code, but with zero rows, that code will never be used. So, the first thing we need to do is to implement the section- and row-count methods in a nontrivial way. This means we need to develop a *model* for the table data; with the RootViewController mediating between the view and this model, we'll have implemented the classic Model View Controller design pattern.

A table model does not have to be anything fancy; it's not a class unto itself as it is in other languages. For a one-section table, it's practical to just use an NSArray, which contains the objects the table represents. The array's length gives you the number of rows, and the contents for a given cell can be looked up with the array's objectAtIndex: method.

At the beginning of this chapter, we said that we would reuse the previous chapter's Movie class as the data for our table. In Groups & Files, Ctrl+click or right-click the Classes folder, and choose Add > Existing Files. Navigate to the previous Movie project,[1] select its Classes folder, use the Command key (⌘) to select Movie.h and Movie.m, and click the Add button. In the next dialog box, make sure the checkbox for "Copy items into destination group's folder (if needed)" is selected, and click Add again to copy the files into this project.

Also add #import "Movie.h" to RootViewController.h, since we'll be using the Movie class in our view controller.

---

1. In the book's downloadable code, for example, this would be ViewControllers/Movie02.

Next, since we're going to be offering an editable list of Movies, we'll want to use an array that we can add to and remove from. So, declare the instance variable NSMutableArray *moviesArray; in the @interface block of RootViewController.h. This array needs to be initialized, and we'll want to provide some data for our table (we'll allow the user to add their own data later), so uncomment the provided viewDidLoad method in RootViewController.m, and add the highlighted code to create one Movie and add it to the array:

TableViews/MovieTable01/Classes/RootViewController.m
```
- (void)viewDidLoad {
    [super viewDidLoad];
▶       moviesArray = [[NSMutableArray alloc] init];
▶       Movie *aMovie = [[Movie alloc] init];
▶       aMovie.title = @"Plaything Anecdote";
▶       aMovie.boxOfficeGross = [NSNumber numberWithInt: 191796233];
▶       aMovie.summary =
▶         @"Did you ever think your dolls were really alive? Well, they are.";
▶       [moviesArray addObject: aMovie];
▶       [aMovie release];
}
```

Now that our model has some genuine data, we need to update the UITableViewDataSource methods to get that data to the on-screen UITableView. The default numberOfSectionsInTableView: returns 1, which is fine as is. However, the tableView:numberOfRowsInSection: returns 0, which is wrong. We want it to return the length of the array:

TableViews/MovieTable01/Classes/RootViewController.m
```
- (NSInteger)tableView:(UITableView *)tableView
                   numberOfRowsInSection:(NSInteger)section {
    return [moviesArray count];
}
```

That will tell the table view that the one section has one row. As a result, when the application runs, it'll call the tableView:cellForRowAtIndexPath: method to get a UITableViewCell for that one row. The template provides a default implementation that creates a cell in code; we just have to customize that cell, immediately after the provided comment // Configure the cell and before return cell;.

What we need to do is to figure out which member of the array—there's only one now, but there will be many later—we want to use for the cell's contents. The key is to use the indexPath variable. An NSIndexPath is an object that specifies a path through a tree structure as a set of zero-based integer indexes. On iPhone OS, this class is extended with properties specifically for use with UITableViews: section and row. In other words, any time you handle tableView:cellForRowAtIndexPath:, the section and row of the cell being requested are indicated as indexPath.section and indexPath.row.

So, right before return cell; in the provided implementation, add the following code:

TableViews/MovieTable01/Classes/RootViewController.m
```
Movie *aMovie = [moviesArray objectAtIndex:indexPath.row];
cell.textLabel.text = aMovie.title;
```

The first gets the member of moviesArray that corresponds to the selected row, which is the value of indexPath.row. Then we just need to present the Movie's title in the cell. The UITableViewCell provides two UILabels as properties: textLabel and detailTextLabel. For this simple case, we set the textLabel's text to the movie title.

That's all that's necessary for a basic table. Build and Go. You'll see the one-row table shown in Figure 5.3, on the next page.

5.4 Table Cells

Thus far, we've provided enough of a data source implementation to get a minimal table on-screen, but there's a lot more we can do with this table, starting with the table cells. After all, while our Movie class has three member properties, we're showing only one of them in the table. Let's look into getting more use from our cells.

Cell Styles

The provided implementation of tableView:cellForRowAtIndexPath: creates a UITableViewCell object called cell that we customize before returning it to the caller. The default cell has three visual properties that can be used to put our data in the cell: textLabel, detailTextLabel, and imageView. In this example, we set the text of the textLabel to get the basic, default appearance. If the Movie class had an NSImage member (like a screenshot or DVD box art), then we could set the imageView's image property to make the image appear on the left side of the cell.

Figure 5.3: A BASIC UITABLEVIEW

To make use of the detailTextLabel, we need to choose a different cell style. The idea of the style is new in iPhone 3.0, and four styles are provided:

- UITableViewCellStyleDefault: Presents the textLabel single block of left-aligned black text. The detailTextLabel is absent. This default is identical in appearance to table cells in iPhone 2.x.

- UITableViewCellStyleSubtitle: Presents a large left-aligned black textLabel and a second line for a smaller, gray detailTextLabel below it, similar to the iPod or Music application.

- UITableViewCellStyleValue1: Presents a large left-aligned black textLabel on the left and a slightly smaller right-aligned detailTextLabel on the right in blue. This layout resembles cells in the Settings application and is intended only for use in group-style tables.

Figure 5.4: UITableViews displaying the four provided UITableViewCell-Styles, in plain and grouped mode

- UITableViewCellStyleValue2: Presents a small right-aligned blue textLabel on the left and a small left-aligned black detailTextLabel on the right, similar to the Contacts application. Again, this button-like style is appropriate for use only in grouped tables.

In Figure 5.4, we can see these four styles in a modified version of the sample application. We've changed the contents of the cells based on their style, because some of the layouts are inappropriate for large strings, particularly UITableViewCellStyleValue2, whose left-side label will truncate after about ten characters. Since the "value" styles are meant for a button-like presentation in grouped tables, the screenshot on the right of the figure puts each cell in its own section, while the left screenshot is a one-section table with four rows.

> **Use Styles for Table Cells, Not CGRect**
>
> Prior to iPhone SDK 3.0, UITableViewCell's designated initializer was initWithFrame:reuseIdentifier:, which took a CGRect (usually the constant CGRectZero, since it wasn't actually used) for the frame argument. The navigation-application template provided a call to this initializer, as did all other table code. However, in iPhone SDK 3.0, this initializer is deprecated in favor of initWithStyle:reuseIdentifier:, which takes one of the style constants instead of the CGRect. It's trivially easy to convert old code to the new standard by just switching to the new call and using the UITableViewCellStyleDefault style.

The provided styles offer some flexibility in presenting your data in the space afforded by a list on a small device like the iPhone. If none of these styles, with or without the optional imageView, suits your needs, then continue to Section 5.7, *Custom Table View Cells*, on page 94, in which we'll look at how to create custom cell layouts.

Cell Reuse

Along with a style, the initializer for a UITableViewCell takes a reuseIdentifier string. Understanding how this object is used is critical to creating tables that perform as expected. Fortunately, the default implementation of tableView:cellForRowAtIndexPath: shows us what this property does and how it is to be used.

A UITableView caches cells for later reuse, which improves performance by recycling cells rather than repeatedly creating them anew. When a cell completely scrolls off the top or bottom of the screen, it becomes available for reuse. So, when you need to create a table cell in tableView:cellForRowAtIndexPath:, you first try to retrieve an existing cell from the table's cache. If it works, you just reset that cell's contents; if it fails, presumably because no cached cells are available, only then do you create a new cell.[2]

---

2. This means that dequeuing occurs only when there is enough data for the table to fill the screen and the user has scrolled far enough for one or more cells to go entirely off-screen.

Here's the default implementation in tableView:cellForRowAtIndexPath::[3]

TableViews/MovieTable01/Classes/RootViewController.m
```
static NSString *CellIdentifier = @"Cell";
UITableViewCell *cell =
        [tableView dequeueReusableCellWithIdentifier:CellIdentifier];
if (cell == nil) {
cell = [[[UITableViewCell alloc] initWithStyle:UITableViewCellStyleDefault
                reuseIdentifier:CellIdentifier] autorelease];
}
```
(Lines 1–7)

Line 1 creates a *cell identifier*, a string that indicates the kind of cell we want. The idea here is that if you use different styles of cells in the same table (either default styles or layouts of your own creation), you will need to distinguish them in the table's cache so you get back the style of cell you need. In the default case, you use only one style, so any arbitrary string like "Cell" will suffice. Next, lines 2–3 attempt to *dequeue* a cell, that is to say, to retrieve a cell from the table's cache, passing in the identifier to indicate what kind of cell is needed. If this fails, then a new cell is allocated and initialized.

5.5 Editing Tables

So now, we've covered how to provide table contents and gain some control over how the contents of a cell are presented. The next step is to make the table editable. What this really means is that we want to make the table serve as an interface for editing the underlying model. When we delete a row in the table, we want to delete the object from the model, and when we add an item to the model, we want the table updated to reflect that.

Let's start with deletes, which are easier. In fact, the commented-out code provided by the navigation-application template includes the basics of what we need to provide deletion. Start with tableView:canEdit-RowAtIndexPath:. The default implementation (and the default behavior, if this UITableViewDataSource method is not implemented at all) is to not permit editing of any row. Uncomment the default implementation, and change it to return YES;.

---

3. We've reformatted the default code to fit the layout of this book.

To implement the delete, we need to implement tableView:commitEditing-Style:forRowAtIndexPath:. The commented-out implementation has an **if-then** block for handling cases where the editing style is UITableViewCellEditingStyleDelete and UITableViewCellEditingStyleInsert. We need to support the latter only. To perform a delete, we need to do two things: remove the indicated object from the moviesArray model, and then refresh the on-screen UITableView. For the former, UITableView provides the method deleteRowsAtIndexPaths:withRowAnimation:, which is exactly what we need. Add the highlighted line to the default implementation, as shown here, and delete the **else** block for UITableViewCellEditingStyleInsert:

TableViews/MovieTable01/Classes/RootViewController.m

```objc
- (void)tableView:(UITableView *)tableView
            commitEditingStyle: (UITableViewCellEditingStyle)editingStyle
            forRowAtIndexPath:(NSIndexPath *)indexPath {
    if (editingStyle == UITableViewCellEditingStyleDelete) {
        // Delete the row from the data source.
        [moviesArray removeObjectAtIndex: indexPath.row];
        [tableView deleteRowsAtIndexPaths:
                       [NSArray arrayWithObject:indexPath]
                   withRowAnimation:UITableViewRowAnimationFade];
    }
}
```

This gives us swipe-to-delete behavior, but some users don't even know it exists. Fortunately, since we're a navigation app, we have a navigation bar at the top of the screen that is well suited to hosting an Edit button. As in other apps, its default behavior when active is to add an "unlock to delete" button to the left side of every table row that allows editing, which brings up the right-side Delete button when tapped.

In the viewDidLoad method you uncommented, you might have noticed the following commented-out code:

TableViews/MovieTable01/Classes/RootViewController.m

```objc
    // Uncomment the following line to display an Edit button in the
    // navigation bar for this view controller.
// self.navigationItem.rightBarButtonItem = self.editButtonItem;
```

You might recall from Section 5.2, *Setting Up Table-Based Navigation*, on page 77 that in MainView.xib, the RootViewController came set up with UINavigationItem as a child element. That represents the blue bar above the table, typically used for forward-back navigation and for editing tables. It has two properties for setting buttons in the bar: leftBarButtonItem and rightBarButtonItem. Then, on the right side of this assignment, notice the reference to self.editButtonItem. Every UIViewController supports this editButtonItem property, which returns a UIBarButtonItem that calls the view controller's setEditing:animated: method and toggles its state between Edit and Done.

The commented-out line is almost what we want, but let's put the Edit button on the left, so we can leave the right side for an Add button that we'll create later. So, here's the line you'll need in viewDidLoad:

TableViews/MovieTable01/Classes/RootViewController.m
```
self.navigationItem.leftBarButtonItem = self.editButtonItem;
```

Once you Build and Go, you should now be able to tap the Edit button and bring up the unlock-to-delete button for all the rows. In Figure 5.5, on the next page, we can see the table in editing mode (with some more sample data to fill out its rows).

5.6 Navigating with Tables

Our next task is to allow the user to add a table row. In the previous chapter, we developed a MovieEditorViewController, and that's perfectly well suited to entering the fields of a new Movie object or editing an existing one. And once created, it would be simple enough to add the new Movie object to the model and update the table.

So, where do we put the editor? In the previous chapter, we used the UIViewController method presentModalViewController:animated: to slide in the editor. In this case, we're going to learn something new: how to use the navigation objects at our disposal. We created the project as a navigation-based application in part because it gave us a good starting point for our table, and navigation also turns out to be a good idiom for switching between our viewing and editing tasks.

Navigation on the iPhone uses a "drill-down" metaphor that you are probably familiar with from the Mail, iPod/Music, and Settings applications. In the SDK, this is managed by a UINavigationController, which maintains the navigation state as a stack of view controllers. Every time you drill down, you push a new UIViewController onto the stack. When you go back, you pop the current view controller off the stack, returning to the previous view. The navigation is handled in code, independent of how it's represented on-screen: whether you navigate by tapping rows in a table or buttons in the navigation bar, the underlying stack management is the same.

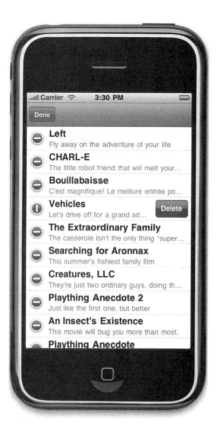

Figure 5.5: USING THE DEFAULT EDITBUTTONITEM TO DELETE ROWS FROM A UITABLEVIEW

Adding the MovieEditorViewController

To try this, let's get to the MovieEditorViewController by means of the navigation API. In fact, we'll use it for two purposes: to edit items already in the table and to create new items.

As with the Movie class, you'll need to copy the MovieEditorViewController.h and MovieEditorViewController.m files to your project's Classes folder and then add those copies to the Xcode project. Also copy over the MovieEditorViewController.xib (with Add > Existing Files as before) to the project's Resources group. In the earlier examples, this editor view was presented modally and took up the whole screen. In this application, it's part of the navigation, and therefore the navigation bar will take up some space above the view. Fortunately, Interface Builder lets us simulate a navigation bar to make sure everything still fits in the view. Open the nib in IB, select the view, and bring up its Property inspector (⌘ 1). Under Simulated Interface Elements, set Top Bar to Navigation Bar to see how the view will look as part of the navigation. In this case, the Done button won't be pushed off-screen, but you might want to adjust its position to get it inside IB's dashed margin.

To bring up the movie editor, our RootViewController needs to push an instance of the MovieEditorViewController on to the navigation stack. We could create the view controller in code, but since we only ever need one instance, it makes sense to create it in Interface Builder. The first step, then, is to create an IBOutlet in RootViewController.h. Add an instance variable MovieEditorViewController* movieEditor; inside the **@interface**'s curly-brace block, and then declare the property as an outlet after the close brace:

TableViews/MovieTable01/Classes/RootViewController.h

@property (nonatomic, retain) IBOutlet MovieEditorViewController *movieEditor;

As usual, you'll need to **@synthesize** this property in the .m file. Also, remember to put #import "MovieEditorViewController.h" in the header.

Now you're ready to create an instance of MovieEditorViewController in Interface Builder. Open RootViewController.xib with IB, and drag a UIViewController from the Library into the nib document window. Select this view controller, and use the Identity inspector (⌘ 4) to set its class to MovieEditorViewController. The last step is to connect this object to the outlet you just created. Ctrl+click or right-click File's Owner (or show its Connections inspector with ⌘ 2), and drag a connection from movieEditor to the view controller object you just created. We're done with IB for now, so save the file.

Editing an Existing Table Item

Let's start by using the MovieEditorViewController to edit an item in the table. When the user selects a row, we'll navigate to the editor and load the current state of the selected Movie object into the editor.

The first thing we need to do is to react to the selection event. The UITableViewDelegate gets this event in the delegate method tableView:didSelectRowAtIndexPath:. The navigation-application template provides a commented-out version of this method in RootViewController, though its sample code creates a new view controller programatically. We don't need to do that, since we already have the next view controller. It's the movieEditor that we just set up in Interface Builder. So, we just need to set up that view controller and navigate to it.

Declare an instance variable of type Movie* named editingMovie in the header file. It remembers which Movie object is being edited, so we'll know what to update in the table when we navigate to the table. Once you've done that, the steps here are pretty simple. Remember what movie we're editing, tell the MovieEditorViewController what movie it's editing, and navigate to that view controller with the UINavigationController's pushViewController:animated: method.

TableViews/MovieTable01/Classes/RootViewController.m
```
- (void)tableView:(UITableView *)tableView
            didSelectRowAtIndexPath:(NSIndexPath *)indexPath {
    editingMovie = [moviesArray objectAtIndex:indexPath.row];
    movieEditor.movie = editingMovie;
    [self.navigationController pushViewController:movieEditor animated:YES];
}
```

What's interesting about the last step is how we get a reference to the navigation controller... remember, we haven't defined an ivar or property for it; in fact, the navigation controller was created for us in MainWindow.xib, and we haven't touched it with IB. The neat trick is the navigationController property defined by the UIViewController class and therefore inherited by RootViewController. This property (also callable as an instance method) looks through the object hierarchy to find a parent or ancestor object that is a UINavigationController. Thanks to this method, you never need to explicitly make connections to your navigation controller. Your root view controller and any view controllers it pushes onto the navigation stack can get to the navigation controller with this property, using it to navigate forward or back or to update the on-screen navigation bar.

This is all we need to do to the movie editor view; now we need a way to get back from the editor to the root. MovieEditorViewController has a done method that's connected in IB to the Done button,[4] but its implementation needs to be updated. Instead of dismissing itself as a modal view controller, it needs to navigate back to the previous view controller:

TableViews/MovieTable01/Classes/MovieEditorViewController.m
```
- (IBAction)done {
    [self.navigationController popViewControllerAnimated:YES];
}
```

---

4. If we didn't already have a Done button in the view, it would be more typical to set up a Done or Back button in the navigation bar. The navigation in the example in Chapter 8, *File I/O*, on page 129 will work like this.

As you can see, the MovieEditorViewController also can use the inherited navigationController property to get the UINavigationController.

This will navigate to and from the movie editor; the only task left to attend to is to update the table when we return from an edit. The RootViewController will get the viewWillAppear: callback when we navigate back to it, so we can use that as a signal to update the table view:

TableViews/MovieTable01/Classes/RootViewController.m
```
- (void)viewWillAppear:(BOOL)animated {
    [super viewWillAppear:animated];
        // update table view if a movie was edited
        if (editingMovie) {
            NSIndexPath *updatedPath = [NSIndexPath
                    indexPathForRow: [moviesArray indexOfObject: editingMovie]
                    inSection: 0];
            NSArray *updatedPaths = [NSArray arrayWithObject:updatedPath];
            [self.tableView reloadRowsAtIndexPaths:updatedPaths
                            withRowAnimation:NO];
            editingMovie = nil;
        }
}
```

We gate our update logic with a check to see whether a movie is being edited, since this method will also be called at other times (at startup, for example). If we are returning from an edit, we need to identify the one table row being updated. We can figure this out by getting the array index that corresponds to editingMovie, constructing an NSIndexPath that goes to that row in section 0 of the table, and pass the path to the table view's reloadRowsAtIndexPaths:withAnimation: method.

Adding an Item to the Table

Another thing we'd like to support is the ability to add new items to the table. We can actually make this a special case of editing. When the user taps an Add button, we quietly add an empty Movie to the table model, insert a table row, and navigate to the editor.

Previously, we used the navigation bar's leftBarButtonItem for the provided editButtonItem, so let's put the Add button on the right side of the navigation bar. We don't inherit an Add button from UIViewController like we did with the Edit button, so we'll create one ourselves.

First, go to RootViewController.h, and set up an IBAction to handle an event from the button we're about to create:

`TableViews/MovieTable01/Classes/RootViewController.h`

```
-(IBAction) handleAddTapped;
```

Now, since we need to work with the navigation objects that Xcode created for us, we'll use Interface Builder to open the MainWindow.xib file, where they're defined. Switch the view mode in the nib document window to list or column view, and double-click the Navigation Controller object. This will bring up a window with the navigation bar at the top and a view placeholder at the bottom that says it's loaded from RootViewController. You'll notice that the Edit button is absent from the left side of the navigation bar, because we add it with code at runtime.

Go to the Library, and find the icon for the Bar Button Item. This is different from the usual Round Rect Button, so make sure the object you've found lists its class as UIBarButtonItem. Drag the bar button to the right side of the navigation bar, where you'll find it automatically finds its way to a highlighted landing spot, making it the navigation bar's rightBarButtonItem. Select the bar button, bring up the Attributes inspector (⌘ 1), and change its identifier to Add. This will change its appearance to a simple plus sign (+).

The next step is to connect this button to the handleAddTapped method. This is a little different from the connections you've made thus far. First, when you bring up the button's Connections inspector (⌘ 2), you won't see the usual battery of touch events like Touch Up Inside. Instead, there's a single Sent Action called selector. This is because the UIBarButtonItem has a different object hierarchy than regular buttons and doesn't have UIControl, UIView, and UIResponder as superclasses. Instead, this object has properties called target and selector; when the bar button is tapped, the method named by selector is called on the target object. You could set both of those properties in code; since we're already here in Interface Builder, let's set it up here.

To set the selector and target, we drag the selector action from the Connections inspector to one of the other objects in the nib. This time, however, we *don't* drag it to the File's Owner. Since this is the MainWindow.xib, the File's Owner proxy object points to a generic UIApplication. The handleAddTapped method that we want the button to call is defined in the RootViewController class, so we drag the connection to the Root View Controller object in the nib window, as shown in Figure 5.6, on the next page. When you release the mouse button at the end of the

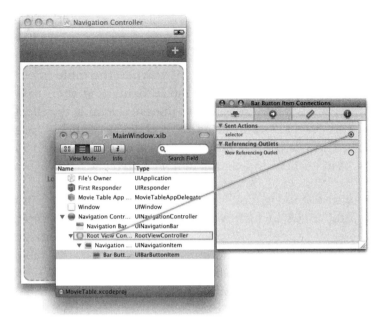

Figure 5.6: CONNECTING A UIBARBUTTONITEM'S SELECTOR TO THE ROOTVIEW-CONTROLLER

drag, the names of the target's IBAction methods will appear, and you'll select the only one: handleAddTapped.

With the connection made, save in IB and return to Xcode. Now we can implement the handleAddTapped method that will be called when the user taps the Add button:

TableViews/MovieTable01/Classes/RootViewController.m

```
-(IBAction) handleAddTapped {
    Movie *newMovie = [[Movie alloc] init];
    editingMovie = newMovie;
    movieEditor.movie = editingMovie;
    [self.navigationController pushViewController:movieEditor animated:YES];
    // update UITableView (in background) with new member
    [moviesArray addObject: newMovie];
    NSIndexPath *newMoviePath =
            [NSIndexPath indexPathForRow: [moviesArray count]-1 inSection:0];
    NSArray *newMoviePaths = [NSArray arrayWithObject:newMoviePath];
    [self.tableView insertRowsAtIndexPaths:newMoviePaths withRowAnimation:NO];
    [newMovie release];
}
```

This method starts by creating an empty Movie object, setting it as the editingMovie, and navigating to the MovieEditorViewController, much like the code to edit an existing Movie did. What's different is that after navigating, it does cleanup work on the table view (while the table is out of sight) by adding the new object to the model array and then calling insertRowsAtIndexPaths:withRowAnimation: to update the table to reflect the new state of the model. The inserted Movie has blank fields, but when the user returns from the editor, the object will be updated in viewWillAppear:, just like when an existing item is edited.

Let's review. We used the navigation-application template to set up an application with a table view, which we backed with a model (a simple NSMutableArray) to provide a list of Movie objects. After looking at the various table cell styles, we added the ability to delete from the table either with horizontal swipes (by implementing tableView:canEditRowAtIndexPath:), or with the Edit button (by adding the default editButtonItem and implementing tableView:commitEditingStyle:forRowAtIndexPath:). Then we looked at how to access the UINavigationControl to navigate between view controllers and used the MovieEditorViewController to edit a Movie indicated by a selected row in the table and then to edit a new Movie in response to the tap of an Add bar button.

5.7 Custom Table View Cells

Back in Section 5.4, *Cell Styles*, on page 81, we looked at the four cell styles provided by iPhone OS. Although they suit a wide range of uses, sometimes you might want something else. If your GUI uses a unique color theme, the default black or blue text on white cells might not suit you. If you need to populate more than two labels, then none of the available styles will work for you.

It is possible, with a little work, to custom design your own table cell in Interface Builder and have your table use this design instead of the built-in styles. In this section, we'll use this technique to create a table that shows all three of the Movie fields.[5]

---

5. Because we will change so much in the project to use custom table cells, the book's downloadable code examples have split this exercise into a separate project. The previous material is represented by MovieTable01, and the custom-cell project is MovieTable02.

Designing a Custom Table Cell

Every UITableViewCell has a contentView, so it's possible to programmatically create subviews and add them to this view; some Apple sample code does this. The problem is that you then have to customize the location, font, size, and other properties of each subview with code, without a visual editor. The second approach is to create a UITableViewCell in a nib file, add the subviews visually, and load that nib when the table needs a new cell. This is what we'll do.

In Xcode, select the Resources group and use File > New File to create a new file, choose User Interface from the iPhone OS section, and create an empty nib file called MovieTableCell.xib. Open this file in Interface Builder. The document will contain just the usual two proxy objects: File's Owner and First Responder. From the Library, drag a table view cell into the nib window. Double-click to edit the object, which will open a small window the size of a typical table view cell, with a gray area designated as the *content view*.

The content view is really just an IB visual artifact, a placeholder for the created-at-runtime view that contains all our subviews, so we'll place our UI elements directly on top of it. The Movie class has three fields, so we'll use three labels to put those fields in a single cell, adjusting the font, color, sizing, and layout appropriate to the items' respective importance. Drag three UILabels from the library into the cell, using the positioning handles and the Attributes inspector (⌘1) to customize their location, bounds, color, and font. For the samples in the book's downloadable example code, here's what we used:

- *Movie Title*: Georgia 17-point font, yellow text, left-aligned near the left side of the cell, toward the top

- *Box Office Gross*: Helvetica 17-point font, green text, right-aligned near the right edge

- *Summary*: Helvetica 10-point font, light blue text, along the entire bottom of the cell

Our cell design in Interface Builder is shown in Figure 5.7, on the next page. We used lighter colors because we plan to use a black background for the table, although this makes the colors harder to see against the background of the gray Content View placeholder. You'll also notice that we've put somewhat plausible data in each of the fields to get a sense of how much space each needs and what they'll look like with real data. Save MovieTableCell.xib, then open RootViewController.xib, find the table,

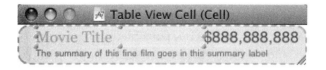

Figure 5.7: DESIGNING A CUSTOM UITABLEVIEWCELL IN INTERFACE BUILDER

and use the Attributes inspector to set its background color to black. We have to do this because parts of the table cell are transparent and we might not have enough cells to fill the table, and we want empty parts of the table to have the same background as populated cells.

Loading and Using a Custom Table Cell

We have a custom table cell, so how do we use it in the table? If you think about it, we really need many table cells. The default behavior of the table is to create a new cell in code each time we try to fail to dequeue a reusable cell from the table's cache. If we're going to use the cell from this nib, then we have to load a new custom cell each time we would have created a cell with code.

There's an interesting trick to how we do this. We can manually load the nib in code and find the cell within the nib. To do this, create an IBOutlet property in RootViewController.h to hold onto a UITableViewCell loaded from the nib.

TableViews/MovieTable02/Classes/RootViewController.h
```
@interface RootViewController : UITableViewController {
            // ... other ivars omitted here for space
            UITableViewCell *nibLoadedCell;
}

@property (nonatomic, retain) IBOutlet MovieEditorViewController *movieEditor;
@property (nonatomic, retain) IBOutlet UITableViewCell *nibLoadedCell;

-(IBAction) handleAddTapped;

@end
```

Now, go back to editing MovieTableCell.xib in Interface Builder. Select File's Owner, bring up its Identity inspector (⌘ 4), and change its class to RootViewController. Having done this, you should be able to switch to the Connections inspector (⌘ 2) and connect the nibLoadedCell outlet to

> **The Secret of File's Owner**
>
> The technique of loading a custom table cell from a nib should also demystify the nature of File's Owner in Interface Builder. In IB, File's Owner is a proxy object that refers to whatever object "owns" the nib file. You can set the class of File's Owner in order to access that class' outlets and actions, but all you're really doing is making an implicit assertion that some object of that class will be the one that owns the nib when it's loaded. Here, you see the other side of that relationship: loadNibNamed:owner:options: loads the nib, specifying an owner object. Any connections to File's Owner get connected to or from this object as part of loading the nib.

the cell object in the nib window. While you're in IB, select the table cell, bring up its Attributes inspector, and change the identifier (the first field) to Cell. You may recognize this as the "reuse identifier" string we used in Section 5.4, *Cell Reuse*, on page 84.

Now for the surprising part. In RootViewController.m, go to the tableView:cellForRowAtIndexPath: method, and rewrite the if (cell==nil) block as follows:

TableViews/MovieTable02/Classes/RootViewController.m
```
Line 1  if (cell == nil) {
     2          [[NSBundle mainBundle] loadNibNamed:@"MovieTableCell"
     3                  owner:self options:NULL];
     4          cell = nibLoadedCell;
     5  }
```

This eliminates the programmatic creation of the table cell, but the means by which cell gets assigned is not necessarily obvious, because the most important step is implicit. On line 2 is where we load the MovieTableCell nib. This returns an NSArray of the nib contents, which we could iterate over to find the table cell object. But we don't have to, because we declared an outlet from that cell to the nibLoadedCell property. The outlets are connected as a consequence of loading the nib, meaning that when loadNibNamed:owner:options: returns, the nibLoadedCell has a reference to the custom cell loaded from the nib, which we can then assign to the local variable, cell, on line 4.

Assigning Values in a Custom Table Cell

Each time a new cell is needed, loadNibNamed:owner:options: will be called again, creating a new cell object in memory. So, at the end of the **if**, we have a cell (either dequeued from the table or loaded from the nib) that we need to customize with the values of a Movie from the model. But with a custom cell, we can no longer use the textLabel or detailTextLabel properties. Instead, we need a way to access the subviews we added in Interface Builder.

One option would be to create a custom UITableViewCell subclass, declare and connect outlets in that class, and then cast the cell to that class when loaded. The only downside is that there are lots more classes to write, one for every kind of table cell in your application. Here's a somewhat more direct technique. Open the cell in Interface Builder, and select the title label. Open the Attributes inspector, and scroll down to the field marked Tag. The tag is a simple, unique integer to identify one view within a view hierarchy. Use the Attributes inspector to set the title label's tag to 1, the box office gross label to 2, and the summary label to 3.

Now, back in tableView:cellForRowAtIndexPath:, you can customize each label's text by looking up the label with the cell's viewWithTag: method.

TableViews/MovieTable02/Classes/RootViewController.m
```
// Configure the cell.
Movie *aMovie = [moviesArray objectAtIndex:indexPath.row];
UILabel *titleLabel = (UILabel*) [cell viewWithTag:1];
titleLabel.text = aMovie.title;
UILabel *boxOfficeLabel = (UILabel*) [cell viewWithTag:2];
boxOfficeLabel.text =   [NSString stringWithFormat: @"%d",
        [aMovie.boxOfficeGross intValue]];
UILabel *summaryLabel = (UILabel*) [cell viewWithTag:3];
summaryLabel.text =   aMovie.summary;
return cell;
```

And now, we're ready to go. We have a custom cell design in a nib, along with new table code to load and populate that cell. Build and Go to see a table like the one shown in Figure 5.8, on the next page.

5.8 Sorting Table Data

Another common task for developers who use tables is to sort the data in the table. Fortunately, Cocoa and Objective-C give us some unique advantages and make this an enviably easy task.

Figure 5.8: A UITableView with custom-designed cells

To add sortability, we'll start by adding a sorting control to our user interface.[6] Open MainWindow.xib in Interface Builder, and double-click the Navigation Controller object to bring up its view. Drag a segmented control from the Library to the center of the navigation bar. The UISegmentedControl is a useful control that allows the user to select one of a small number of preset values. Even though it automatically adjusts its size for the limited space of the navigation bar, this one won't have room for many options, so let's just use three. Select the segmented control, and open the Attributes inspector. Set the number of segments to 3, using the Title field to set the titles of the segments to A-Z, Z-A, and $ (or whatever monetary symbol makes the most sense for your locale).

---

6. Once again, we're making enough changes to merit a separate project in the book's downloadable code. The sorting version of the project is MovieTable03.

We'll need to access this segmented control from code, so we'll need an outlet to it. In RootViewController.h, declare the instance variable UISegmentedControl* sortControl; and set up a property for it with the usual **@property** and **@synthesize** statements, as with the other properties you've already created. You'll also need to declare this method to handle the event when the user taps the sort control:

TableViews/MovieTable03/Classes/RootViewController.h
```
-(IBAction) handleSortChanged;
```

In IB, still in MainWindow.xib, you should now be able to connect the Root View Controller object's sortControl outlet to the segmented control, as well as connect the segmented control's Value Changed event to the Root View Controller's handleSortChanged method.

We'll need to sort the array both in response to the user changing the segmented control and to other causes: adding or editing an item will require a re-sort, plus we'll need to sort the array when the application first starts up. So, let's plan on writing a sortMoviesArray method, which we'll get to in a minute. We can now implement handleSortChanged pretty trivially:

TableViews/MovieTable03/Classes/RootViewController.m
```
-(IBAction) handleSortChanged {
        [self sortMoviesArray];
        [self.tableView reloadData];
}
```

Whenever the sort type changes, we sort the array and then tell the UITableView to reload all its data. This could be expensive, but a sort may well change every row of the table, making the update of individual rows impractical. We also need to add these two lines of code to the bottom of viewWillAppear: so that we re-sort and update the table when the application starts up, when an item is edited, and when an item is added.

How do we perform the sort? It's actually pretty easy. The sortMoviesArray method needs to appear in the implementation before any call to it (or else you can put the method signature in the header file, although that exposes it publicly). To perform the sort, we'll rely on the fact that the NSArray provides a number of methods to return a sorted copy of an array, and NSMutableArray offers these as methods to sort the mutable array itself. Some of these take function pointers or Objective-C selectors, allowing you to write a custom sorting function. But the easiest option is to use *sort descriptors*.

The NSSortDescriptor is a class that describes a sorting criteria, consisting simply of a key and a **BOOL** to indicate whether the sort is ascending. The way this works is to use Key-Value Coding to access the field to sort by. The key is a string that defines a *key path*, which is a dot-separated path of "getable" properties of an object. For each step of the path, the path segment is retrieved by attempting to access a property, accessor method, or instance variable with the key name. The sort descriptor then uses a default selector defined for many Cocoa objects, compare:, to actually perform the sort.[7]

Our Movie objects are very simple, having three properties. To sort alphabetically by the title, we just create a sort descriptor whose key is title, the name of the property.

With that description in mind, look how simple it is to implement all three of our sorts:

TableViews/MovieTable03/Classes/RootViewController.m

```
-(void) sortMoviesArray {
    NSSortDescriptor *sorter;
    switch (sortControl.selectedSegmentIndex) {
        case 0: // sort alpha ascending
            sorter = [[NSSortDescriptor alloc]
                    initWithKey:@"title" ascending:YES];
            break;
        case 1:  // sort alpha descending
            sorter = [[NSSortDescriptor alloc]
                    initWithKey:@"title" ascending:NO];
            break;
        case 2:
        default: // sort $$ ascending
            sorter = [[NSSortDescriptor alloc]
                    initWithKey:@"boxOfficeGross" ascending:YES];
            break;
    }
    NSArray *sortDescriptors = [NSArray arrayWithObject: sorter];
    [moviesArray sortUsingDescriptors:sortDescriptors];
    [sorter release];
}
```

This implementation sets up a single NSSortDescriptor appropriate to the selected sort type and puts it into an array for use by NSMutableArray's sortUsingDescriptors:. The reason this takes an array is that you could

---

7. If your properties were custom classes that didn't respond to compare:, you could change the sorting selector, but you'll usually be sorting Cocoa classes like NSString and NSNumber, which have sensible implementations of compare:.

Figure 5.9: SORTING THE TABLE ALPHABETICALLY BY TITLE

provide multiple descriptors to perform secondary sorts; two objects determined to be equal by the first sort descriptor would then be sorted by the second descriptor in the array, and so on.

With these changes, our sorting behavior is ready to go. In fact, you will see it as soon as you launch the application, since viewWillAppear: makes a call to sortMoviesArray at startup. The new startup view of the table is shown in Figure 5.9.

Chapter 6
Navigation

In this chapter, we'll look at how to manage a hierarchy of data with a navigation controller. It's important that you keep in mind all you learned in Chapter 5, *Table Views*, on page 75 for two reasons—first because you were introduced to an application built from the navigation controller template and second because we will be doing most of our navigation from one table filled with data to another. You don't have to use a table view to use a nav controller; it just makes a lot of sense for us to build on what you know. Let's get started with an example.

6.1 Navigating Through Mail

The best way to get a feel for what you can do with a navigation controller is to take a look at how Apple's Mail app on your iPhone works. At the top of the screen just below the system status bar (where the carrier is and what network you are connected to, and so on), you will find the navigation bar. In the Mail app this is the grayish blue bar just below the white status bar. In the center of the navigation bar you'll find the word "Accounts" if you have multiple email accounts or "Mailboxes" if you don't. We'll assume you have only one account. Below this is a table view of your mailboxes.

Let's drill down a little bit into a mailbox and pay attention to the navigation bar as we switch to the next view. Select your Inbox. The navigation bar will change in three ways. First, there is a new button on the left side that links back to the Mailboxes view we just came from. The title in the center has changed to say "Inbox," and it contains the count of unread messages. And lastly, there is a new Edit button on the right side. If you press it, it puts your table view of messages into edit mode, adding red delete links to the left side of each table cell. The Edit button has changed to a Cancel button. Press Cancel to leave the edit mode.

Select one of your messages. The navigation bar has changed again. The Back button that said "Mailboxes" on the previous view now says "Inbox" and has the count of unread messages. In other words, the Back button displays the title of the view we were just on. Once again, the title of the current view has changed. This time it tells you the number of the current message and the count of all messages. On the right of the navigation bar is a segmented control with one arrow pointing up and another arrow pointing down, which lets you navigate up and down your list of messages.

The Title and Back buttons combine to help users understand where they are in your application's hierarchy. Adding functional buttons like the Edit button makes it easier for your users to perform the most common tasks for that particular view.

6.2 The Navigation Controller

The navigation controller captures all the functionality you need to manage a hierarchy of data. The cool thing is that by using the nav controller, applications can get the navigation from general to specific information, and back, for free.

A nav controller maintains a stack of view controllers. Each view controller represents one node in the hierarchy. As the user drills down into more detailed information, another view controller that manages that detailed information is pushed onto the stack. When the user pops back up from the detailed to the more general information, the current detailed view controller is popped off the stack, and the view controller for the more general information becomes the top view controller again.

Here's a basic diagram of the stack of view controllers for a DVD library application. In the remainder of this chapter, we are going to build out this application so we can see how navigation-based applications are built.

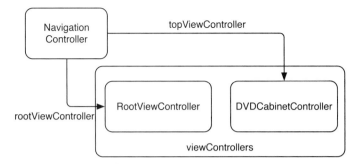

This navigation controller has two view controllers in the viewControllers stack. Each entry is ordered; the one on the left is the most general, and the one on the right is the most specific. The most general in this case is the RootViewController, and the most specific is the DVDCabinetController.

6.3 Navigation-Based Applications

Fire up Xcode if it is not already running, and create a new navigation-based application named DVDCase. Our new project contains a couple of classes and a couple of nib files. We are going to start by examining the MainWindow.xib nib file. Open it by double-clicking the file. This nib file contains the genesis of our application, so let's look at it in a bit of detail.

Change the view mode in the nib document window to show the list mode of objects. You should see something that looks like this:

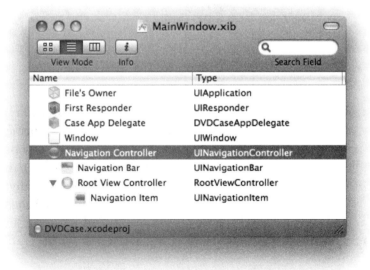

Select and expand the navigation controller as you see in the figure. Notice there are two objects under the nav controller. The first is the navigation bar and is responsible for showing the contextual information on the top of the screen. The second is the root view controller. Let's look at the nav bar first.

If you double-click the navigation controller, you should see a window that looks like this:

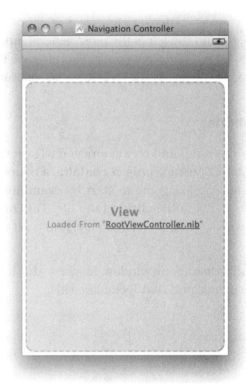

The bar on the top of the window is the navigation bar. This bar is what holds the contextual information that keeps the user informed about where they are in the hierarchy of data. The center of the nav bar displays the title of the currently active view controller. We can set the title in code or in Interface Builder. On the left side, there is typically a Back button, unless we are on the root view controller, in which case the left side is empty. The right side is usually occupied by an action button; often this button is the Edit button and is used to put the table view into edit mode.

The remainder of the window is where the content of the RootViewController.xib will go when the application is running. Before heading to the code, let's look at this nib file. Open it by clicking the blue link in the window. Take a look at the view by double-clicking the object named Table View in the nib document. You should see a picture that looks like this. Notice that this view has the nav bar at the top to let us see what it will look like when we run the application.

Notice also that the table view's dataSource and delegate outlets are set to the File's Owner. Setting these two outlets allows the table view to get its data and react to selections and such. The downloadable code bundle for RootViewController.m has the tableView:numberOfRowsInSection: and tableView:cellForRowAtIndexPath: methods implemented, but they are included here for reference. These methods populate the initial table view with Home and Work rows. For more information on how that works, look at Section 5.3, *Modeling Table Data*, on page 78.

Navigation/DVDCase/Classes/RootViewController.m
```
- (NSInteger)tableView:(UITableView *)tableView
 numberOfRowsInSection:(NSInteger)section {
   return 2;
}
```

```objc
- (UITableViewCell *)tableView:(UITableView *)tableView
        cellForRowAtIndexPath:(NSIndexPath *)indexPath {

    static NSString *CellIdentifier = @"Cell";

    UITableViewCell *cell =
            [tableView dequeueReusableCellWithIdentifier:CellIdentifier];
    if (cell == nil) {
            cell = [[[UITableViewCell alloc] initWithStyle:UITableViewStylePlain
                            reuseIdentifier:CellIdentifier]
                autorelease];
    }

    cell.accessoryType = UITableViewCellAccessoryDisclosureIndicator;
    switch (indexPath.row) {
      case 0:
        cell.textLabel.text = @"Home";
        break;
      case 1:
        cell.textLabel.text = @"Work";
        break;
      default:
        break;
    }
    return cell;
}
```

In this code we are setting the cell's accessoryType to UITableViewCellAccessoryDisclosureIndicator. This indicates to the user that selecting this row will cause a new view controller with more detail about the selection to navigate in.

There are three options for the accessory type. In our example we want to use the Disclosure Indicator. The other two options are Detail Disclosure Button and Checkmark. The Detail Disclosure Button shows the blue circle with a chevron in it and should be used to indicate that the user has two options when clicking the cell. Click in the cell to invoke an action, and then click the blue chevron and navigate to the detail. The Checkmark option should be used when the item can be selected, without navigation. And finally, the indicator we are using in this example should be used when clicking the row will take you to the details of the item represented in the cell.

With these two nib files and the table view data code in place, we have the makings of a navigational application. Now that we have seen the configuration of the nib files, let's look at the code that does navigation.

6.4 Pushing View Controllers

Next we need to do the actual navigation. To do that, we need to implement the tableView:didSelectRowAtIndexPath: and push a new view controller onto the navigation controller's stack. Here is the code to do that:

Navigation/DVDCase/Classes/RootViewController.m

```
- (void)tableView:(UITableView *)tableView
didSelectRowAtIndexPath:(NSIndexPath *)indexPath {
  if(0 == indexPath.row) {
    self.cabinetController.key = @"home";
    self.cabinetController.title = @"Home";
  } else {
    self.cabinetController.key = @"work";
    self.cabinetController.title = @"Work";
  }
  [self.navigationController pushViewController:self.cabinetController
                                       animated:YES];
}
```

We have not created the cabinetController yet, but we will shortly. This controller will be responsible for displaying the contents of the cabinet in another table view.

The conditional setting of the view controller's key property is required so that it will know which list of DVDs to display. We will look at the data layout shortly to understand what is going on there. We are also conditionally setting the title of the view controller to either Home or Work.

Apart from getting the cabinetController ready to display the contents of the cabinet, it takes only a single line of code to get a new view controller onto the stack. That's pretty cool—calling the pushViewController:animated: does all the hard work of making a view controller's view active as well as animating its arrival.

In this method, we use the cabinetController property that needs to be added to the RootViewController interface along with an instance variable and a @synthesize statement in the implementation. We will look at this in more detail in Section 6.5, *Customizing the Navigation Bar*, on the next page.

The pushViewController:animated: method is also responsible for managing the navigation bar. As the new view controller is pushed onto the stack, the navigation bar is updated with the new view controller's title, and the Back button's title is replaced with the previous view controller's title. Since the title of the RootViewController is set to *Cases*, that is the title that shows up in the Back button. Here is the code that sets the title:

Navigation/DVDCase/Classes/RootViewController.m
```
- (void)viewDidLoad {
  [super viewDidLoad];
  self.title = @"Cases";
}
```

Next we need to create the cabinetController view controller, but before we do, let's talk a bit more about the navigation bar and how it is customized. It's not just the title that we can customize.

6.5 Customizing the Navigation Bar

We can control what shows up in the navigation bar via the properties of your view controllers.

This picture shows how the stack of view controllers sets the titles of the various pieces of the nav bar. The topViewController's title is placed in the center as the title of the nav bar. The title of the view controller just behind the topViewController in the stack is placed into the Back button.

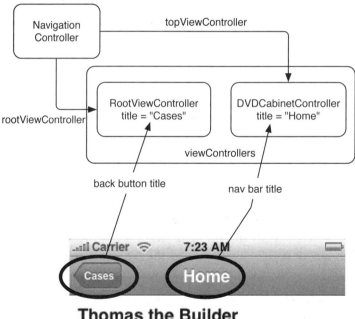

The title in the center of the navigation bar is set to the title of the current view controller. The right button is typically left blank, but it is often set to be an Edit button.

The right button can be replaced with a custom instance of UIBarButtonItem. With a custom instance of the Bar Button Item, you can specify your own behavior by supplying the target and action. You can even go so far as to fully replace the view used to draw the button by setting the customView property.

You have a lot of flexibility in customizing the way the navigation bar works in your application. In our DVDCase application, we customize only by setting the title of the view controllers. In a more sophisticated application, you could fully customize the nav bar. We have not talked about custom views yet, but we will in Chapter 19, *Drawing in Custom Views*, on page 385. Basically, you can replace the UILabel that is used by default to draw the titles with a view of your own that draws anything you want into the title area.

To see one example of the way you'd want to use the ability to customize the nav bar, open the YouTube application, and click the Most Viewed tab. The title view of the nav bar has been replaced with a segmented control. In your app, you can use any UIControl or, as we said earlier, any view, even a custom one of your own design.

Our new view controller will manage the list of DVDs in each cabinet. So, we need to create a new table view controller and its associated user interface. To accomplish this task, we need to do the following things:

1. Create a new subclass of UITableViewController.
2. Create a new nib file that will contain the UI for our new table view controller.
3. Configure the nib file to have a table view, and connect the table view to our new view controller.
4. Add an outlet to the RootViewController so it knows the new table view controller.
5. Update RootViewController.xib to set this outlet.

Right-click the Classes group in Xcode, and choose Add > New File. Choose to make a subclass of UITableViewController (select Cocoa Touch Classes -> Objective-C class and then UITableViewController from the pulldown), click Next, and then name your new controller DVDCabinetController. Now we need to add the new controller outlet to the RootView-

Controller. Here is the updated code in the header. Remember to add the corresponding import and synthesize to the implementation file.

`Navigation/DVDCase/Classes/RootViewController.h`

```
@class DVDCabinetController;

@interface RootViewController : UITableViewController {
  DVDCabinetController *cabinetController;
}

@property (nonatomic, retain) IBOutlet DVDCabinetController *cabinetController;

@end
```

Now that we have an outlet, the next thing we need to do is connect it in IB. Open the RootViewController.xib file, and add a new view controller. Change its class to DVDCabinetController in the Identity inspector (⌘-4). Now make the connection from the File's Owner to the new view controller by Ctrl+dragging from the File's Owner to the new view controller, and choose the cabinetController outlet to make the connection. Set the NIB name (Attributes inspector, ⌘-1) to DVDCabinetController. Save your work.

Now we need to create the DVDCabinetController.xib nib file. In Xcode, Ctrl+click the Resources group, and choose Add > New File. Choose the View XIB template, click Next, and name your new nib file DVDCabinetController.xib. Open this file so we can change it to be the UI we need.

The initial view placed in this file is a generic UIView, but we need a table view. So, delete the View object, and replace it with a UITableView. Next we need to set the class of the File's Owner object to DVDCabinetController. Select the File's Owner, open the Identity inspector (⌘-4), and change the Class. Now we can make the necessary connections. Ctrl+drag from the File's Owner's view outlet to the newly placed Table View object, and choose the view outlet. Now connect the dataSource and delegate from the Table View to the File's Owner. Building any table view–based UI typically involves these steps: creating a new subclass of UITableViewController, creating a new View XIB file, switching the view for a table view, and setting the data source and delegate.

Now that we have our UI set up, we need to go back and fill in the code for our DVDCabinetController. For the most part, this table view controller is like the others we have built. This class is responsible for displaying the list of DVDs contained in the selected case. To show the data, you will need to implement the tableView:numberOfRowsInSection: and table-

View:cellForRowAtIndexPath: methods, as you have in all the table view controllers that you have implemented.

To meet its responsibility, this class needs two instance variables: a marker to know which cabinet was selected and a container to hold the data for each cabinet. We use an NSString object called key to keep track of the marker. We use an NSDictionary to keep track of the data. Remember that we set the key property in the tableView:didSelectRowAtIndexPath: method on RootViewController when the user selects the Home or Work row. The data is purely internal, so we don't need a property for it. Don't forget to @synthesize the key property.

Here is the source for DVDCabinetController header file:

Navigation/DVDCase/Classes/DVDCabinetController.h
```
@interface DVDCabinetController : UITableViewController {
  NSDictionary *data;
  NSString *key;
}

@property(nonatomic, retain) NSString *key;

@end
```

Here we see the declaration of a dictionary to hold our data. We will see the initialization of that data in the viewDidLoad method in just a moment. The second instance variable is of more interest just now. This key, if you recall from the discussion of RootViewController's implementation of tableView:didSelectRowAtIndexPath:, is the way we communicate which of the cabinets was selected. We are going to see how it's used as soon as we look at how our data is arranged. Here is the code for initializing the data object:

Navigation/DVDCase/Classes/DVDCabinetController.m
```
- (void)viewDidLoad {
  [super viewDidLoad];
  NSArray *keys = [NSArray arrayWithObjects:@"home", @"work", nil];
  NSArray *homeDVDs = [NSArray arrayWithObjects:@"Thomas the Builder", nil];
  NSArray *workDVDs = [NSArray arrayWithObjects:@"Intro to Blender", nil];
  NSArray *values = [NSArray arrayWithObjects:homeDVDs, workDVDs, nil];
  data = [[NSDictionary alloc] initWithObjects:values forKeys:keys];
}
```

This code is creating a dictionary keyed on the two values we used in the RootViewController implementation. Although it would be better to use symbolic constants here for the keys, I wanted the code to be very clear. So, if you were doing this in a more sophisticated project, make sure that you used some kind of constant instead of hard-coding the value.

Now that we have the data layout in our minds, let's look at how we implement the tableView:numberOfRowsInSection: method:

Navigation/DVDCase/Classes/DVDCabinetController.m
```
- (NSInteger)tableView:(UITableView *)tableView
 numberOfRowsInSection:(NSInteger)section {
    return [[data valueForKey:self.key] count];
}
```

All we need to do is grab the array for the particular cabinet and return the count of that array. When the user clicks Home in the UI of the first table view, the key is set to home so we get the proper array, and vice versa, when the work key is set. And here is the code for the tableView:cellForRowAtIndexPath::

Navigation/DVDCase/Classes/DVDCabinetController.m
```
- (UITableViewCell *)tableView:(UITableView *)tableView
         cellForRowAtIndexPath:(NSIndexPath *)indexPath {

  static NSString *CellIdentifier = @"Cell";

  UITableViewCell *cell = [tableView
                     dequeueReusableCellWithIdentifier:CellIdentifier];
  if (cell == nil) {
    cell = [[[UITableViewCell alloc] initWithStyle:UITableViewCellStyleDefault
                               reuseIdentifier:CellIdentifier]
           autorelease];
  }

  cell.textLabel.text = [[data valueForKey:self.key] objectAtIndex:indexPath.row];

  return cell;
}
```

Again we grab the proper array from the dictionary of data, and we have our text for the cell.

One last bit of code to make sure our table view is always showing the correct information. Here is the code for the viewWillAppear: method:

Navigation/DVDCase/Classes/DVDCabinetController.m
```
- (void)viewWillAppear:(BOOL)animated {
  [super viewWillAppear:animated];
  [self.tableView reloadData];
}
```

Remember from Section 4.3, *Adding Outlets and Actions to the Controller*, on page 58 that the viewWillAppear: method gets called each time the view that the controller manages is about to become visible. This makes the perfect spot for us to reload the data so that when the user is switching between her home and office cabinets, the list of movies will get updated. Calling reloadData causes the table view to reload its entire contents. As we said before, this can be inefficient, but for this small example, it is no big deal.

This is a very typical pattern used to tie together two view controllers. The parent VC knows about the child VC through a connection made in Interface Builder (IB). The parent sets a value on the child VC, which it uses as a key to find the data it should display. If it turns out that you need two-way communication (from child to parent), you can make that connection in IB as well using the File's Owner.

6.6 Popping View Controllers

Finally, let's look at how to remove the child view controller when the user is done. For our example application we are going to just dismiss the child view controller. Here is the implementation of the tableView:didSelectRowAtIndexPath: in our DVDCabinetController class:

Navigation/DVDCase/Classes/DVDCabinetController.m

```
- (void)tableView:(UITableView *)tableView
didSelectRowAtIndexPath:(NSIndexPath *)indexPath {
  [self.navigationController popViewControllerAnimated:YES];
}
```

Here we call the popViewControllerAnimated: method to dismiss our view controller. Since the view controllers are arranged in a stack, the navigation controller will replace the DVDCabinetController's view with the RootViewController's view, and the user will be brought back to the top of the stack. Build and Go to see your work.

If you are building an application with a deep hierarchy and you want to provide your users with the ability to pop back to the root of the hierarchy at any time, you can do that by invoking the popToRootViewControllerAnimated: method. You can also pop to any of the view controllers in the hierarchy via popToViewController:animated:.

Navigation is a fundamental piece of many iPhone applications because the data in our lives is so often hierarchical. If you are building an application that manages data like that, then you will find the navigation controller a natural fit.

Chapter 7

Tab Bar Controllers

You've used view controllers to help the user get more out of your application. The table view and friends help you display data in lists. The navigation controller helps your users move from generic data to specific data. In this chapter, we will use the tab controller to organize our application around areas of functionality.

7.1 When to Use Tabs

The problem as always is to neatly arrange your user interface so that your users have just what they need when they need it. However, when you have several different interaction modes or different ways to look at the same data, the navigation paradigm just does not work. After all, navigation is supposed to be moving from less specific information to more specific information with each new view. When your user interface does not fit into the navigation paradigm, consider the tab controller.

The tab controller is particularly good at organizing or grouping several different areas of functionality. Each area of functionality has its own tab bar item and is active only when the user clicks that item. Whenever your users interact with information in several different ways, a tab controller can help you organize your user interface to focus on the task the user is trying to accomplish.

Each view controller gets its own tab along the bottom of the screen, and when the user taps on the tab, the view controller associated with that tab becomes active. Contrast this with the navigation controller. When the user taps on an entry in a table, a new view controller is made active and is focused on the more detailed information about that item.

A tab bar controller does not reveal more detailed information when the user taps on a different tab; instead, it reveals different information.

Consider the clock application, which is a great example of each tab containing different information and different functionality. As you click between World Clock and Alarm, you don't expect to see more detailed information; you expect completely different data.

The app we build in this chapter will have six different view controllers that each display our data in different sort orders. You can see a screenshot here that uses five U.S. states as our data:

As the user taps on each different tab bar item, the new sorting is displayed. Let's get started building an application using a tab controller.

If you are having trouble with a section, refer to the earlier material to help bolster that intuition.

7.2 Creating a Tab Bar Controller

The initial example application is going to provide two different ways to sort our state data. Because we have two ways to look at the data, the example will have two view controllers. The first view controller will display the state data sorted by population, and the second will sort the data by area.

Like the rest of the view controllers you have seen, Xcode has a template for tab controller–based applications that will create a bunch of resources for you. In Xcode, create a new project with the Tab Bar Application template, and name it States. Then click the Save button.

We can start to understand this new project by looking at the details of two of the objects created for us by the template. Open the MainWindow.xib file by double-clicking the file in Xcode. The document window should have five objects in it; the one we will be working with initially is called Tab Bar Controller. Select the tab bar controller, and inspect it with the Attributes inspector (⌘-1). You should see something like this:

The Attributes inspector lists the view controllers that the tab controller contains and allows you to edit that list by adding, removing, or reordering the view controllers. This tab controller has two controllers that it is managing titled First and Second. Both of these controllers are UIViewControllers (instead of something more specific like a navigation controller). Soon you are going to replace these two view controllers with your state data–sorting view controllers.

The next object we look at is the States App Delegate. Open the Connections inspector for this object by selecting it and hitting ⌘-[2]. Notice that this object has two outlets, tabBarController and window. Recall that the application delegate gets notification when the application is finished loading the main nib file and is about to start processing events. We use that delegate hook to finalize any initialization and to place our user interface onto the screen. The delegate adds the view defined by the tabBarController onto the window. Here is the code:

`TabBarControllers/States/Classes/StatesAppDelegate.m`
```
- (void)applicationDidFinishLaunching:(UIApplication *)application {
    [window addSubview:tabBarController.view];
    [self createData];
}
```

This code is straightforward; we just ask the tab controller for its view and then place that as a subview of the window. You have seen something similar in each of the view controller examples, but what is interesting here is where the tab controller gets the view to display. Instead of a stack of controllers like the nav controller, the tab controller has a list of controllers and an explicit "selected controller." The view that the tab controller returns is the view for the selectedController. We also create the data for our application with a call to the method createData, which we will talk more about shortly.

Now that we have covered the basics of how a simple tab controller application works, let's start customizing it with our particular view controllers that will sort our data in different ways.

7.3 View Controllers in Tab Controllers

In this example, we need two view controllers—one to present a list of the five most populous states and the other to present the five largest states by acres of land. The data for this example is hard-coded in the application delegate. In a real-world application, you'd build a set of model classes to represent your data and then store that data in a database or file (we prefer a database). For this simple example, though, we will just use NSDictionary objects to hold the data. Here is the code for the California data extracted from the createData method. The remainder of the data for the other states is created in the same way and added to the same data array but not shown.[1]

`TabBarControllers/States/Classes/StatesAppDelegate.m`
```
NSMutableArray *data = [NSMutableArray array];

[data addObject:[NSDictionary dictionaryWithObjectsAndKeys:
            [NSNumber numberWithInt:36553215], @"population",
            @"California", @"name",
            [NSNumber numberWithInt:163770], @"area", nil]];
```

After you create all the test data, you then need to set the states property to the array you have just created. The code should look like this:

`TabBarControllers/States/Classes/StatesAppDelegate.m`
```
self.states = [NSArray arrayWithArray:data];
```

Don't forget to add the states instance variable and property to the StatesAppDelegate interface and synthesize the property inside the implementation of the StatesAppDelegate class.

We create two methods to retrieve this data, one to return it in population order and the second to return it in acre order. The first method is here:

`TabBarControllers/States/Classes/StatesAppDelegate.m`
```
- (NSArray *)statesByPopulation {
  NSRange range = NSMakeRange(0, 5);
  return [self.states subarrayWithRange:range];
}
```

Since our data is created in population order, there is not much to do here except to limit the data to the first five states (from the requirements). The next method to get the data by acre order is a bit more interesting:

`TabBarControllers/States/Classes/StatesAppDelegate.m`
```
- (NSArray *)statesByArea {
  NSSortDescriptor *sorter = [[[NSSortDescriptor alloc]
                      initWithKey:@"area" ascending:NO] autorelease];
  NSArray *sorted = [self.states sortedArrayUsingDescriptors:
                [NSArray arrayWithObject:sorter]];
  NSRange range = NSMakeRange(0, 5);
  return [sorted subarrayWithRange:range];
}
```

---

1. Grab the code from the downloadable bundle, though, because the order is important.

In this method, we have to re-sort the data using an NSSortDescriptor. A sort descriptor lets us specify a property to sort on and whether we want the results ordered ascending or descending. Then we invoke the sortedArrayUsingDescriptors: method on our states array and get back an array that is re-sorted based on the area instead of population. Then we simply subset the array to get the first five elements.

Now that we have our data, the rest of the work is to get that data into the respective table views. The two view controllers that we are going to build have essentially the same code. They differ only by the method used to get the list of states they display. We need to create two classes then; we called ours ByPopulationViewController and ByAreaViewController. Select the Classes group in Xcode, Ctrl+click and choose Add New File > Objective-C Class, and in the "Subclass of" pull-down choose UITableViewController. Then click Next, set the name to ByPopulationViewController, and click Finish. We need to add three methods to this class. The first one will get the state data from the application delegate:

TabBarControllers/States/Classes/ByPopulationViewController.m
```
- (NSArray *)states {
  return [(StatesAppDelegate *)[[UIApplication sharedApplication] delegate]
       statesByPopulation];
}
```

This view controller is going to be showing the data sorted by population, so it invokes the statesByPopulation method. Please keep in mind that we are doing this to keep the example focused on tab controllers. In a more sophisticated example, we'd use something like Core Data (see Chapter 11, *Core Data*, on page 203) to manage the data instead of getting it from the application's delegate.

Next up we need to provide the proper data to the table view (that we will create and configure shortly). Recall from Section 5.3, *Modeling Table Data*, on page 78 that we need to provide the row count for a section and then fill the cells for each cell in the section. The code to do that is here:

TabBarControllers/States/Classes/ByPopulationViewController.m
```
- (NSInteger)tableView:(UITableView *)tableView
 numberOfRowsInSection:(NSInteger)section {
   return self.states.count;
}
```

```
- (UITableViewCell *)tableView:(UITableView *)tableView
         cellForRowAtIndexPath:(NSIndexPath *)indexPath {
  static NSString *CellIdentifier = @"Cell";
  UITableViewCell *cell = [tableView
                          dequeueReusableCellWithIdentifier:CellIdentifier];
  if(nil == cell) {
    cell = [[[UITableViewCell alloc] initWithStyle:UITableViewCellStyleDefault
                                  reuseIdentifier:CellIdentifier] autorelease];
  }
  cell.textLabel.text = [[self.states objectAtIndex:indexPath.row]
                         objectForKey:@"name"];
  return cell;
}
```

The code is straightforward, and we have seen several instances of this before in Chapter 5, *Table Views*, on page 75. The only new bit here is the use of the method objectForKey:, which is just getting the object from the dictionary using the key.

Next we need to configure this view controller in Interface Builder. If you don't still have it open, double-click the MainWindow.xib file in the Xcode project to fire up IB and load the file. Switch to the list mode on the document viewer in IB.

In this view, you can see two view controllers that are part of this tab controller. This view is well suited for manipulating tab controllers because you can select the view controllers that the tab controller contains and you can manipulate them.

For example, select the first view controller, and then hit ⌘-4 to bring up the Identity inspector. Set the class name to ByPopulationViewController, hit ⌘-1 to bring up the Attributes inspector, and change the title to By Population. Now we have built and configured our "by population" view controller.

Repeat these steps for the "by area" view controller, remembering to replace the call to statesByPopulation with a call to statesByArea. Also make the same adjustments in IB to the second view controller. However, with this second controller, its view will be loaded out of another nib file (more for demonstration purposes than out of necessity).

When you are done, you should have something like this:

Before we move on to creating the interface for the "by area" view controller, let's add the table view for the population controller first. In IB delete the view associated with the ByPopulationViewController class by selecting the view and hitting Delete. Next click ⌘-⇧-L to bring up the Library, and choose Data Views from the top list. Then select a table view, and place it as a child of the ByPopulationViewController.[2] Save your work. Now let's move on to making the interface for the ByAreaViewController.

In IB, select the second view controller for the tab controller, and then in the Attributes inspector (⌘-1) set the nib filename to ByAreaView.[3]

Now we need to create the nib file. In Xcode ^+click the Resources group, and choose New File. Choose User Interfaces from the list on the left of the wizard, then choose View XIB from the list on the right, and click Next. Then specify the name as ByAreaView, and click Finish.

Open your new nib file, and change the class of the File's Owner object to ByAreaViewController. Next we need to do the same thing we did for the

---

2. Placing the table view as a child of the ByPopulationViewController automatically makes this table view the view for that controller, so we don't have to do any connections.
3. The template project creates a nib file called SecondView.xib. We could rename this file and then reuse it, but instead we are going to ignore it. You can delete it at this point because the project won't need it after we create the new nib file.

ByPopulationViewController. First delete the existing view, and then add a table view to the nib file. Once the new table view is added, connect the ByAreaViewController's view outlet to the table view. Next you need to connect the table view's dataSource and delegate outlets to the File's Owner.[4] Save your work.

Now your project is very nearly finished. Back in MainWindow.xib, we need to set the title for both of the view controllers. This title is where the tab controller gets the title for the buttons. So, select the tab controller, hit ⌘-1, and then double-click each title in the list of view controllers for this tab controller. Change their titles to By Population and By Area appropriately.

Run the application now, and you should see the first tab button selected and the list of states ordered from largest population at the top to smaller populations at the bottom. Click the By Area list, and you should see the five largest states in terms of area.

This application is fairly simple and has only two tabs. In many cases, though, you might want to have more tabs. The good news is that the UITabBarController makes it super easy to handle this case. Let's look at that next.

7.4 Many Controllers

One of the great things about the tab controller is how much work it does for us automatically. If you give it more than five controllers, it automatically puts a button called More into the tab. When the user clicks the More button, the tab controller displays a navigation controller that lists the remainder of the controllers along with an Edit button. If the user clicks the Edit button, they are allowed to rearrange the list of controllers so that their favorites show up in the main bar and the others are in the navigation controller.

Tab controllers default to doing all this stuff for you. When you add your list of view controllers (either in Interface Builder or via code), the tab controller of course knows when there are five or more and thus when to add this new functionality. The tab controller also allows us to control much of what it does. First let's look at controlling which view controllers are able to be reordered.

---

4. We have to perform the connections here because we are editing the view in a separate nib file.

The customizableViewControllers property is where the tab controller looks to see what can be reordered. If a view controller is not in this list, then the tab controller won't allow its location in the list to be changed. By default, all the view controllers the tab controller is managing will be in this list, so if you don't want some of them moved around, you will need to remove them. Let's look at an example to make these concepts clearer.

In the previous section, we had only two controllers that displayed a list of five states in two different orders. In this example, we expand on these listings to provide six different orders that we can view the states in: ordered alphabetically by name, ordered by population, or ordered by area. Each order has an ascending or descending controller for a total of six. Now, in the real world, you wouldn't make a UI like this, but this example provides a minimal addition to the previous example that will teach you about the tab controller without us getting bogged down in UI design.

You saw a screenshot of the states application running in the simulator at the beginning of this chapter. The items along the bottom are in the order defined in Interface Builder.

The code for all this is very similar to what we saw in the previous example. One method was added for each of the four additional sorting controllers, and each of the controllers invokes its respective methods to get the state data. Here is an example of one of the sorting methods:

TabBarControllers/StatesMore/Classes/StatesAppDelegate.m

```
- (NSArray *)statesAscendingByName {
    NSSortDescriptor *sorter = [[[NSSortDescriptor alloc]
                                 initWithKey:@"name" ascending:YES] autorelease];
    NSArray *sorted = [self.states sortedArrayUsingDescriptors:
                       [NSArray arrayWithObject:sorter]];
    NSRange range = NSMakeRange(0, 5);
    return [sorted subarrayWithRange:range];
}
```

This code is very similar to what we saw earlier. It is invoked by the ByNameAscendingViewController controller. The code for that controller is again very similar to the controllers from the previous example, so we won't go into any more detail here.

The interesting part of this is the way the tab bar controller automatically manages the extra controllers that won't fit onto the tab bar. If you have more than five controllers, the tab controller places the first four and replaces whatever would have been in the fifth place with the More button. Here's the configuration for our example:

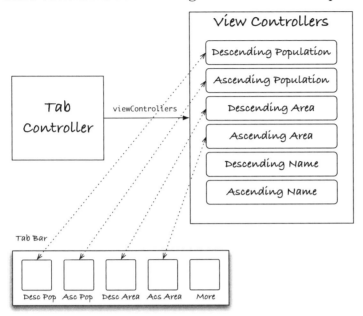

The way the tab controller does this magic is by implementing the UITabBarDelegate protocol and becoming the delegate of the tab bar. There are several methods in this protocol, but the big picture is that the tab bar lets its delegate decide what to do when the More button is pressed. The tab controller decides to put up a navigation controller with the list of additional view controllers in it. The cool thing about this, though, is that the magic is not really magic; you can do the same thing by writing your own implementation of the UITabBarDelegate, or you could do something completely different. For example, you could display an icon for each of the view controllers instead of their titles.

The same basic principles apply for editing the order of the view controllers. The tab bar's delegate decides what to do, and since the tab controller is the default delegate, it displays a modal view with the icons and titles for each of the controllers. But again, you are free to replace that with whatever you'd like in your implementation.

Congratulations. You have been through all the base view controllers that come with Cocoa Touch. You now have in your tool belt all the tools you need to make just about any type of user interface you need on the iPhone. Next up we will start to see some of the lower-level APIs to interact with the iPhone's filesystem.

Chapter 8
File I/O

As an iPhone application developer, you're entrusted by the user with the care of their data. As a result, you'll frequently need to save (or "persist," if you must) their data to long-term storage. After all, any time your application delegate gets the applicationWillTerminate: message, you may need to quickly save program state before your application ends, and you'll also want to be able to restore your state the next time you come up. Beyond that, your application is also responsible for maintaining and honoring the user's preferences. And in some cases, your app may be well served by keeping its information in a database for speedy search and retrieval of needed data.

In this chapter, we'll look at the filesystem. This is the most basic and fundamental system for long-term data storage. It will often be your first choice for data storage, but keep in mind there are some other systems that you may choose instead. Your application may use the system's facilities for saving and loading preferences (covered in Chapter 9, *Preferences*, on page 163), or you might opt for the performance of a relational database, either via the SQLite3 API (covered in Chapter 10, *The SQLite Database*, on page 185) or Core Data (covered in Chapter 11, *Core Data*, on page 203). Two other APIs facilitate long-term storage, but they are different because they share data with other applications: the image APIs (covered in Chapter 20, *Drawing Images and Photos*, on page 397) and the Address Book API (covered in Chapter 23, *Address Book*, on page 441).

iPhone OS sits atop a filesystem that's more or less identical to what you'd find on a Mac. Files and directories exist in the same form you'd

Figure 8.1: NAVIGATION FLOW OF FILESYSTEMEXPLORER SAMPLE APPLICATION

expect, with names, hierarchy, POSIX-style attributes, and so on.[1] As you'll see in this chapter, you have access to the filesystem equivalent to what you'd have in the Mac version of Cocoa: you can navigate through the directory hierarchies, read from and write to files, create and delete directories, get attributes, and more.

The one catch is that you can't see the *whole* filesystem, just the contents of your application's own home directory. This is part of iPhone OS's *sandbox* security model. Your application can do as it pleases within its own reserved portion of the filesystem (its *sandbox*) but cannot access any other part of the filesystem. This means that other applications' code and data, the user's music and videos, the system data, and everything else outside your home folder is off-limits, and attempts to read from or write to these parts of the filesystem will fail.

8.1 Exploring Your Filesystem

To explore common file I/O activities, let's develop an application to browse the filesystem, starting in your home directory. This Filesystem-Explorer application will need to support the following tasks:

- List the contents of a directory in a UITableView
- Navigate to items of that directory by showing either:
    - Another UITableView of a selected subdirectory's contents
    - An overview of a selected file's contents

---

1. One significant difference: the iPhone filesystem is case sensitive, not just case observant like the Mac.

- Read and display the contents of a file
- Create a new directory with a user-supplied name
- Create a new file with user-supplied name and contents
- Delete a selected file

A preview of the sample application is shown in Figure 8.1, on the facing page. As you navigate through directories, you see their contents in table views. When you select a file, you see an overview of the file, and from there you can read its contents into another view. We'll develop a navigation-based application that moves between the following view controllers (which have corresponding IB-developed views):

- DirectoryViewController: Shows contents of a directory as a table. Selecting a row takes you to either another DirectoryViewController (if selection is a directory) or an overview of a flat file. It also has an Add button for adding new files and directories and deletes files or directories with the horizontal swipe gesture.
- FileOverviewViewController: Displays metadata about a file: size, dates, and so on.
- FileContentsViewController.: Reads contents of a file into a UITextView, a graphic component for presenting (and optionally editing) multiple lines of scrollable text.
- CreateDirectoryViewController: Displays a UITextField prompt for a new directory name.
- CreateFileViewController: Offers a text field for a new filename and a UITextView for its contents.

About Your Application's Directories

Before we start navigating into files and subdirectories, take a look at the home directory. This set of directories will exist in every iPhone OS application you create. Four directories are available to you, though it's highly likely you'll never look into two of them.

- Documents: This is the primary directory for you to store your user's flat files. It is empty by default, and you're free to make whatever use of it you like.
- *Application-Name*: This directory is your application bundle, containing nibs, localizations, executable code, and other resources. You can see FilesystemExplorer showing its own application bundle in Figure 8.2, on the next page. In all likelihood, you won't do a lot of direct file I/O in this directory. Delete a .nib file while your

Figure 8.2: DISPLAYING THE CONTENTS OF AN APPLICATION'S BUNDLE

app is running in the simulator, and you're probably heading for a crash. On the actual device, you won't even be permitted to try. On the other hand, when you retrieve images, sounds, and other resources from your application bundle, the paths will point into this directory.

- Library: This directory exists solely as a parent to its Preferences directory. You don't need to write files to Preferences directly; you'll be using the preferences APIs, like NSUserDefaults, instead. Preferences are discussed in more detail in Chapter 9, *Preferences*, on page 163.
- tmp—The "temporary" directory is meant for short-lived, temporary files that have no long-term relevance. You can use it as your application's "scratch pad" if you need to, say, write out a file incrementally before uploading it over the network. If you use this directory, you should delete its contents on a regular basis (such as at startup or shutdown) in order to conserve space. Note that iTunes' backup of your device will ignore anything in tmp.

8.2 Creating Our Project

As you might expect from the flow diagram, you should create a new Xcode project named FilesystemExplorer using the navigation-based application template, as introduced in Chapter 6, *Navigation*, on page 103. For this application, we will be applying some pretty serious customization.

Because this is the third time we've seen a navigation-based application, we hope you're getting the hang of it. With this chapter, we start to move away from UIKit and the Xcode tools and more into the rest of the APIs. So that we can focus on new material, we'll increasingly take some things for granted and assume you know the following:

- To create a property, you need a backing instance variable, a **@property** declaration in the .h file, and (usually) a **@synthesize** in the .m file.
- When you refer to another class, you need to **import** its .h.
- You connect IBOutlets and IBActions with Interface Builder to wire your GUI to your code.
- You implement a protocol by putting its name in angle braces (<...>) as part of your **@interface** declaration and then implementing its defined methods. You do this a lot for delegates.

Ideally, you're starting to think conceptually about the parts of your app. As you build more apps, you'll find you worry less about Ctrl+ dragging this to that and more about how you're expressing relationships between the objects you declare in your code and those that you create in IB.

Refactoring Your Code

Let's start with the default table view. The class that's generated for us is called RootViewController. In this application, that's not a very descriptive view controller name. Since we'll not only start with a given directory but be able to drill down repeatedly into other directories, this isn't so much a "root" view as a "directory" view. This gives us an opportunity to exercise Xcode's refactoring abilities as promised.

To refactor something in your code—a class, method, or member—you just double-click its name in any source file and select Refactor either from the Edit menu or by right-clicking the name to bring up a contextual menu. So, open RootViewController.h, select the RootViewController class name from the **@interface** declaration, and select Edit > Refactor. The refactoring window that comes up lets you type in a new name and

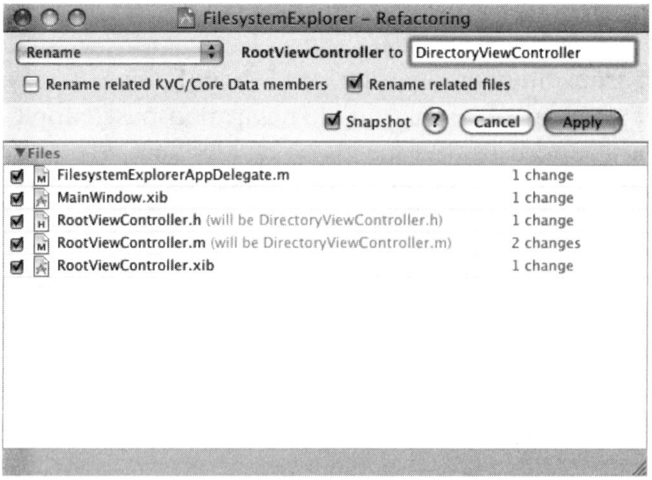

Figure 8.3: REFACTORING THE ROOTVIEWCONTROLLER CLASS

then preview all the code (and possibly filenames) that will be changed. In Figure 8.3, we can see what happens when we prepare to refactor RootViewController into DirectoryViewController.[2]

Applying this changes the source files (and renames them), but the RootViewController.xib file that contains the view still has its old name. If you want to change that too—and you don't have to, this is just so you'll know how—then you have to make changes in a few different places. You can begin by selecting RootViewController.xib in Resources folder of the Groups & Files list and selecting Rename from the right-click menu. That will rename the file, but it will also break your application.

To understand why, double-click MainWindow.xib to launch Interface Builder. If you open the navigation controller object, you'll see its view: a gray navigation bar at the top and a large dash space that says "View: Loaded from RootViewController.xib." And this would be *bad*, because we just renamed RootViewController.xib, but apparently MainWindow.xib didn't get the memo. Fortunately, the fix is pretty simple. In the navigation controller preview window, click the view, and open the Attributes inspector (⌘ 1). As you can see in Figure 8.4, on the next page, all you need to do is reset the nib name to your renamed view controller.

---

2. Be sure to check your imports too; in updating the book for iPhone 3.0, we found that

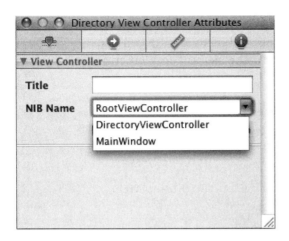

Figure 8.4: REFACTORING THE MAINWINDOW.XIB FILE TO USE A DIFFERENT NIB FOR THE FIRST VIEW

Exploring Directories

To show the contents of a directory, your DirectoryViewController will need an NSArray instance variable called directoryContents, which will just contain the contents of a given directory as NSString paths. Cocoa doesn't have any special class to represent files and often just uses NSStrings, for which there are a variety of path-related instance methods.[3] Add this ivar to DirectoryViewController.h, and then switch to the implementation file. With the array in place, it's simple enough to provide the two needed UITableViewDataSource methods to populate the table that lists the directory contents. First, the number of rows in the table is simply the length of the directoryContents array, so replace the provided tableView:numberOfRowsInSection: with a one-line implementation that returns the array length:

FileIO/FilesystemExplorer/Classes/DirectoryViewController.m

```
- (NSInteger)tableView:(UITableView *)tableView
            numberOfRowsInSection:(NSInteger)section {
    return [directoryContents count];
}
```

---

the refactor failed to change #import "RootViewController.h" in FilesystemExplorerAppDelegate.m to #import "DirectoryViewController.h", and the project wouldn't build until we fixed that.
3. Files can also be represented as file://-type NSURLs, and there are methods to convert file references between these string and URL representations.

Similarly, a table cell is constructed with the name of the file, which is just an indexed array element. Use that to set the cell's text, as shown in the highlighted lines added to the default implementation of tableView:cellForRowAtIndexPath:::

FileIO/FilesystemExplorer/Classes/DirectoryViewController.m

```
- (UITableViewCell *)tableView:(UITableView *)table
            cellForRowAtIndexPath:(NSIndexPath *)indexPath {
    static NSString *CellIdentifier = @"DirectoryViewCell";
        UITableViewCell *cell =
            [table dequeueReusableCellWithIdentifier:
            CellIdentifier];
        if (cell == nil) {
        cell = [[[UITableViewCell alloc]
                    initWithStyle:UITableViewCellStyleDefault
                    reuseIdentifier:CellIdentifier] autorelease];
        }
        cell.textLabel.text = (NSString*)
                [directoryContents objectAtIndex: indexPath.row];
        return cell;
}
```

But how do we get a chance to load these into the table? At some point, we have to tell the view controller which directory it's presenting. One option is to make the directoryPath a property of the view controller. Create the instance variable NSString *directoryPath, and then declare it as a property:

FileIO/FilesystemExplorer/Classes/DirectoryViewController.h

```
@property (nonatomic, retain) NSString *directoryPath;
```

When the property is set, we can load the directory contents into the array and refresh the table view. But to do this, we have to write our own implementation of the property getter and setter methods, since a @synthesized property won't know to do the extra work of populating the array. Instead, we implement the property ourselves by providing the directoryPath and setDirectoryPath: methods. Providing our own implementation lets us load the directory contents when the path is set and, while we're at it, set the navigation title to the current directory's name.[4]

---

4. Since we're doing nothing interesting in the getter, we could use the usual **@synthesize** and have it created for us, and then we'd hand-code just the setDirectoryPath:. We just thought it was useful here to show both to emphasize that there's nothing magical going on in properties.

FileIO/FilesystemExplorer/Classes/DirectoryViewController.m

```
-(NSString*) directoryPath {
    return directoryPath;
}

-(void) setDirectoryPath: (NSString*) p {
    [p retain];
    [directoryPath release];
    directoryPath = p;
    [self loadDirectoryContents];
    // also set title of nav controller with last path element
    NSString *pathTitle= [directoryPath lastPathComponent];
    self.title = pathTitle;
}
```

Since this retains the directoryPath, be sure to release it in dealloc, just as you would do with any other property that uses the retain attribute. The attributes represent your contract with any code that uses your property, and your implementation should match the declared writability, atomicity, and memory-management attributes.[5]

Now that we have a means of calling a loadDirectoryContents method, we need to go ahead and discover all the contents of a given directory using Cocoa's file APIs. NSFileManager is the primary class you'll use to work with the filesystem. Since there isn't a "file" class that you call methods on, you instead get an instance of the manager—a shared default instance retrieved with the class method defaultManager—and call methods on that. NSFileManager has a method called directoryContentsAtPath:, which returns an NSArray of NSStrings representing files or directories within the given path, so it's a snap for our DirectoryViewController to hang on to the directory contents (remember that you'll also need to provide a header for this method in DirectoryViewController.h):

FileIO/FilesystemExplorer/Classes/DirectoryViewController.m

```
- (void) loadDirectoryContents {
    [directoryContents release];
    directoryContents = [[NSFileManager defaultManager]
        directoryContentsAtPath: directoryPath];
    [directoryContents retain];
}
```

Since this retains the new directory contents, be sure to add a [directoryContents release]; to your dealloc method.

---

5. Note that property contracts are *not* enforced by the compiler or the Objective-C runtime; the **@synthesize** directive represents the only programmatic use of the attributes.

The first DirectoryViewController has already been created by the template as part of MainWindow.xib. So, how do we tell it the first directoryPath to use? Let's address this problem when the application starts. In FilesystemExplorerAppDelegate.h, declare an instance variable called directoryViewController for an IBOutlet to the DirectoryViewController, and then open MainWindow.xib with IB, examine the FilesystemExplorerAppDelegate's connections (with the Connections inspector, ⌘ 2), and connect directoryViewController to the DirectoryViewController object that is a child of the navigation controller (you'll probably have to put IB in list or columns mode to find it).

Now that we can see the first DirectoryViewController, let's give it a starting directoryPath. You can do this with a one-line call in FilesystemExplorerAppDelegate's applicationDidFinishLaunching: method, after it has finished its work setting up the main application window with makeKeyAndVisible:

FileIO/FilesystemExplorer/Classes/FilesystemExplorerAppDelegate.m
```
// populate the first view
directoryViewController.directoryPath = NSHomeDirectory();
```

If you want, you can now build and run the application. It will show you a four-line table with the contents of your home directory: Documents, FilesystemExplorer.app, Library, and tmp. You'll also notice the navigation title is the home directory name, the long application ID string. Now let's start navigating into these directories.

As we can see, we can get the application's home directory with a single call to NSHomeDirectory(). In Mac Cocoa, this method is defined as returning the path to the user's home directory, but on the iPhone, it returns the *application's* home directory, which is a directory that contains the application bundle, along with the "helper" directories Documents, Library, and tmp.

This directory has a long unique identifier string as its name, meaning that the entire path, when run in the simulator, might be something such as /Users/cadamson/Library/Application Support/iPhone Simulator/User/Applications/5C73EBC6-BDB3-46E7-B7EC-203A0BA6417B, where the last path element is an arbitrary *application ID* that is regenerated every time you build your application. Of course, since the iPhone security model won't let you navigate any higher than this in the filesystem hierarchy, the name of the directory or what's above it is pretty much irrelevant.

It's worth noting that we won't always need to use the application directory returned by NSHomeDirectory(), and we often want a path to one of its known subdirectories instead, such as Documents. There's a special technique for searching system paths for known special directories, illustrated when we copy a database file to Documents in Section 10.3, *Copying to the Documents Directory*, on page 192.

8.3 Getting File Attributes

The sample application needs to know which directory items are files and which are directories in order to handle the user tapping on a cell and knowing whether to take the user to a file view or a directory view. As it turns out, there are several ways to figure out whether a given path represents a directory or a file; one of the easiest is to just use NSFileManager's fileExistsAtPath:isDirectory: method. The second parameter is a pointer to a BOOL, and when the method returns, its value is set to YES if the path represents a directory. So, a typical (and hypothetical) use looks like this:

```
BOOL myPathIsDir;
BOOL fileExists = [[NSFileManager defaultManager]
          fileExistsAtPath: selectedPath
          isDirectory: &myPathIsDir];
NSLog (myPathIsDir ? @"My path is a directory" :
          @"My path is a file");
```

Notice how you pass the *address* of the BOOL to the method and check its value after the method call. Yes, it's very C-like. You have to do this because the method's return value is already being used to indicate whether a file exists at the specified path. For what it's worth, if you care *only* about checking whether something exists at a given path and don't care whether it's a directory or a file, you can simply pass NULL for isDirectory or, better yet, use the one-argument fileExistsAtPath:.

Now that we can tell a directory from a file, our filesystem explorer GUI can figure out what to do when the user taps on a row, which calls the event-handling delegate method tableView:didSelectRowAtIndexPath:. If the selected item is a directory, then you create another DirectoryViewController with initWithNibName:bundle:, set its directoryPath property, and push the view controller to the UINavigationController. To come back to this directory listing, we don't have to do anything; the user will just use the Back button provided by the navigation controller.

FileIO/FilesystemExplorer/Classes/DirectoryViewController.m

```
- (void)tableView:(UITableView *)tableView
        didSelectRowAtIndexPath:(NSIndexPath *)indexPath {
    NSString *selectedFile = (NSString*)
    [directoryContents objectAtIndex: indexPath.row];
    BOOL isDir;
    NSString *selectedPath =
    [directoryPath stringByAppendingPathComponent: selectedFile];
    if ([[NSFileManager defaultManager]
            fileExistsAtPath:selectedPath isDirectory:&isDir] && isDir) {
        DirectoryViewController *directoryViewController =
                [[DirectoryViewController alloc]
                        initWithNibName: @"DirectoryViewController"
                        bundle:nil];
        [[self navigationController]
                pushViewController:directoryViewController animated:YES];
        directoryViewController.directoryPath = selectedPath;
        [directoryViewController release];
    }
}
```

With just this first half of the **if** in place (closing curly braces as necessary), you can Build and Go to navigate forward and backward through the directories. For example, you can go to Library and then Preferences and then use the navigation bar's Back button to come back.

If the user taps on a row corresponding to a file, we'll need a new view, which is managed by a class we'll call a FileOverviewViewController. Use New File to create this new view controller, and then **import** its header in DirectoryViewController.m. Like the DirectoryViewController, this is a view controller with a filePath property, whose setter method we implement ourselves (as we did with the DirectoryViewController) and use as a signal to update the GUI.

FileIO/FilesystemExplorer/Classes/FileOverviewViewController.m

```
-(NSString*) filePath {
    return filePath;
}

-(void) setFilePath: (NSString*) p {
    [p retain];
    [filePath release];
    filePath = p;
    [self updateFileOverview];
    // also set title of nav controller with last path element
    NSString *pathTitle= [filePath lastPathComponent];
    self.title = pathTitle;
}
```

> **Joe Asks...**
> **Why Create View Controllers in Code?**
>
> Back in Chapter 6, *Navigation*, on page 103, we created the second view controller for the DVDCase navigation example in Interface Builder, but for FilesystemExplorer, we're suggesting you create the view controllers programmatically. The difference is that DVDCase's navigation is deterministic. It always goes from the RootViewController to the DVDCabinetController, so we could create both objects at build time in IB. When we explore the filesystem, we don't know the order of view controllers we need. We could visit one directory after another, putting more and more DirectoryViewController instances on the navigation stack, before visiting a file (which would require a FileOverviewController). Since we have to create the DirectoryViewControllers programatically, we think it's easier in this case to create all the navigation view controllers programatically, except for the first one, which Xcode provides automatically as part of the NavigationController in MainWindow.xib.

So, to navigate to the file overview in DirectoryViewController's tableView: didSelectRowAtIndexPath:, we create a FileOverviewViewController, push it to the navigation controller, and set the filePath to make the new controller update its views (which we do last to ensure the view has been loaded from the nib).

FileIO/FilesystemExplorer/Classes/DirectoryViewController.m
```
    else {
        FileOverviewViewController *fileOverviewViewController =
            [[FileOverviewViewController alloc]
                initWithNibName: @"FileOverviewView"
                bundle:nil];
        [[self navigationController]
            pushViewController:fileOverviewViewController animated:YES];
        fileOverviewViewController.filePath = selectedPath;
        [fileOverviewViewController release];
    }
}
```

To keep things simple, the file overview just shows the file's name, size, and last-modified date. In Figure 8.5, on page 143, we see what this view looks like when viewing the actual FilesystemExplorer binary file inside the application bundle. You may be used to associating your custom view controllers to IB-created views at this point, but let's go through the steps again:

1. Create a new UIViewController subclass, adding IBOutlets and IBActions in the header file for connections you'll want to establish in IB. In this case, we need three UILabel outlets. Remember to back these with instance variables and to **@synthesize** them in FileOverviewViewController.m.

   FileIO/FilesystemExplorer/Classes/FileOverviewViewController.h
   ```
   @property (nonatomic, retain) IBOutlet UILabel *fileNameLabel;
   @property (nonatomic, retain) IBOutlet UILabel *fileSizeLabel;
   @property (nonatomic, retain) IBOutlet UILabel *fileModifiedLabel;
   ```

   We also need two methods: one for a method to update the labels when the property is set and one IBAction to handle a button that will let the user read the file's contents.

   FileIO/FilesystemExplorer/Classes/FileOverviewViewController.h
   ```
   -(void) updateFileOverview;
   -(IBAction) readFileContents;
   ```

2. Create a new view nib in Xcode named FileOverviewView.xib, and open it in IB to build the user interface. Use the Identity inspector (⌘ 4) to set the File's Owner to your custom view controller class.

3. Ctrl+drag to connect your class's outlets to the GUI controls and events (like any button's Touch Up Inside event) to your actions. Also be sure to connect the File's Owner's view outlet to the main View in the IB window, or you'll crash when you load this nib (this is something you have to always remember to do when you create nibs with the View XIB template).

4. Add this custom view controller to your navigation controller with pushViewController:animated:, which we did earlier in DirectoryViewController's tableView:didSelectRowAtIndexPath:.

As mentioned earlier, providing your own setter for the filePath property lets you take extra action when the property is set, such as updating the labels, which we'll do by calling an updateFileOverview method. The first thing we want to do in this method is to create some data formatting objects so that when we populate the labels, we'll get more human-readable number and date formats.

Figure 8.5: DISPLAYING FILE ATTRIBUTES

FileIO/FilesystemExplorer/Classes/FileOverviewViewController.m
```
-(void) updateFileOverview {
        if (self.filePath != NULL) {
                NSString *fileName = [self.filePath lastPathComponent];
                fileNameLabel.text = fileName;

                NSDateFormatter *dateFormatter = [[NSDateFormatter alloc] init];
                [dateFormatter setDateStyle:NSDateFormatterMediumStyle];
                [dateFormatter setTimeStyle:NSDateFormatterNoStyle];

                NSNumberFormatter *numberFormatter =
                [[NSNumberFormatter alloc] init];
                [numberFormatter setPositiveFormat: @"#,##0.## bytes"];
```

Now, we once again turn to the NSFileManager class. It provides the method fileAttributesAtPath:traverseLink:, which returns an NSDictionary of the file's attributes as key-value pairs (the traverseLink parameter indicates whether symbolic links should be followed or whether the caller wants attributes for the link itself; this is largely irrelevant in this app). The possible key values are described in NSFileManager's documentation. For now, the ones we want are NSFileSize and NSFileModificationDate, so continuing where we left off...

```
FileIO/FilesystemExplorer/Classes/FileOverviewViewController.m
            NSDictionary *fileAttributes =
            [[NSFileManager defaultManager]
                    fileAttributesAtPath: self.filePath
                    traverseLink: YES];
            NSDate *modificationDate = (NSDate*)
                    [fileAttributes objectForKey: NSFileModificationDate];
            NSNumber *fileSize = (NSNumber*)
                    [fileAttributes objectForKey: NSFileSize];
            fileSizeLabel.text =
                    [numberFormatter stringFromNumber: fileSize];
            fileModifiedLabel.text =
                    [dateFormatter stringFromDate: modificationDate];

            [numberFormatter release];
            [dateFormatter release];

        }
    }
```

As we can see, although all the keys are NSStrings, the value types vary by the attribute requested. In this case, the modification date is an NSDate, while the file size is an NSNumber. The NSFileManager documentation for each attribute indicates the type of its value. Having said that, there aren't actually that many attributes you'll need on an iPhone. Many of the attributes are holdovers either from the Mac (for example, NSFileHFSTypeCode and NSFileHFSTypeCreator) or from ownership and permissions-oriented attributes from the world of Unix, neither of which is relevant to a file on an iPhone.[6]

8.4 Reading Data from Files

The next, and probably most useful, step is to read the contents of a file. As it turns out, there are a lot of ways to do this, more than is practical to describe here.

You may have noticed a Read Contents button in Figure 8.5, on the previous page. Tapping this should navigate to a new view to show the contents of the file. To show the file's contents, you'll need to create the class FileContentsViewController. Give it a filePath property, which the FileOverviewViewController class can set to tell it which file to read.

---

[6]. Surprisingly, although the seemingly useful NSFileCreationDate is defined, the attribute is always **NULL** on the device and in the simulator as of iPhone SDK 3.0.

Now you can implement the readFileContents event handler in FileOverviewViewController.m:

FileIO/FilesystemExplorer/Classes/FileOverviewViewController.m
```
- (IBAction) readFileContents {
        FileContentsViewController *fileContentsViewController =
                [[FileContentsViewController alloc]
                        initWithNibName: @"FileContentsView"
                        bundle:nil];
        fileContentsViewController.filePath = filePath;
        fileContentsViewController.title =
                [NSString stringWithFormat: @"%@ contents",
                        [filePath lastPathComponent]];
        [[self navigationController] pushViewController:
                fileContentsViewController animated:YES];
        [fileContentsViewController release];
}
```

Create a new nib for the file contents overview, FileContentsView.xib, which needs just a UITextView of the file's contents. Declare this outlet in FileContentsViewController.h, and then look ahead to Figure 8.6, on page 148. That's where you'll design the view in IB, and where you'll also connect the outlet and set File's Owner to FileOverviewViewController. Since this view will be part of a navigation, be sure to turn on the navigation bar "simulated interface element" in the view's property inspector.

If you build and run now, you can navigate to files, look at their overviews, and then tap the Read Contents button to navigate to the contents view, though we haven't yet populated its members.

Since the text view's contents are maintained as an NSString, loading the file into the view can be reduced to a single line of code:[7]

```
myString =
        [NSString stringWithContentsOfFile: filePath
                                usedEncoding: NULL
                                error: NULL];
```

We can't really use that here, since the file you want to load might contain something other than text (most of the files you can explore at the moment are binaries). In that case, you can use another one-line call to load the file into an NSData object, an object-oriented wrapper around a byte buffer.

---

[7]. Note that the simpler initWithContentsOfFile: and stringWithContentsOfFile: shown in earlier betas are deprecated in favor of new methods that take or return a text encoding parameter and that return an error.

```
NSData *fileData = [NSData dataWithContentsOfFile: filePath];
```

Of course, this gives you an NSData, while the text area needs a string. Still, these two methods are convenient options if your file is small enough to fit into memory, which is often the case. So, keep them in mind. But what if you have a big file that you don't necessarily want to load all at once? What if you want to parse through a file or simply upload over the network, reading and writing chunks as you go? In cases like these, it's appropriate to go the more traditional route of working with *streams*.

In the stream metaphor, the bytes of a file are a "stream" of data that flows in one direction, from the beginning of the file to the end. To read from a file, you open a stream from a given file and keep reading bytes until the stream runs out.

The essential class for reading from a file is NSInputStream, a subclass of NSStream. To read and write from a file, you need to do the following:

1. Open an input stream from the file.

2. Allocate a buffer of memory to hold bytes read from the stream.

3. Repeatedly call read:maxLength: to read the next few bytes from the stream. The first parameter is the buffer you want the bytes copied into, and maxLength specifies how many bytes you can handle on each read (typically the size of your buffer). This method will return the number of bytes read. It may be less than maxlength. If it's 0, the end of stream has been reached, and if it's -1, a read error has occurred.

4. Do something with the bytes you've read into the buffer.

5. Once you reach the end of the stream, close it.

Here's how FileContentsViewController can use an NSInputStream to read the contents of a file into a UITextView.[8] You call this loadFileContentsIntoTextView from viewDidLoad, so the file contents are loaded into the text view before we navigate to it.

---

[8]. Code this if you like, but we're going to provide a preferred alternative in the next section.

```
FileIO/FilesystemExplorer/Classes/FileContentsViewController.m
```
```objc
- (void) loadFileContentsIntoTextView {
        // open a stream to filePath
        NSInputStream *inputStream = [[NSInputStream alloc]
                initWithFileAtPath: filePath];
        [inputStream open];
        // read and dump to NSTextView
        NSInteger maxLength = 128;
        uint8_t readBuffer [maxLength];
        BOOL endOfStreamReached = NO;
        // NOTE: this tight loop will block until stream ends
        while (! endOfStreamReached) {
                NSInteger bytesRead = [inputStream read: readBuffer
                        maxLength:maxLength];
                if (bytesRead == 0) {
                        endOfStreamReached = YES;
                } else if (bytesRead == -1) {
                        // TODO - should have an error dialog
                        endOfStreamReached = YES;
                } else {
                        NSString *readBufferString =
                        [[NSString alloc]
                         initWithBytesNoCopy: readBuffer
                         length: bytesRead
                         encoding: NSUTF8StringEncoding
                         freeWhenDone: NO];
                        [self appendTextToView: readBufferString];
                        [readBufferString release];
                }
        } // while ! endOfStreamReached
        [inputStream close];
        [inputStream release];
}
```

For convenience, the example code has a method called appendText-ToView:, which copies an NSString to the end of the text view's text string. This way, with each buffer full of data you read, you incrementally fill in the text view until the stream ends and you've displayed the file contents to the user.

```
FileIO/FilesystemExplorer/Classes/FileContentsViewController.m
```
```objc
-(void) appendTextToView: (NSString*) textToAppend
{
        fileContentsTextView.text = [NSString stringWithFormat:
                @"%@%@", fileContentsTextView.text, textToAppend];
}
```

As you can see in the stream-reading block, each time a block of bytes is read into the buffer, it's converted into an NSString and appended to

Figure 8.6: FILE CONTENTS

the view. The example's hard-coded use of an NSUTF8StringEncoding is appropriate for simple text files, but not for binary files.[9] In Figure 8.6, we can see what Info.plist, a binary settings file, looks like when read in this way.

8.5 Asynchronous File Reading

There's actually a pretty significant problem with this code. The tight loop will tie up the main application thread while reading the stream—meaning the application can't do any work rendering the GUI, handling input, or doing anything else until you reach the end of the stream. As files get bigger and the time to read them gets longer, this becomes more and more of a problem.

The key problem is that the read:maxLength: method will block until it has bytes available to read. This is a bigger deal when we're dealing with

---

9. Most of the default application contents are binary, so we won't see many meaningful contents until we add the ability to create new files.

network streams and their inherent unreliability, but taking a more fault-tolerant approach is a good habit to adopt.

An NSStream can be made to work *asynchronously*. You can register for intermittent callbacks whenever the stream has bytes available, and if there aren't any, then you don't block. To do this, you send the stream a delegate (for example, the FileContentsViewController itself) that implements the stream:handleEvent: method. Then you schedule the stream to periodically check whether bytes are available and, if so, to call your delegate method. So, here's the setup that you call in viewDidLoad instead of the earlier loadFileContentsIntoTextView:

FileIO/FilesystemExplorer/Classes/FileContentsViewController.m

```
Line 1   - (void) setUpAsynchronousContentLoad {
     2       // open a stream to filePath
     3       NSInputStream *inputStream =
     4           [[NSInputStream alloc] initWithFileAtPath: filePath];
     5       [inputStream setDelegate: self];
     6       [inputStream scheduleInRunLoop:[NSRunLoop currentRunLoop]
     7               forMode:NSDefaultRunLoopMode];
     8       [inputStream open];
     9       [inputStream release];
    10   }
```

The scheduleInRunLoop:forMode: call in line 6 may look like a mystic incantation, but it's actually pretty much boilerplate. What this call indicates is that you want the current thread's *run loop* to be the one that handles checking on the stream—this is absolutely the typical option—and that you want the run loop to operate in its default mode, as opposed to one of several esoteric event-processing modes (most of which are defined for Mac Cocoa and don't exist on the iPhone). In all likelihood, you'll never call this method with options other than these defaults.

The delegate is required to implement a single stream:handleEvent: method. The first argument is the stream to read that produced the event, and the second is an NSStreamEvent indicating the event that occurred. These events are constants, enumerated in the NSStream class (and described in its documentation), that indicate the following:

- The opening of a stream
- The availability of a stream for reading or writing
- An error from the stream
- The end of the stream

So, the method implementation typically involves doing a **switch** on the NSStreamEvent and deciding what to do in that case. For the FilesystemExplorer example, all we need to do is the following:

- When bytes are available, read from the stream.
- If an error occurs, display a dialog box, close the stream, and unschedule the callbacks.
- If the end of stream is reached, close the stream, and unschedule the callbacks.

Here's a basic implementation of stream:handleEvent::

FileIO/FilesystemExplorer/Classes/FileContentsViewController.m

```
- (void)stream:(NSStream *)theStream handleEvent:(NSStreamEvent)streamEvent {
    NSInputStream *inputStream = (NSInputStream*) theStream;
    switch (streamEvent) {
        case NSStreamEventHasBytesAvailable: {
            NSInteger maxLength = 128;
            uint8_t readBuffer [maxLength];
            NSInteger bytesRead = [inputStream read: readBuffer
                            maxLength:maxLength];
            if (bytesRead > 0) {
                NSString *bufferString = [[NSString alloc]
                    initWithBytesNoCopy: readBuffer
                    length: bytesRead
                    encoding: NSUTF8StringEncoding
                    freeWhenDone: NO];
                [self appendTextToView: bufferString];
                [bufferString release];
            }
            break;
        } // case: bytes available
        case NSStreamEventErrorOccurred: {
            // dialog the error
            NSError *error = [theStream streamError];
            if (error != NULL) {
                UIAlertView *errorAlert = [[UIAlertView alloc]
                    initWithTitle: [error localizedDescription]
                    message: [error localizedFailureReason]
                    delegate:nil
                    cancelButtonTitle:@"OK"
                    otherButtonTitles:nil];
                [errorAlert show];
                [errorAlert release];
            }

            [inputStream removeFromRunLoop: [NSRunLoop currentRunLoop]
                    forMode:NSDefaultRunLoopMode];
            [theStream close];
            break;
        }
```

```
                case NSStreamEventEndEncountered: {
40                  [inputStream removeFromRunLoop: [NSRunLoop currentRunLoop]
                                    forMode:NSDefaultRunLoopMode];
                    [theStream close];
                }
            }
45      }
```

Lines 4 to 19 handle the "bytes available for reading" event, which is where we read from the stream. NSStreams read and write bytes as type **uint8_t**, so we use a buffer of this type (line 6) and then attempt to read from the stream into the buffer (line 7). If we succeed in reading any bytes, we use them to create a new string (line 10), using the NSUTF8StringEncoding, and append this string to the text view.

There's one significant hazard to watch out for with this approach. By making the loading asychronous, we've decoupled the streaming code from the state of the GUI. In general, this is a good thing; it's why the GUI doesn't block while loading a big file. But it also means that the user could navigate back and out of the file contents view before the stream is done loading. There's no point letting the stream continue working if that happens, and it could even lead to a crash.[10]

So, if the user navigates away from this view, we want to stop the loading process. We can handle this by getting the viewWillDisappear: message. When that happens, we'll want to unschedule the stream from the run loop and close it, the same things we do in a normal end-of-stream or error case. But to do any of that, the stream will need to be an instance variable. So, declare NSInputStream *asyncInputStream in the header file, and rewrite the setUpAsynchronousContentLoad method:

FileIO/FilesystemExplorer/Classes/FileContentsViewController.m

```
- (void) setUpAsynchronousContentLoad {
    [asyncInputStream release];
    // open a stream to filePath
    asyncInputStream =
            [[NSInputStream alloc] initWithFileAtPath: filePath];
    [asyncInputStream setDelegate: self];
    [asyncInputStream scheduleInRunLoop:[NSRunLoop currentRunLoop]
            forMode:NSDefaultRunLoopMode];
    [asyncInputStream open];
}
```

---

10. In earlier versions of this chapter, that's exactly what happened because of an over-released object. A reader helped us find this bug by reporting it on the errata page.

Notice that the stream code is the same; the only difference is how we manage the instance variable. Now we can stop the stream if the user hits the Back button and dismisses the view.

FileIO/FilesystemExplorer/Classes/FileContentsViewController.m
```
- (void) viewWillDisappear: (BOOL) animated {
      [asyncInputStream removeFromRunLoop:[NSRunLoop mainRunLoop]
            forMode:NSDefaultRunLoopMode];
      [asyncInputStream close];
      [super viewWillDisappear: animated];
}
```

8.6 Creating and Deleting Files and Directories

Now that we've drilled down through directories to files and their contents, let's look at a few more common file-related tasks. We'll provide the user with the ability to create a new directory, to create a new file, or to delete the selected item from a directory view.

To provide the add actions, we'll add an Add button to the DirectoryViewController. We can do this in the viewDidLoad method using @selector and the name of a method to indicate what method will be called when the button is tapped:

FileIO/FilesystemExplorer/Classes/DirectoryViewController.m
```
- (void)viewDidLoad {
   [super viewDidLoad];
      UIBarButtonItem *addButton = [[[UIBarButtonItem alloc]
            initWithBarButtonSystemItem:UIBarButtonSystemItemAdd
            target:self
            action:@selector(showAddOptions)] autorelease];
      self.navigationItem.rightBarButtonItem = addButton;
}
```

When the button is tapped, we'll insert an UIActionSheet, which provides a modal set of buttons, as shown in Figure 8.7, on the facing page. The sheet is initialized with a title and names for the buttons, including a Cancel button that simply dismisses the sheet. There's also an optional red button for "destructive" actions, but we don't need that in this sheet. The method takes a variable number of arguments for the "normal" buttons and requires you to provide a final NULL argument to end the list of button titles.

CREATING AND DELETING FILES AND DIRECTORIES ◀ 153

Figure 8.7: DISPLAYING A UIACTIONSHEET FOR FILE ACTIONS

FileIO/FilesystemExplorer/Classes/DirectoryViewController.m
```
-(void) showAddOptions {
    NSString *sheetTitle = [[NSString alloc]
                    initWithFormat: @"Edit \"%@\"",
                    [directoryPath lastPathComponent]];
    UIActionSheet *actionSheet = [[UIActionSheet alloc]
                    initWithTitle: sheetTitle
                    delegate: self
                    cancelButtonTitle: @"Cancel"
                    destructiveButtonTitle: NULL
                    otherButtonTitles: @"New File", @"New Directory", NULL];
    [actionSheet showInView: self.view];
    [sheetTitle release];
    [actionSheet release];
}
```

This requires us to declare that our class implements UIActionSheetDelegate and to provide a callback method for when one of the buttons is tapped.

FileIO/FilesystemExplorer/Classes/DirectoryViewController.m

```
- (void)actionSheet:(UIActionSheet *)actionSheet
        clickedButtonAtIndex:(NSInteger)buttonIndex {
    if (buttonIndex == 0)
        [self createNewFile];
    else if (buttonIndex == 1)
        [self createNewDirectory];
}
```

Obviously, we'll need to write two methods, createNewFile and createNewDirectory, before this will compile.

Let's start by creating new directories. Once again, the NSFileManager is the class you'll need for basic file-related functionality. First, though, it's worth checking ahead of time that you'll be able to write to the directory that you want to add to. This is a simple call to isWritableFileAtPath:, which, if successful, leads us to push to a new CreateDirectoryViewController class that we'll write next.

FileIO/FilesystemExplorer/Classes/DirectoryViewController.m

```
- (void)createNewDirectory {
    BOOL canWrite = [[NSFileManager defaultManager]
                     isWritableFileAtPath: self.directoryPath];
    if (! canWrite) {
        NSString *alertMessage = @"Cannot write to this directory";
        UIAlertView *cantWriteAlert =
            [[UIAlertView alloc] initWithTitle:@"Not permitted:"
                    message:alertMessage
                    delegate:nil
                    cancelButtonTitle:@"OK"
                    otherButtonTitles:nil];
        [cantWriteAlert show];
        [cantWriteAlert release];
        return;
    }
    CreateDirectoryViewController *createDirectoryViewController =
        [[CreateDirectoryViewController alloc]
            initWithNibName: @"CreateDirectoryView"
            bundle:nil];
    createDirectoryViewController.parentDirectoryPath = directoryPath;
    createDirectoryViewController.directoryViewController = self;
    createDirectoryViewController.title = @"Create directory";
    [[self navigationController]
            pushViewController:createDirectoryViewController animated:YES];
    [createDirectoryViewController release];
}
```

We'll need another view to let the user type in the new directory's name (shown in Figure 8.8, on page 156). The CreateDirectoryViewController will need a property for the parentDirectoryPath and another for the directoryViewController that created it, for reasons we'll explain shortly. For the GUI, define an IBOutlet for the directory name field. You may also want to implement UITextFieldDelegate to dismiss the keyboard (with resignFirstResponder) if the user taps Return, though it isn't really necessary for such a simple form. Next, use viewDidLoad to set up a Save button in the navigation bar that calls the createNewDirectory method:

`FileIO/FilesystemExplorer/Classes/CreateDirectoryViewController.m`

```
- (void)viewDidLoad {
    [super viewDidLoad];
    UIBarButtonItem *saveButton =
            [[UIBarButtonItem alloc]
                    initWithBarButtonSystemItem: UIBarButtonSystemItemSave
                    target: self
                    action: @selector(createNewDirectory)];
    self.navigationItem.rightBarButtonItem = saveButton;
    [saveButton release];
}
```

In createNewDirectory, create the directory by building a path string and using the NSFileManager:

`FileIO/FilesystemExplorer/Classes/CreateDirectoryViewController.m`

```
- (void) createNewDirectory {
    [directoryNameField resignFirstResponder];
    NSString *newDirectoryPath =
            [parentDirectoryPath stringByAppendingPathComponent:
                    directoryNameField.text];
    [[NSFileManager defaultManager]
            createDirectoryAtPath:newDirectoryPath
            attributes: nil];

    [directoryViewController loadDirectoryContents];
    [directoryViewController.tableView reloadData];
    [self.navigationController popViewControllerAnimated:YES];
}
```

The second parameter in createDirectoryAtPath:attributes: is an NSDictionary whose keys are the attribute constants that you saw earlier. A lot of them are irrelevant on the iPhone, particularly for directories, so it's common to just use a **NULL** dictionary. The last thing we need to do here is to alert the previous DirectoryViewController (a property in this class) to reload the directory contents and update its table view before we navigate back to it.

Once you've created some directories, you may well want to remove them. We can handle this back in DirectoryViewController by catching the

Figure 8.8: CREATING A NEW DIRECTORY

"swipe" gesture, via tableView:commitEditingStyle:forRowAtIndexPath:. First, when we get the delete-swipe, we get the path to delete and make sure it's writable.

FileIO/FilesystemExplorer/Classes/DirectoryViewController.m
```
- (void)tableView:(UITableView *)tableView
           commitEditingStyle:(UITableViewCellEditingStyle)editingStyle
           forRowAtIndexPath:(NSIndexPath *)indexPath {
    // handle a delete swipe
    if (editingStyle == UITableViewCellEditingStyleDelete) {
        NSString *selectedFile = (NSString*)
                [directoryContents objectAtIndex: indexPath.row];
        NSString *selectedPath =
                [directoryPath stringByAppendingPathComponent:
                        selectedFile];
        BOOL canWrite =
                [[NSFileManager defaultManager]
                        isWritableFileAtPath: selectedPath];
        if (! canWrite) {
            // show a UIAlert saying path isn't writable
```

If it appears we can delete, then we use NSFileManager, which provides a simple method for deletion, removeItemAtPath:error:.

FileIO/FilesystemExplorer/Classes/DirectoryViewController.m
```
NSError *err = nil;
if (! [[NSFileManager defaultManager] removeItemAtPath: selectedPath error:&err]) {
    // show a UIAlert saying cannot delete
```

Notice that this method does two things to indicate the results of the attempted deletion. It returns a BOOL that indicates whether the action succeeded, and if you pass in the address of an NSError object, that object will be populated with a more descriptive error if the delete action fails. You can then use this error object to provide feedback to the user.

Assuming the delete succeeds, we then refresh the list of files and update the table view:

FileIO/FilesystemExplorer/Classes/DirectoryViewController.m
```
            NSArray *deletedPaths = [NSArray arrayWithObject: indexPath];
            [self loadDirectoryContents];
            [self.tableView deleteRowsAtIndexPaths: deletedPaths
                    withRowAnimation: YES];
        }
}
```

Now let's turn to our other Add option in the DirectoryViewController, adding files for which we use another view controller, CreateFileViewController, paired with the view shown in Figure 8.9, on the following page. This class needs two IBOutlets for the filename UITextView and the file contents UITextArea. It also needs properties for the DirectoryViewController that created it and the parentDirectoryPath in which it will create new files.

FileIO/FilesystemExplorer/Classes/CreateFileViewController.h
```
@property (nonatomic,retain) IBOutlet UITextField *fileNameField;
@property (nonatomic,retain) IBOutlet UITextView *fileContentsView;
@property (nonatomic,retain) NSString *parentDirectoryPath;
@property (nonatomic,retain) DirectoryViewController *directoryViewController;
```

Once you've created CreateFileViewController, and its corresponding CreateFileView.xib, you can navigate to it in the createNewFile method called by DirectoryViewController's actionSheet:clickedButtonAtIndex:

To copy the data from the text field to a new file, there are two actions involved: creating the empty file and writing data to it.

As it turns out, there are one-line methods available to dump an entire NSString or NSData to a given path. Check out NSString's writeToFile:atomically:encoding:error: and NSData's writeToFile:atomically: and writeToFile:options:error: methods.

Figure 8.9: CREATING A NEW FILE

You can also supply an NSData as the contents argument to the NSFileManager class's createFileAtPath:contents:attributes: method.

However, to ensure that we don't block the user interface while we write, we can again use the asynchronous nature of NSStreams.

8.7 Writing Data to Files

The stream metaphor for writing to a file is almost identical to that used for reading a file:

1. Open a stream to a file.
2. Repeatedly write blocks of data to the stream from a memory buffer.
3. When you've written all your data, close the stream.

The stream-writing subclass of NSStream is NSOutputStream, and its essential method is write:maxLength:. As a subclass of NSStream, we inherit the scheduleInRunLoop:forMode method used earlier for asynchronous, nonblocking I/O. Let's see how that works.

First, when we bring up the CreateFileViewController, we want to add a navigation bar button called Save, which will set up our asynchronous stream.

FileIO/FilesystemExplorer/Classes/CreateFileViewController.m

```
- (void)viewDidLoad {
    [super viewDidLoad];
        UIBarButtonItem *saveButton =
                [[UIBarButtonItem alloc]
                        initWithBarButtonSystemItem: UIBarButtonSystemItemSave
                        target: self
                        action: @selector(setUpAsynchronousContentSave)];
        self.navigationItem.rightBarButtonItem = saveButton;
}
```

Next, we need to write the setUpAsynchronousContentSave that this button will call. To coordinate between this method and the callback, we'll need three instance variables in the header file: the output stream, a buffer to write, and a pointer to what part of the buffer we're writing.

FileIO/FilesystemExplorer/Classes/CreateFileViewController.h

```
NSOutputStream *asyncOutputStream;
NSData *outputData;
NSRange outputRange;
```

Now we can write setUpAsynchronousContentSave. Its tasks will be to call NSFileManager's createFileAtPath: to create a file to write into, to set up the stream, and to schedule it with the run loop.

FileIO/FilesystemExplorer/Classes/CreateFileViewController.m

```
- (void) setUpAsynchronousContentSave {
        [asyncOutputStream release];
        [outputData release];
        outputData = [[fileContentsView.text
                dataUsingEncoding: NSUTF8StringEncoding] retain];
        outputRange.location = 0;
        NSString *newFilePath = [parentDirectoryPath
                        stringByAppendingPathComponent: fileNameField.text];
        [[NSFileManager defaultManager] createFileAtPath:newFilePath
                        contents:nil attributes:nil];
        asyncOutputStream =     [[NSOutputStream alloc]
                                initToFileAtPath: newFilePath append: NO];
        [asyncOutputStream setDelegate: self];
        [asyncOutputStream scheduleInRunLoop:[NSRunLoop currentRunLoop]
                                forMode:NSDefaultRunLoopMode];
        [asyncOutputStream open];
}
```

On line 4, this converts the text view's NSString representation into an NSData block of bytes that we can use with the stream APIs. We also set up an NSRange to keep track of the next byte of that data that needs to be written to the stream. After creating the file, we create the stream (line 11), set the delegate to **self** so this class gets the callbacks (line 13), schedule the stream with the run loop (line 14), and open the stream (line 16).

Because the asynchronous API is inherited from NSStream, the callback is the same as we saw for the file-reading case: stream:handleEvent:. Of course, when we're writing to the stream, we'll get different events, such as NSStreamEventHasSpaceAvailable to indicate the stream's readiness to accept a write command.

FileIO/FilesystemExplorer/Classes/CreateFileViewController.m

```
Line 1  - (void)stream:(NSStream *)theStream handleEvent:(NSStreamEvent)streamEvent {
     -       NSOutputStream *outputStream = (NSOutputStream*) theStream;
     -       BOOL shouldClose = NO;
     -       switch (streamEvent) {
     5             case NSStreamEventHasSpaceAvailable: {
     -                   uint8_t outputBuf [1];
     -                   outputRange.length = 1;
     -                   [outputData getBytes:&outputBuf range:outputRange];
     -                   [outputStream write: outputBuf maxLength: 1];
    10                   if (++outputRange.location == [outputData length]) {
     -                       shouldClose = YES;
     -                   }
     -                   break;
     -             }
    15             case NSStreamEventErrorOccurred: {
     -                   // dialog the error
     -                   NSError *error = [theStream streamError];
     -                   if (error != NULL) {
     -                       UIAlertView *errorAlert = [[UIAlertView alloc]
    20                           initWithTitle: [error localizedDescription]
     -                           message: [error localizedFailureReason]
     -                           delegate:nil
     -                           cancelButtonTitle:@"OK"
     -                           otherButtonTitles:nil];
    25                       [errorAlert show];
     -                       [errorAlert release];
     -                   }
     -                   shouldClose = YES;
     -                   break;
    30             }
     -             case NSStreamEventEndEncountered:
     -                   shouldClose = YES;
     -       }
```

```
                if (shouldClose) {
35                      [outputStream removeFromRunLoop: [NSRunLoop currentRunLoop]
                                forMode:NSDefaultRunLoopMode];
                        [theStream close];

                        // force update of previous page and dismiss view
40                      [directoryViewController loadDirectoryContents];
                        [directoryViewController.tableView reloadData];
                        [self.navigationController popViewControllerAnimated:YES];
                }
        }
```

When the stream is ready to accept bytes, line 6 sets up a buffer of type **uint8_t**, the type needed for NSOutputStream's write:maxLength:. We've used a trivial length of one byte to simplify the code (fewer bounds checks). We copy a byte from the NSData into this buffer on line 8, and we write that byte to the stream on line 9. If we've reached the end of the data to write, line 10 sets a cleanup flag, which is also used for handling error and end-of-stream events. When any of these things happen, we unschedule the stream from the run loop (line 35) and close the stream (line 37). Finally, we clean up the GUI by making the DirectoryViewController that created this CreateFileViewController reload the directory contents, refresh its GUI, and then navigate back to the DirectoryViewController (lines 40–42).

With this, our long effort to implement DirectoryViewController's Add button is finally complete. You can create directories and files, see them appear in the directory views' tables, and tap on the files you create to read their contents. You may also notice different behavior with the simulator than on the device. The simulator will let you create files and directories inside the application bundle, while the actual device will not. In fact, you shouldn't plan on writing data anywhere other than the Documents and tmp folders. But within those folders, you're free to set up whatever directory structures and file contents suit you and your app.

8.8 Property Lists and NSCoding

Having discussed the conveniences for dumping NSString and NSData objects to disk and writing out the file manually with an NSOutputStream, you might wonder whether that covers all the ways you might want to persist your data with flat files. Not entirely. There are two more options to consider for writing structured data to disk that we'll consider briefly.

You've seen .plist files since way back in Section 3.4, *Working with Xcode and Interface Builder*, on page 28 when we introduced the Info.plist file that contains your application's basic settings. These *property lists* are serialized representations of structured application data.[11] Cocoa supports property lists consisting of the following classes:

- NSNumber
- NSString
- NSData
- NSDate
- NSArray
- NSDictionary

Property list structures can be arbitrarily deep, so you could have an array of dictionaries of string-to-date mappings and save the whole thing to disk with just two lines of code:

```
xmlData = [NSPropertyListSerialization dataFromPropertyList: somePlistObject
                format:NSPropertyListXMLFormat_v1_0
                errorDescription:&error];
[xmlData writeToFile:somePath atomically:YES]
```

Similarly, you can read this structure back into memory with propertyListFromData:mutabilityOption:format:errorDescription:.

And if you have objects that don't consist of the property list types, you can implement the NSCoding protocol, which allows the various NSCoder subclasses to persist the object's state to disk. This kind of approach might be more typical of enterprise applications, but the classes are there if you want to use them on iPhone. For more information on this approach, see Apple's *Archives and Serialization Programming Guide for Cocoa* [App08a]. In fact, we'll use this technique with the NSCoder subclass NSKeyedArchiver to pack name-value pairs into an NSData and send it over the network in Section 13.7, *Network Game Logic*, on page 284.

---

11. In iPhone OS 2.x, the application's Info.plist was human-readable XML, but in 3.0, the format has changed.

Chapter 9
Preferences

Your iPhone application will be started and shut down frequently. Any time the user hits the home key or stops to take an incoming call, your application will be terminated immediately. Your users will appreciate you leaving things the way they were when they last ran the application—and that includes saving and restoring their preferences.

There are effectively two approaches for managing preferences on the iPhone, not technically mutually exclusive, but so different in their approach and results that they might as well be. It's very important to pick one approach and stick to it, because if you don't, you'll have preferences in two totally different places, making yourself and your user do twice as much work.

We'll start with an approach that's more work for you but less for your user: having your application manage its own preferences and offering a user interface for the preferences.

9.1 Displaying a Flippable Preference View in Your Application

In a sense, the idea of handling your own preferences should be pretty straightforward after reading Chapter 8, *File I/O*, on page 129 and mastering the various forms of working with the iPhone's filesystem. You choose some sort of strategy for writing and reading your data to and from a file, and there's your preference-persistence strategy. If you choose to roll your own like this, it's particularly appealing to store your preferences as an NSDictionary of name-value pairs, which you'd write with writeToFile:atomically: and read with initWithContentOfFile:. One huge advantage of this approach is that the saved files are property lists (see Section 8.8, *Property Lists and NSCoding*, on page 161), so the

Figure 9.1: FLIPPING THE MAIN VIEW TO SHOW A PREFERENCES VIEW

values of your preferences could be NSArrays or embedded NSDictionary objects, in addition to the simpler property list member types (NSString, NSDate, NSNumber, or NSData).

In fact, managing preferences within your own application is not a question of storage so much as it is one of presentation. If you're going to manage the user's preferences, then you'll need a GUI to let them configure those preferences. And in the limited space of an iPhone screen, where does that go?

If you've used all the applications included with the phone, you know the answer: on the "back" of your view. The simpler applications that come with the iPhone and the iPod touch—Weather and Stocks—have a small button at the bottom right that brings up a preferences GUI with an animated "flip," as if the settings were located on the back side of the application's main view.

The sample application called FlippingPreferableClock illustrates this technique, as you can see in Figure 9.1, on the preceding page. This application simply shows the current time and displays the name of a time zone. The time zone used by the clock and the format for showing the time in 12- or 24-hour format are the preferences exposed by the sample application. The main view consists simply of these two UILabels, along with a UIButton to flip to the preferences view.

Xcode provides a template to create this kind of utility application, which is what we'll use for this example. When you create a utility application, Xcode's sets you up with a large number of classes and nibs:

- A typical application delegate class, with the MainViewController as a property.

- A "main view" for the primary application view, the one that comes up at launch and that presents the main functionality, implemented with a custom MainView class, a MainViewController, and a MainView.xib nib file. The view contains an "info" button at the bottom right, connected to a showInfo action. This method loads the FlipsideViewController from a nib, sets the flip-side view controller's delegate to the main view controller, sets the new-in-3.0 modalTransitionStyle property to set up the "flip" animation, and shows the flip-side view controller modally.

- A "flip-side view" for the configuration GUI, again comprised of a custom FlipsideView class, a FlipsideViewController, and a FlipsideView.xib. The view contains a UINavigationButton with a Done button that calls a done method. This method gets the delegate (that is, the main view controller) and calls its flipsideViewControllerDidFinish to let the main view controller know to update itself in response to changed settings and to dismiss the flip-side view controller.

- Other typical boilerplate files, such as main.m, which launches the application.

Without writing any code, you can create a utility application project called FlippableClock and click Build and Go to try flipping the view with the Info and Done buttons. Once you've tried that, let's start providing the functionality.

> **Caution: Utility Application Changes in iPhone 3.0**
>
> The earliest public betas of the iPhone SDK didn't include code for flippable applications and expected developers to set up their own animation. Fortunately, the final 2.0 SDK included a Utility Application template that used a RootViewController that mediated between the "main" and "flip-side" view controllers. Both of these approaches were covered in earlier betas of this book.
>
> In iPhone SDK 3.0, the UIViewController gained a new property, modalTransitionStyle, which allows the Utility Application template to flip directly between two view controllers via presentModalViewController:animated: and dismissModalViewControllerAnimated:, without needing the RootViewController class. We cover this new, simpler approach in this chapter, and you might find the new transition styles useful any time you need to bring in a modal view controller.
>
> However, apps using modalTransitionStyle, including any apps built on the new utility application template, will *not* compile for or run on iPhone OS 2.x. We've left the old example in the download code as FlippingPreferableClockFor2.0, in case you need it.

9.2 Managing Preferences

We said that the I/O part of managing the preferences would be the easy part, easier than managing the flipping views, so let's take care of that right away. In MainViewController.h, we'll **#define** keys for a preferences dictionary and for some default values, and in the **@interface**, we'll declare an NSMutableDictionary to hold the preference values at runtime and a path to the preferences file:

FileIO/FlippingPreferableClockFor3.0/Classes/MainViewController.h

```
#define TWENTY_FOUR_HOUR_PREF_KEY @"24HourDisplay"
#define TIME_ZONE_PREF_KEY @"TimeZone"
#define DEFAULT_TWENTY_FOUR_HOUR_PREF @"NO"
#define DEFAULT_TIME_ZONE_PREF @"America/Detroit"

@interface MainViewController : UIViewController <FlipsideViewControllerDelegate> {

    NSMutableDictionary *clockPrefs;
    NSString *prefsFilePath;
```

And with that, we can find a path to a prefs file in the application bundle and attempt to load its contents as an NSMutableDictionary, building a dictionary from the default values if the file doesn't already exist:

FileIO/FlippingPreferableClockFor3.0/Classes/MainViewController.m

```
- (void) initPrefsFilePath {
        NSString *documentsDirectory =
        [NSHomeDirectory() stringByAppendingPathComponent:@"Documents"];
        prefsFilePath = [documentsDirectory stringByAppendingPathComponent:
                        @"flippingprefs.plist"];
        [prefsFilePath retain];
}

- (void) loadPrefs {
        if (prefsFilePath == nil)
                [self initPrefsFilePath];
        if ([[NSFileManager defaultManager] fileExistsAtPath: prefsFilePath]) {
                clockPrefs = [[NSMutableDictionary alloc]
                        initWithContentsOfFile: prefsFilePath];
        }
        else {
                clockPrefs = [[NSMutableDictionary alloc] initWithCapacity: 2];
                [clockPrefs setObject: DEFAULT_TIME_ZONE_PREF
                        forKey: TIME_ZONE_PREF_KEY];
                [clockPrefs setObject: DEFAULT_TWENTY_FOUR_HOUR_PREF
                        forKey: TWENTY_FOUR_HOUR_PREF_KEY];
        }
        NSString *prefTimeZone = [clockPrefs objectForKey: TIME_ZONE_PREF_KEY];
        BOOL uses24Hour = [(NSString*)
                [clockPrefs objectForKey: TWENTY_FOUR_HOUR_PREF_KEY] boolValue];
        [self setClockToTimeZoneName: prefTimeZone uses24Hour: uses24Hour];
}
```

You'll notice this ends with a call to a convenience method, setClockToTimeZoneName:uses24Hour:. We'll call that to update the GUI both at startup (which means you'll want to add [self loadPrefs]; to your viewDidLoad method) and when the flip-side view changes the preferences. Of course, we haven't even created the GUI yet, so let's work on that next.

The template provides .h and .m files for MainView and FlipsideView, allowing you to do custom rendering of each, but for this example we need to edit only the view controller and the .xib files. In MainViewController.h, create IBOutlets for UILabels called timeLabel and timeZoneLabel. Double-click MainView.xib (not MainWindow.xib, which is just boilerplate) to open it in IB. Drag two labels into the view, customize their style and size as you see fit, and connect them to the outlets you just defined.

Now that we have IBOutlets to the timeLabel and timeZoneLabel, we can update them with the setClockToTimeZoneName:uses24Hour: method mentioned earlier. For this, add the instance variables NSString* timeZoneName and NSDateFormatter *clockFormatter to the header file, and code the method in the implementation file before loadPrefs, which calls it:

FileIO/FlippingPreferableClockFor3.0/Classes/MainViewController.m

```
-(void) setClockToTimeZoneName: (NSString*) tz uses24Hour: (BOOL) u24h {
        [timeZoneName release];
        [tz retain];
        timeZoneName = tz;
        // set time formatter with 24 hour preference and time zone
        if (clockFormatter == nil) {
                clockFormatter = [[NSDateFormatter alloc] init];
        }
        // see formats at
        // http://unicode.org/reports/tr35/tr35-4.html#Date_Format_Patterns
        [clockFormatter setTimeZone: [NSTimeZone timeZoneWithName: tz]];
        if (u24h)
                [clockFormatter setDateFormat: @"HH:mm:ss"];
        else
                [clockFormatter setDateFormat: @"h:mm:ss a"];
}
```

The NSDateFormatter is doing the interesting work here, taking a time zone name and a formatting string to produce the appropriate text for the timeLabel.[1]

The last thing we need to do to display the time is to use the formatter to help set the value of the timeLabel. Add the following updateClockView method:

FileIO/FlippingPreferableClockFor3.0/Classes/MainViewController.m

```
-(void) updateClockView {
        if (! clockFormatter) {
                timeLabel.text = @"";
                timeZoneLabel.text = @"";
                return;
        }
        NSDate *dateNow = [NSDate date];
        timeLabel.text = [clockFormatter stringFromDate: dateNow];
        timeZoneLabel.text = timeZoneName;
}
```

---

1. Note that the NSDateFormatter will pick up some of its behavior from the locale and the system settings. In particular, if you have set 24 Hour Time in the Settings application, then the NSDateFormatter will ignore any format string set by your application and will *only* produce 24-hour times.

Call this method from the end of viewDidLoad with [self updateClockView];. If you build and run your code at this point, you should see the current time in the default time zone displayed on your screen. Granted, we haven't done anything with preferences yet. Let's go to the flip side, where we allow the user to configure the preferences.

9.3 Changing and Updating Preferences

For the flip side, your FlipsideViewController.h will need IBOutlet properties for a UISwitch called twentyFourHourSwitch and a UIPickerView called timeZonePicker (don't forget to synthesize them as well). You'll also need an NSArray instance variable called timeZoneNames to populate the picker. Double-click FlipsideView.xib to edit it in IB, and add the switch and the picker, wire them to the outlets, and add some appropriate labels.

A picker is similar in many ways to a table, so you also need to connect the picker's delegate and datasource outlets to File's Owner. We'll use these connections to allow the flip side to populate its picker with known time zones. In FlipsideViewController.h, add the protocol declaration <UIPickerViewDataSource, UIPickerViewDelegate> to the end of the @interface statement. Then in the implementation file, you'll need to get a list of known time zones from the NSTimeZone class method knownTimeZoneNames via a loadTimeZoneNames method that you can call from viewDidLoad.

FileIO/FlippingPreferableClockFor3.0/Classes/FlipsideViewController.m

```
-(void) loadTimeZoneNames {
    if (timeZoneNames)
        return;
    NSArray *unsortedTimeZoneNames = [NSTimeZone knownTimeZoneNames];
    timeZoneNames =
        [unsortedTimeZoneNames sortedArrayUsingSelector:
            @selector(caseInsensitiveCompare:)];
    [timeZoneNames retain];
}
```

With this array populated, you can implement the needed data source and delegate methods for the time zone picker:

FileIO/FlippingPreferableClockFor3.0/Classes/FlipsideViewController.m

```
-(NSInteger) pickerView: (UIPickerView*) pickerView
        numberOfRowsInComponent: (NSInteger) component {
    return [timeZoneNames count];
}
```

```
- (NSInteger) numberOfComponentsInPickerView:(UIPickerView *)pickerView {
    return 1;
}

- (NSString *)pickerView:(UIPickerView *)pickerView
              titleForRow:(NSInteger)row
             forComponent:(NSInteger)component {
    return (NSString*) [timeZoneNames objectAtIndex: row];
}
```

Now we have a picker that can show all the known time zones. Of course, when we navigate to the flip side, we should set the picker to the main view's current time zone and set the 24-hour switch to the state currently being displayed. To do this, make the MainViewController's clockPrefs a property, and let the flip side read it to initialize its own state. Set this up in FlipsideViewController's viewDidLoad, after the [self loadTimeZoneNames]; that you added earlier. You'll also have to #import "MainViewController.h", since you're casting delegate to a MainViewController and accessing its clockPrefs property.

FileIO/FlippingPreferableClockFor3.0/Classes/FlipsideViewController.m

```
[self loadTimeZoneNames];
// init to values from main view
MainViewController *mainVC = (MainViewController*) delegate;
NSString *timeZone =
     [mainVC.clockPrefs objectForKey:TIME_ZONE_PREF_KEY];
NSString *twentyFourHourPref =
     [mainVC.clockPrefs objectForKey:TWENTY_FOUR_HOUR_PREF_KEY];
[timeZonePicker
     selectRow: [timeZoneNames indexOfObject:timeZone]
     inComponent: 0 animated:NO];
twentyFourHourSwitch.on = [twentyFourHourPref boolValue];
```

To review, the user starts up, and the time zone and 24-hour settings, which are instance variables in MainViewController, are loaded from the filesystem or set to defaults. When they navigate to the flip-side view, the current values are used for the initial state of the time-zone picker and 24-hour switch. Now let's suppose the user changes one or both of these. When he or she taps Done, the new values need to be communicated to the MainViewController. Where should we do that?

The template provides the FlipsideViewController with a done method, which calls the MainViewController's flipsideViewControllerDidFinish:. So, we could write our update code in either class. That leads to an interesting design decision: should the MainViewController make its time zone and 24-hour values settable by the flip side (or, for that matter, any code), or should we pull values from the flip side into the main? In the interest of minimizing what we expose, we'll use the latter approach and have MainViewController's flipsideViewControllerDidFinish: fetch the values from the flip side when it's called.

Still, instead of exposing the switch and picker views, let's declare two "getter" methods in FlipsideViewController.h, called selectedTimeZone and uses24Hour (using the following signatures), and implement them in FlipsideViewController.h.

FileIO/FlippingPreferableClockFor3.0/Classes/FlipsideViewController.m

```
-(NSString*) selectedTimeZone {
        return [timeZoneNames objectAtIndex:
                [timeZonePicker selectedRowInComponent: 0]];
}
-(BOOL) uses24Hour {
        return twentyFourHourSwitch.on;
}
```

Now we can have the MainViewController update its display by retrieving the selected values from the flip side, updating the time and time-zone labels in the main view, and saving the updated preferences to disk. By default, flipsideViewControllerDidFinish: contains a single line to dismiss the flip-side view controller. We'll do our work right before that.

FileIO/FlippingPreferableClockFor3.0/Classes/MainViewController.m

```
- (void)flipsideViewControllerDidFinish:
                (FlipsideViewController *)controller {
        timeZoneName = [controller selectedTimeZone];
        BOOL uses24Hour = [controller uses24Hour];
        NSString *selected24HourDisplayS = uses24Hour ? @"YES" : @"NO";
        [clockPrefs setObject: timeZoneName forKey: TIME_ZONE_PREF_KEY];
        [clockPrefs setObject: selected24HourDisplayS
                forKey: TWENTY_FOUR_HOUR_PREF_KEY];
        // save prefs to documents folder
        [self savePrefs];

        // update display to changed prefs
        [self setClockToTimeZoneName: timeZoneName uses24Hour: uses24Hour];
        [self updateClockView];

        // from template
        [self dismissModalViewControllerAnimated:YES];
}
```

This method uses the selectedTimeZone and uses24Hour methods we wrote for the FlipsideViewController and puts their values into the clockPrefs. Then, before updating the clock's time zone and performing the flip, it calls a savePrefs method to actually save the preferences to the filesystem. Since the preferences are stored in an NSDictionary, we can use a one-line call to save them as a plist.

FileIO/FlippingPreferableClockFor3.0/Classes/MainViewController.m
```
- (void) savePrefs {
        [clockPrefs writeToFile: prefsFilePath atomically: YES];
}
```

And with that, our locally managed preference settings make a round-trip: from the file into the MainViewController where they're used to format the timeLabel and timeZoneLabel... from there to the FlipsideViewController where they set the initial state of the picker and switch in the flip-side view... then back to the MainViewController, where they're used to update the display and are saved back to the filesystem.

Granted, by using an easy-to-save NSDictionary, we didn't have to do much work to save or load the preferences and spent most of our time working with the unique arrangement of the flippable utility application. If you wanted to use some other scheme for persisting your preferences, like your own file format, the NSUserDefaults (see Section 9.6, *A Hybrid Approach*, on page 183), or even a database (see Chapter 10, *The SQLite Database*, on page 185), then the only difference would be in your implementations of savePrefs and loadPrefs.

9.4 Side Trip: Updating the Clock Label Every Second

It's not necessarily germane to the discussion of preferences, but the technique for keeping the clock label constantly updated is a good one to know. The trick is to create an NSTimer—an object that makes repeated, scheduled calls to a given method—to periodically regenerate and redisplay the clock label's content.

You begin by creating a method that will be called periodically. We've already done that, because we can just repeatedly call updateClockView.

Next, you create an NSTimer to call that method at some given frequency (in seconds). Declare the instance variable clockViewUpdateTimer in MainViewController.h, and then add a startClock method to the implementation file, before viewDidLoad, which will need to call it.

FileIO/FlippingPreferableClockFor3.0/Classes/MainViewController.m
```
-(void) startClock {
        // since timer's first callback will occur after interval,
        // do one up-front refresh
        [self updateClockView];
```

```
        // now set up timer to repeatedly call updateClockView
        clockViewUpdateTimer = [NSTimer
                scheduledTimerWithTimeInterval:0.2
                target:self
                selector:@selector (updateClockView)
                userInfo:nil
                repeats:YES];
}
```

This makes an immediate call to update the labels and then sets up an NSTimer to repeatedly do the same thing. The class method scheduledTimerWithTimeInterval:target:selector:userInfo:repeats: is a mouthful but lets us spell out everything we need for periodic callbacks: how often to call back, what method is to be called (a combination of the target object and the selector method signature), an object to pass to the callback (we don't need one), and whether the NSTimer should keep making this callback or stop after the first call. Notice how you can refer to a method by means of the @selector keyword, which gives you the selector type needed for the selector parameter.

Once you've added this method, replace [self updateClockView]; in viewDidLoad with a call to start the clock: [self startClock];. This will update the view right away and then begin the regular updates. Build and run the app again, and you should see the clock update every second by changing its display as you change its preferences on the flip side.

9.5 Using the System Settings Application for Preferences

The second option for handling preferences is to not manage them in your own application at all. Instead, you let the iPhone OS's Settings application manage the preferences for you.

With this approach, all you do is provide a bundle file that describes your user-configurable preferences: their types, possible values, user-readable strings, and so on. The Settings application provides a GUI for this, and you simply read in the values in your application. In Figure 9.2, on the following page, we can see the Settings application offering the two preferences for the clock application, as exposed by a settings bundle.

We'll apply this approach in a rewrite of the clock application. In Xcode, create a view-based application project called BundlePreferableClock. Again, this application will show a constantly updating clock, allowing the user to set a 12/24-hour display preference and to choose the

Figure 9.2: PREFERENCES AS EXPOSED BY SETTINGS.BUNDLE TO THE SETTINGS APPLICATION

time zone to display. Since our application won't be responsible for presenting a preferences GUI, it won't need to flip and will instead have just a single main view.

For this application, we'll start the preferences data and build the user interface only at the end.

Let's begin by providing a 29x29 icon in PNG format for the Settings application to display in its table of configurable applications. This file *must* be named Icon-Settings.png; if it isn't found, iPhone OS will scale your app's main icon (Icon.png) and use it instead. Ctrl+click (or right-click) the Resources group, choose Add > Existing Files, and choose Icon-Settings.png, being sure to select "Copy item into destination group's folder (if needed)" in the second dialog box. In the book's downloadable

Figure 9.3: CREATING A NEW SETTINGS BUNDLE

sample code, we've put a blue tint on the background of our icon to distinguish it from the main application icon.

Creating a Settings.bundle File

Next, we need to specify what settings the System Preferences application will manage. Begin by selecting the top-level project icon in Xcode's Groups & Files panel, and use the New File menu item. In the New File window, choose the Settings Bundle template, as shown in Figure 9.3.

This adds Settings.bundle, which looks like a little white construction toy block, to your Groups & Files display. You'll notice it has a disclosure triangle, because it's actually a folder that contains two items by default: a Root.plist file and an en.lproj localization folder. These are the minimal contents for a settings bundle.

Take a look at Root.plist by double-clicking it, which brings up an editor like that shown in Figure 9.4, on the following page. The starter entries provide a "Strings Table" reference for finding localizations. and an array called PreferenceSpecifiers, with four entries. These four are

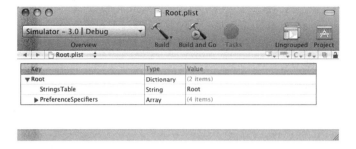

Figure 9.4: INITIAL CONTENTS OF SETTINGS BUNDLE'S ROOT.PLIST

examples that you can inspect to get the hang of the preference bundle format but that you'll want to delete in favor of your own preferences.

Each element of the PreferenceSpecifiers array is a dictionary of name-value pairs that specify one preference. The contents of the directory vary somewhat by type of preference, but most of them contain the following keys:

- Type: One of several constants indicating the type of the preference and therefore how the user configures it: a switch for boolean values, a text field for strings, and so on
- Key: The string that the application will use as a key to read the preference at runtime
- DefaultValue: The value that should be presented by default

Beyond these, other key-value pairs depend on the type of preference being offered. The available types are as follows:

- PSTextFieldSpecifier displays an editable text field with an optional title.
- PSTitleValueSpecifier shows a noneditable title for read-only preference values.
- PSToggleSwitchSpecifier presents an on/off switch for boolean preferences.
- PSSliderSpecifier uses a slider control for a preference whose value is a number between a set minimum and maximum.
- PSMultiValueSpecifier shows a table of possible values, from which the user can select one item.

> ### Joe Asks...
> #### When Do I Use a Flip, and When Do I Use a Bundle?
>
> For obvious reasons, the two preferences paradigms in this chapter should be considered mutually exclusive. Having some of your preferences in a flippable view while others are managed by the Settings application would mean twice as much work for you and your user. So, which approach is right for you?
>
> The precedent set by Apple's default apps suggests that only the simplest apps manage their own preferences. There's a good reason for this. With so little screen space—five buttons at the bottom, maybe two at the top—apps like Mail and Safari just can't afford the room for a preferences button.
>
> Furthermore, it seems like the simple apps tend to merit more frequent preference changes. If you travel, you'll set and reset your Weather cities frequently, but are you *ever* going to mess with your Safari settings?
>
> Putting preferences in a completely different place might seem strange at first, but users will be used to it by the time they run your app, so unless your app is really simple or perhaps is a game with its own distinctive GUI, you probably want your preferences in a bundle.

- PSGroupSpecifier groups other preferences entries together under a common title.
- PSChildPaneSpecifier offers a link to a child page of preferences, which is defined in a separate .plist file.

So, setting up a preferences bundle for the clock will require just two entries in the PreferenceSpecifiers array: one of type PSToggleSwitchSpecifier to present the 12/24-hour preference and one of type PSMultiValueSpecifier to choose a time zone. We can start with the toggle switch. Delete the four default array items by right-clicking them and choosing the Cut menu item. Now click the PreferenceSpecifiers array, and notice that a button pops up on the far right side of the row. This is the Add button and will create a new table row. Click it to create a new row, and then set the type of that row to Dictionary. You can now click this row to add values to the dictionary. To specify the 12/24-hour preference,

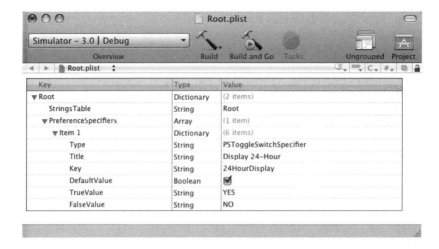

Figure 9.5: CREATING A BOOLEAN PREFERENCE

you'll need to add the following key-value pairs to the dictionary, also shown in Figure 9.5.

- Type: The string PSToggleSwitchSpecifier
- Title: A short user-readable string, like "Display 24 hour"
- Key: A string, like "24HourDisplay," for your application to use as a dictionary key for finding this preference value
- DefaultValue: A boolean default value
- TrueValue: The string to be provided to your application when the user has switched this preference on
- FalseValue: The string to be provided to your application when the user has switched this preference off

Providing the time zones is potentially more burdensome. The time zones provided to the picker in the FlippingPreferableClock (in the earlier Section 9.1, *Displaying a Flippable Preference View in Your Application*, on page 163) came from NSTimeZone's knownTimeZoneNames, which generates several hundred time zone strings. As you might imagine, using Xcode's point-and-click interface to create all those table cells and enter all those strings would be really burdensome.

Fortunately, the property list GUI is not the only way to edit .plist files. If you Ctrl+click or right-click Root.plist, you'll see that the Open As submenu gives you four options: XML Property List, Text Property List,

Figure 9.6: EDITING ROOT.PLIST AS SOURCE

Source Code File, and Plain Text File. Use either of the last two to open a textual view of the property list, as shown in Figure 9.6, which will be much more amenable to pasting and quickly reformatting large blocks of text, like the time zone names.

At the top of the file, we can see the XML representation of the 12/24-hour clock preference identifier. Now we need to add one for the time zone names. This will be a specifier of type PSMultiValueSpecifier, which takes two arrays, one called Titles and the other Values, to define the preferences as the user sees them and the values provided to the application.[2]

---

2. Yes, there is a maintenance hazard keeping the arrays in sync with each other. We don't know why they didn't use an array of title-value dictionaries.

So, enter the following XML as a new dict child of the PreferenceSpecifiers array:[3]

```
<dict>
        <key>Type</key>
        <string>PSMultiValueSpecifier</string>
        <key>Title</key>
        <string>Time Zone</string>
        <key>Key</key>
        <string>TimeZone</string>
        <key>DefaultValue</key>
        <string>America/Detroit</string>
        <key>Titles</key>
        <array>
                <string>Chicago</string>
                <string>Detroit</string>
                <string>Vancouver</string>
        </array>
        <key>Values</key>
        <array>
                <string>America/Chicago</string>
                <string>America/Detroit</string>
                <string>America/Vancouver</string>
        </array>
</dict>
```

With the bundle created, you can now view and edit your preferences through the Settings application, just as soon as you install the app to the simulator or device with a Build and Go (yes, even though your application doesn't yet do anything).

To try it, run the example program and quit out of it with the home button. Launch the Settings application, and you should see your application with the name you provided in the bundle and the Icon-Settings.png like in Figure 9.7, on the facing page.

Even without writing any code in your application, the preferences GUI is done... a significant win when compared to having to develop your own GUI and persistence scheme, as we did in the first half of the chapter. What's left now is for the application to actually read in and use the preferences.

---

[3]. The downloadable sample code uses many more cities. We show just a few here to keep the book's code more readable.

Figure 9.7: SETTINGS APPLICATION DISPLAYING CONFIGURABLE APPLICATIONS

9.6 Loading Preferences Configured in the Settings Application

Having set up the bundle, reading preferences in your application is quite simple. You'll want to start by defining both the keys for your preferences (which of course must match the keys you defined in the preference specifiers in Root.plist) and the default values:

FileIO/BundlePrefereableClock/Classes/BundlePrefereableClockViewController.m
```
NSString *TwentyFourHourPrefKey = @"24HourDisplay";
NSString *TimeZonePrefKey = @"TimeZone";
NSString *DefaultTimeZonePref = @"America/Detroit";
```

Next, to read in the settings, you get an NSUserDefaults object by means of the class method standardUserDefaults. The NSUserDefaults lets you retrieve preferences as strings, dictionaries, BOOLs, and so on, by means of methods named stringForKey:, dictionaryForKey, boolForKey, and so on. Define the instance variables BOOL show24Hour and NSString *timeZoneName in BundlePreferableClockViewController.h, and then add the following method to the .m file to read the preferences into these ivars:

FileIO/BundlePrefereableClock/Classes/BundlePrefereableClockViewController.m

```
- (void) loadPrefs {
        // set app defaults
        timeZoneName = DefaultTimeZonePref;
        // read user prefs
        NSUserDefaults *defaults = [NSUserDefaults standardUserDefaults];
        NSString *userTimeZone = [defaults stringForKey:TimeZonePrefKey];
        if (userTimeZone != NULL)
                timeZoneName = userTimeZone;
        [userTimeZone release];
        show24Hour = [defaults boolForKey:TwentyFourHourPrefKey];
}
```

As you can see, once the NSUserDefaults object is loaded, it's pretty simple to pull out the preferences values one by one, though you could also get the entire set of preferences in a single call with the method dictionaryRepresentation. It's also worth noting that the stringForKey: may return NULL if the user hasn't set that preference, which is why you want to always be able to fall back on the defaults in your code.

You'll call loadPrefs in viewDidLoad, so the preferences will be read in when the view comes up.

Speaking of the view, we still haven't built the clock GUI that uses these preferences. As with the first clock application, you'll need to create labels in IB for the time and the time zone name connected to IBOutlets called timeLabel and timeZoneLabel. Also declare an NSTimer *clockViewUpdateTimer instance variable for the label-updating timer, and an NSDateFormatter *clockFormatter to format it.

FileIO/BundlePrefereableClock/Classes/BundlePrefereableClockViewController.m

```
-(void) setClockFormatter {
        if (clockFormatter == nil) {
                clockFormatter = [[NSDateFormatter alloc] init];
        }
        if (show24Hour)
                [clockFormatter setDateFormat: @"H:mm:ss"];
        else
                [clockFormatter setDateFormat: @"h:mm:ss a"];
        // also update time zone
        [clockFormatter setTimeZone: [NSTimeZone timeZoneWithName: timeZoneName]];
}
```

```
-(void) updateClockView {
        NSDate *dateNow = [NSDate date];
        timeLabel.text = [clockFormatter stringFromDate: dateNow];
        timeZoneLabel.text = timeZoneName;
}

-(void) startClock {
        [self updateClockView];
        if (!clockViewUpdateTimer) {
                clockViewUpdateTimer = [NSTimer scheduledTimerWithTimeInterval:0.2
                                target:self
                                selector:@selector (updateClockView)
                                userInfo:NULL
                                repeats:YES];
        }
}

- (void)viewDidLoad {
        [super viewDidLoad];
        [self loadPrefs];
        [self setClockFormatter];
        [self startClock];
}
```

Click the application's title or icon, and you'll be taken to the first page of its settings, as shown in Figure 9.2, on page 174.

Try changing the settings in this interface, and then go back and rerun the BundlePreferableClock in the simulator or on the device. You should see your preferences honored in the display. One important thing to keep in mind is that when you do a Build and Go in Xcode, the new settings bundle will be copied over to the simulator or device, eliminating the preferences you had set before. So, when you test your preference handling, you'll want to be sure that you stay in the simulator and don't go back to Xcode until you've tested everything.

A Hybrid Approach

In the first section of this chapter, we didn't spend a lot of time on the idea of how to store your properties, noting that it's easy enough to just store .plists in your application's Documents directory, in whatever file format suits you. But even if you want to provide your own preferences GUI, through a flippable interface or perhaps some other presentation, you don't *have* to work exclusively with the file APIs.

Take a look at the documentation for the NSUserDefaults that we use to load the preferences set in the system Settings application. Along with the preference "getters" like boolForKey, integerForKey, and stringForKey:, there are corresponding "setter" methods (setBool:forKey:, setInteger:forKey:) and a catchall setObject:forKey: that works with any of the property list types.

What this means is that if you choose to provide your own preferences GUI, you have a choice: manage the files yourself if that makes more sense for your data, or use the main preferences database by using NSUserDefaults to both read and write the properties.

Chapter 10
The SQLite Database

Have you written database code before? If you have, let us guess: it ran on the server, it was part of some business application, and the license for the database cost more than your house. Any hands up for that one?

Let's consider another vision: a powerful database... in your pocket... for whatever apps you care to use it for... for *free*. That's what you get with the iPhone's built-in SQLite database.

And for readers who haven't used a database in an application before, you're in luck, because the power afforded by a relational database often makes managing your application faster and more reliable than you could ever achieve with flat files. Consider the iPod application, for example. If you relied on the metadata inside the MP3 and AAC files, then finding and playing a given song by its title, or all the songs by an artist, would become a woefully expensive operation as you opened and read thousands of files. With a database, it's a matter of composing sensible queries like select * from music where artist like "%Zappa%"[1] and getting results in a fraction of a second.

The iPhone OS includes the SQLite database, a simple, lightweight, database that bills itself as "the most widely deployed SQL database engine in the world." You can learn more about SQLite at its home page at http://sqlite.org.

In this section, we'll learn the techniques for building and using SQLite databases in your application.

---

1. As always, our relative tastes in music may vary.

> ### Joe Asks...
> #### Should I Be Using Core Data Instead?
>
> Maybe you should. Take a look at this chapter and Chapter 11, *Core Data*, on page 203, and you'll immediately realize they're radically different. Directly using the SQLite API is low-level, manipulating database tables directly. Sometimes that's good for straightforward persistence, but it starts to get tricky when you have to model complex relationships, like one-to-many and many-to-many. Core Data lets you work at the object level, which makes life easier when each object is related to several others (which in turn may have their own object relationships) and you need to maintain those relationships. Core Data is also designed to present its managed objects in a UITableView, whereas apps that use the database directly will have to provide their own table data models (as this chapter's example does).
>
> If you are porting existing SQL code or you already have SQL expertise on your team, you may prefer to work with SQLite. That said, in bringing Core Data to the phone, Apple has made it clear that it is the preferred framework for persisting data on the device or desktop.

10.1 Creating Your Database

Conveniently, your Mac also has SQLite, version 3, as revealed by a simple which on the command line:

```
⇐  Yuna:~ cadamson$ which sqlite3
⇒  /usr/bin/sqlite3
```

We'll use the sqlite3 to create a database that you can then distribute with your application.

For this section's example, we'll build a database-backed "shopping list" that allows the user to add new items and list those that have already been entered.

Before we create the iPhone application itself, let's build the database format and seed it with some data. Like most databases, SQLite works with Structured Query Language (SQL). This is a widely supported standard for expressing database commands as human-readable command strings, with the added benefit that it's a simple matter to create SQL commands in code at runtime. Like most databases, SQLite has an interactive mode that allows you to enter SQL commands and execute them immediately. So, open Terminal (or xterm or whatever), and cd to the DatabaseShoppingList directory. A shopping.db database file already exists; delete or rename it so you can create a new one from scratch.

Now, begin your SQLite session by calling the sqlite3 command, along with a database file:

```
Yuna:DatabaseShoppingList cadamson$ sqlite3 shopping.db
SQLite version 3.4.0
Enter ".help" for instructions
sqlite>
```

This opens the shopping.db file (if it doesn't exist, the file will be created once you write some data to it). Now you can immediately begin entering SQL commands. Since there's no data in the empty database, we'll begin by creating a table for the shopping list, where each row will contain an item name, its price, a group code, and the date the item was added. All it takes is a create table command, followed by a go.[2]

```
sqlite> create table shoppinglist (key integer primary key,
   ...> item text,
   ...> price double,
   ...> groupid integer,
   ...> dateadded date);
sqlite> go
```

This creates the structure of the shoppinglist table, but it doesn't yet have any contents. Add an item with a SQL insert command, like this:

```
sqlite> insert into "shoppinglist" (item, price, groupid, dateadded)
   ...> values ("iPhone 3G", "199", 1, DATETIME('NOW'));
sqlite> go
```

Now you can do a SQL query to see the item you've added:

```
sqlite> select * from shoppinglist
   ...> go
1|iPhone 3G|199.0|1|2008-06-28 12:35:24
```

Having done that, you can seed more data for the application with a script file that we've included with the downloadable sample code. Quit sqlite3 with the .quit command to return to the command line. Delete

---

2. Note that the ...> in the output listing is an indentation inserted by sqlite3 where we've hit the Return key while entering the command. Since the semicolon, and not the Return key, completes the command, you can break lines as you please.

the shopping.db file with rm shopping.db. The download contains a script called make-table-script to insert more items into the shopping list. You can run this script by providing it as input to the sqlite3 command:

⇐ `Yuna:DatabaseShoppingList cadamson$ sqlite3 shopping.db < make-table-script`

Now, go back into SQLite's interactive mode to examine the data that's been created:

```
⇐ Yuna:DatabaseShoppingList cadamson$ sqlite3 shopping.db
⇒ SQLite version 3.4.0
  Enter ".help" for instructions
⇐ sqlite> select * from shoppinglist
⇐    ...> go
⇒ 1|Pop Zero 12-pack|3.99|0|2008-06-28 12:42:23
  2|Mac Pro|2499.99|1|2008-06-28 12:42:23
  3|iPhone 3G|199.0|1|2008-06-28 12:42:23
  4|Potato chips|2.49|0|2008-06-28 12:42:23
  5|Frozen pizza|3.75|0|2008-06-28 12:42:23
  6|Final Cut keyboard|169.95|1|2008-06-28 12:42:23
  7|Mints|0.98|0|2008-06-28 12:42:23
  8|ADC select membership|500.0|1|2008-06-28 12:42:23
  9|iPhone individual membership|99.0|1|2008-06-28 12:42:23
  sqlite>
```

For purposes of the example, we tried to provide the kinds of things that geeks would typically shop for. That's why there are only two group codes: 0 represents groceries, and 1 represents tech. After all, when you have munchies and gear, what else do you really need? [3]

If you haven't done a lot of database work, you might wonder about this key field we've created and marked as a primary key, without ever actually populating. It is a common practice in database programming to give each row a unique ID, often created by the database itself, in order to make unambiguous references to the row later. For example if you had multiple entries for "Frozen pizza," all effectively identical, and wanted to update or delete one, you'd want a unique ID to tell the database what to delete (for example, delete from shoppinglist where key=5). It's also crucial if you later want to combine rows from multiple tables, for example, if you wanted to have a second table that listed a variable number of stores where you might buy items from the shopping list. In such a case, each row of the itemsatstores table would refer to the id of a row in the shoppinglist table.

---

[3]. Actually, we considered having a third group for action figures but figured it would be too much of a hassle coming up with authentic-sounding superhero and RPG character names that wouldn't get us sued by the trademark police.

10.2 Creating the Sample Application

To exercise some database code, our iPhone application will support two functions: adding to the shopping list and viewing its contents. We'll do this with a tab bar application, with one tab each (and therefore one view controller each) for the adding and viewing functions. In Xcode, create a new tab bar application, which we'll call DatabaseShoppingList. By default, the app has two tabs, which is just what we need, though we've chosen to use the Refactor command to rename FirstViewController to AddItemViewController.

By default, the MainWindow.xib has two view controllers to handle the two tabs, which is what we want, though they're named First and Second by default. Find the Selected First View Controller in Interface Builder (you'll have to change IB's view mode to list or column to see it), expand it to get to its Tab Bar Item, bring up the Identity inspector, and change its Title to Add Item. Similarly, change the second view controller's tab bar item's Title to List by Price. We also added small PNG icons to the downloadable example at this point.

MainWindow.xib includes a view for the first tab by default, and rather than create a separate nib, it's fine to just customize this. Delete any current contents and add text fields for an item name and price, a picker for the item type, an Add Item button, and a label at the bottom that we'll use to confirm each add.

You'll then need to edit AddItemViewController.h to add outlets for the text fields, picker, and label, and you'll need to provide an action to handle the button click. Also, implement the picker delegate and data source methods, as well as UITextFieldDelegate, so you can dismiss the keyboard when the user clicks Return. Then connect these outlets and headers from the GUI components to the AddItemViewController. Also connect the text fields' delegates to this view controller, as well as the picker's delegate and dataSource.

SQLite/DatabaseShoppingListComplete/Classes/AddItemViewController.h

```objc
@interface AddItemViewController : UIViewController <UIPickerViewDelegate,
                    UIPickerViewDataSource, UITextFieldDelegate> {

    UITextField *itemNameField;
    UITextField *priceField;
    UIPickerView *groupPicker;
    UILabel *statusLabel;

}
- (IBAction) addShoppingListItem: (id) sender;

@property (nonatomic, retain) IBOutlet UITextField *itemNameField;
@property (nonatomic, retain) IBOutlet UITextField *priceField;
@property (nonatomic, retain) IBOutlet UIPickerView *groupPicker;
@property (nonatomic, retain) IBOutlet UILabel *statusLabel;

@end
```

The picker's delegate and data source methods have nothing to do with the database and can be written now. numberOfComponentsInPickerView: always returns 1, pickerView:numberOfRowsInComponent: returns 2, and pickerView:titleForRow:forComponent: returns the string Groceries if the row is 0 or Tech if it's 1.

The second view is just a table, so it's easier to set up. In Xcode, use File > New File to create a new UITableViewController subclass, called ListByPriceViewController. If you like, you can rename the SecondView.xib file to ListByPriceView.xib. Either way, edit this file in IB, and delete the default view object, replacing it with a table view. Be sure to check the File's Owner object's outlets, reconnecting its view outlet to the table view. Over in MainWindow.xib, select the second view controller, and use the Identity inspector to change its class to ListByPriceViewController. If you renamed the corresponding nib, you'll also need to use the Attributes inspector to change the nib file from the no longer existent SecondView.xib to ListByPriceView.xib.

We now have the basic outline of the app, so let's integrate the database functionality. If you want to skip setting up the GUI and go straight to the database stuff, we've put just the previous GUI setup in the downloadable code bundle as DatabaseShoppingListStarter, while the finished project is DatabaseShoppingListComplete.

10.3 Putting Your Database on the Device

You have a database file and the basics of an application. Now you're ready to access the contents of the database from your code. To do this, you'll be using SQLite's C API, which is not included by default in Xcode projects. That means you'll need to do two things with any iPhone application that uses SQLite:

1. First, add libsqlite3.dylib to the Xcode project. In Groups & Files, expand the Targets item, select DatabaseShoppingList, and bring up its Inspector with ⌘I or the toolbar Info button. Click General, and look at the bottom half of the window: you'll see the frameworks currently linked into the project. Click the + button to add a new framework, and choose libsqlite3.dylib from the sheet. You'll notice we've already done this for the project in the downloadable sample code.

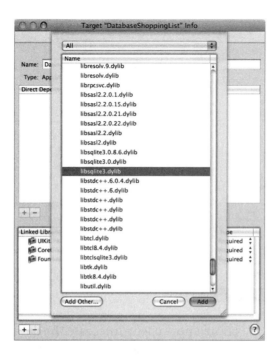

2. In any class that uses the SQLite API, you'll need to include the header file:

   ```
   #include <sqlite3.h>
   ```

You'll also need to copy your database files to the application bundle. You can do this by dragging them from the Finder into the Resources folder in your Xcode project or by Ctrl+clicking the Resources directory and choosing Add > Existing File.

While your database will be in the application bundle, this isn't the file you'll be writing new records to. The reason is simple. You don't have write access to your application's own bundle. Instead, you should copy the database file to another folder—almost certainly the Documents folder—and then work with that file.

Copying to the Documents Directory

So, before using any SQLite functions, copy the database file to a known location (which the DatabaseShoppingListAppDelegate class will refer to with an NSString property called dbFilePath). In the example code, this is a file in the Documents directory whose name is indicated by the constant value DATABASE_FILE_NAME. To set this path when the application starts up, we begin by locating the Documents directory.

>  **Joe Asks...**
>
> ### Do I Have to Create My Database on the Desktop First?
>
> Technically, no. As you'll see in the next few sections, the SQL statements you create in code are no different from those used with the sqlite3 interactive command-line application, so you could do everything at runtime with the programmatic API, including the CREATE TABLE command. When you do this, the initial call to sqlite3_open( ) will create the database file if it doesn't exist already, just as the command-line sqlite3 does.
>
> There might be highly dynamic cases where this is a useful approach, but you'll often find it's more convenient and performant to create your database at development time and include it as part of your application bundle, making a copy to hold the user's data. This approach also lets you revert to the original database easily. Just recopy the original database file from the bundle.

SQLite/DatabaseShoppingListComplete/Classes/DatabaseShoppingListAppDelegate.m
```
NSArray *searchPaths =
      NSSearchPathForDirectoriesInDomains
      (NSDocumentDirectory, NSUserDomainMask, YES);
NSString *documentFolderPath = [searchPaths objectAtIndex: 0];
dbFilePath = [documentFolderPath stringByAppendingPathComponent:
                    DATABASE_FILE_NAME];
```

You'll notice this is a different approach than the NSHomeDirectory( ) function that was shown earlier (in Section 8.2, *Exploring Directories*, on page 135) for finding the application's home directory. This code shown here—Apple's preferred technique for finding the Documents folder—searches known paths for predefined items of interest. In this case, you indicate your interest in the Documents folder by passing the constant NSDocumentDirectory and then pulling out the first path from the returned array.

Next, it's a pretty simple matter to find the database file in the application bundle and have NSFileManager copy it to the Documents directory. In this example, the constants DATABASE_RESOURCE_NAME and DATABASE_RESOURCE_TYPE are just NSStrings for the database filename and extension: shopping and db, respectively.

```
SQLite/DatabaseShoppingListComplete/Classes/DatabaseShoppingListAppDelegate.m
if (! [[NSFileManager defaultManager] fileExistsAtPath: dbFilePath]) {
        // didn't find db, need to copy
        NSString *backupDbPath = [[NSBundle mainBundle]
                pathForResource:DATABASE_RESOURCE_NAME
                ofType:DATABASE_RESOURCE_TYPE];
        if (backupDbPath == nil) {
                // couldn't find backup db to copy, bail
                return NO;
        } else {
                BOOL copiedBackupDb = [[NSFileManager defaultManager]
                        copyItemAtPath:backupDbPath
                        toPath:dbFilePath
                        error:nil];
                if (! copiedBackupDb) {
                        // copying backup db failed, bail
                        return NO;
                }
        }
}
return YES;
```

10.4 Using Your Database on the Device

With the database file now copied to your application's Documents directory, you're now ready to start making calls on it via the SQLite C API. The entire API is quite small, and you'll typically follow a consistent pattern of function calls when using it:

1. sqlite3_open()—to prepare the database for use.
2. sqlite3_prepare()—to set up a prepared SQL statement.[4]
3. sqlite3_step()—to execute the statement, which either writes to the database (in the case of writes, deletes, and updates) or prepares one row of a result set for reading.
4. sqlite3_column_NNN()—to retrieve typed data from the current result set, where *NNN* indicates a data type, such as bytes(), int(), text(), and so on.
5. sqlite3_finalize()—to free resources allocated for use by the database.[5]
6. sqlite3_close()—to close the database.

---

[4] Technically, sqlite3_prepare_v2() is preferred for all new SQLite development.
[5] One of our tech reviewers reminded us that unfinalized statements can result in data loss or corruption, so if you keep a statement around in memory, you might need to finalize it in applicationWillTerminate.

Inserting Values into the Database

So, let's consider the needs of the sample application. It needs to list the items currently in the database—which it can do by querying the shoppinglist table for all its contents and building a GUI from the results—and to allow the user to add new items. We can start with the "add items" functionality, because it's a little easier.

When the user taps the Add Item button, the addShoppingListItem: method begins by sanity-checking the item and price fields. If this succeeds, then we can get the dbFilePath property from the app delegate and use it to open the database with sqlite3_open().

SQLite/DatabaseShoppingListComplete/Classes/AddItemViewController.m

```
if ((([itemNameField.text length] == 0) ||
     ([priceField.text length] == 0) ||
     ([priceField.text doubleValue] == 0.0))
    return;

sqlite3 *db;
int dbrc; // database return code
DatabaseShoppingListAppDelegate *appDelegate = (DatabaseShoppingListAppDelegate*)
    [UIApplication sharedApplication].delegate;
const char* dbFilePathUTF8 = [appDelegate.dbFilePath UTF8String];
dbrc = sqlite3_open (dbFilePathUTF8, &db);
if (dbrc) {
    NSLog (@"couldn't open db:");
    return;
}
```

As a procedural C API, using SQLite means you'll be working with habits different from what you're used to from Cocoa and Objective-C. Specifically, you don't get to work with objects per se, and instead you'll need to create structures that you'll pass into various function calls. You can see that here with db, a pointer to an sqlite3 structure. Rather than allocing it, you call the sqlite3_open() function and pass in an address where you'd like to receive a pointer to a newly created sqlite3 **struct**.

You might also notice that as a C API, you have to pass in the database path as a pointer to a C string: a const char*, terminated by a nul character. Fortunately, NSString's UTF8String method returns exactly what we need here and autoreleases it as an added benefit. Also note that all SQLite functions return a "result code" value, whose values are constants defined in sqlite3.h. In most cases, 0, the value for the constant SQLITE_OK, indicates success, and nonzero results indicate errors or other exceptional conditions.

With the database now open, you are ready to construct an INSERT statement to add your row of data. Here's how the shopping list application does it: [6]

SQLite/DatabaseShoppingListComplete/Classes/AddItemViewController.m
```
sqlite3_stmt *dbps; // database prepared statement
NSString *insertStatementNS = [NSString stringWithFormat:
   @"insert into \"shoppinglist\"\
   (item, price, groupid, dateadded)\
   values (\"%@\", %@, %d, DATETIME('NOW'))",
   itemNameField.text,
   priceField.text,
   [groupPicker selectedRowInComponent: 0]];
const char *insertStatement = [insertStatementNS UTF8String];
dbrc = sqlite3_prepare_v2 (db, insertStatement, -1, &dbps, NULL);
dbrc = sqlite3_step (dbps);
```

In this block, we begin by reserving a pointer to a prepared statement **struct**, that is, a sqlite3_stmt, and then assembling a SQL INSERT statement by filling in values for a format string with the content of the UITextFields and the selected row of the group picker. The resulting string will be an INSERT statement just like those you used earlier in the command-line interactive mode. You then use the sqlite3_prepare_v2() function to create a prepared statement from the C-friendly null-terminated version of this string, passing -1 as the string length to tell SQLite to figure out the length by finding the null terminator (which spares us from having to account for the possibility of multibyte characters in the UTF-8 string).

To actually execute the insert, you call sqlite3_step(). If this method returns SQLITE_DONE, your row has been added to the database.

With the statement having served its purpose, you should now clean up and release any resources it has allocated. You can reuse the sqlite3 struct to create more prepared statements, but in the sample code, I've chosen to just always clean up and close everything after each insert.

---

6. The backslash characters at the end of each line in the insertStatementNS string are not necessarily part of the syntax; they're line-wrap characters needed only for formatting the string for the book's code example. You don't have to break your strings like this, though it can make your code more readable to do so.

$SQLite/DatabaseShoppingListComplete/Classes/AddItemViewController.m
```
// done with the db.  finalize the statement and close
sqlite3_finalize (dbps);
sqlite3_close(db);
```

One final task to attend to in AddItemViewController is to dismiss the virtual keyboard when the user taps Return, which we have do to make sure the user can see and tap the Add Item button. We already set both text fields' delegates to the view controller, so we just need a boilerplate method to resign the first responder:

$SQLite/DatabaseShoppingListComplete/Classes/AddItemViewController.m
```
- (BOOL)textFieldShouldReturn:(UITextField *)textField {
    [textField resignFirstResponder];
    return YES;
}
```

> **Care and Feeding of SQL**
>
> The SQL statements we're using in this example are unrealistically simple, so we can focus on the SQLite API. In particular, we're not doing anything to clean up the input, such as escaping quotes or ensuring that the user input isn't trying to perform its own nefarious SQL (a so-called SQL injection attack).
>
> It's not that we don't think such techniques aren't important—we do—it's just that offering a proper introduction to SQL is way beyond our scope. There are tens of thousands of books on SQL. (Literally. We counted.) If you're at all serious about using the iPhone's database, then either you've already read at least one or you're going to want to do so.

Reading Values from the Database

Having inserted a row into the database, reading rows turns out to be a nearly identical process. The biggest difference in doing a query, other than the content of the prepared statement, is that after you do the sqlite3_step(), you iterate over the results to pull out the values returned by your query.

In the DatabaseShoppingList example, the entire contents of the shoppinglist table can be read into memory each time the view controller for the shopping list gets the viewWillAppear: message. A loadDataFromDb method queries the database and caches the results for use as the data source of a UITableView, the table view that we set up as the only view in the second nib. Specifically, loadDataFromDb creates an NSArray of NSDictionaries, where each dictionary is one row of the database table, with its fields (item name, kind, and price) maintained as name/value pairs. This way, we can implement the tableView:numberOfRowsInSection: to just return the length of the array.

Opening the database and preparing a statement, with sqlite3_open() and sqlite3_prepare_v2(), work exactly as they did earlier, with the obvious exception of the content of the SQL string:

SQLite/DatabaseShoppingListComplete/Classes/ListByPriceViewController.m

```
NSString *queryStatementNS =
@"select key, item, price, groupid, dateadded\
from shoppinglist order by price";
```

Notice how this query indicates the fields we want the query to return. The result set will contain *only* these fields, in this order. The order becomes important soon when you indicate which columns you want to retrieve.

Next, repeatedly call sqlite3_step(), checking to see whether the return code is SQLITE_ROW. If so, a row of result data is ready for you to retrieve with SQLite's "column" functions, as defined in sqlite3.h:

```
const void *sqlite3_column_blob(sqlite3_stmt*, int iCol);
int sqlite3_column_bytes(sqlite3_stmt*, int iCol);
int sqlite3_column_bytes16(sqlite3_stmt*, int iCol);
double sqlite3_column_double(sqlite3_stmt*, int iCol);
int sqlite3_column_int(sqlite3_stmt*, int iCol);
sqlite_int64 sqlite3_column_int64(sqlite3_stmt*, int iCol);
const unsigned char *sqlite3_column_text(sqlite3_stmt*, int iCol);
const void *sqlite3_column_text16(sqlite3_stmt*, int iCol);
int sqlite3_column_type(sqlite3_stmt*, int iCol);
sqlite3_value *sqlite3_column_value(sqlite3_stmt*, int iCol);
```

As you can see, functions are available for returning BLOBs (binary large objects, which are returned as untyped pointers), **byte**s, **double**s, **int**s, strings (as pointers to null-terminated UTF-8 strings[7]), along with metadata functions for getting the names and types of columns. Each function takes as parameters the statement that you executed with sqlite3_step(), as well as a column index, which refers to the order you specified in your SQL SELECT statement (and not necessarily the order of the columns in the actual database table).

So, here's how we can iterate over the query results, converting from SQLite's C-friendly representations into Cocoa objects, which we then put into an NSDictionary for use later by the table.

SQLite/DatabaseShoppingListComplete/Classes/ListByPriceViewController.m

```
while ((dbrc = sqlite3_step (dbps)) == SQLITE_ROW) {
        int primaryKeyValueI = sqlite3_column_int(dbps, 0);
        NSNumber *primaryKeyValue = [[NSNumber alloc]
                initWithInt: primaryKeyValueI];
        NSString *itemValue = [[NSString alloc]
                initWithUTF8String: (char*) sqlite3_column_text (dbps, 1)];
        double priceValueD = sqlite3_column_double (dbps, 2);
        NSNumber *priceValue = [[NSNumber alloc]
                initWithDouble: priceValueD];
        int groupValueI = sqlite3_column_int(dbps, 3);
        NSNumber *groupValue = [[NSNumber alloc]
                initWithInt: groupValueI];
```

---

7. As of SQLite 3, the database can handle UTF-16 text encodings too.

```objc
            NSString *dateValueS = [[NSString alloc]
                    initWithUTF8String: (char*) sqlite3_column_text (dbps, 4)];
            NSDate *dateValue = [dateFormatter dateFromString: dateValueS];

            NSMutableDictionary *rowDict =
                    [[NSMutableDictionary alloc] initWithCapacity: 5];
            [rowDict setObject: primaryKeyValue forKey: PRIMARY_ID_KEY];
            [rowDict setObject: itemValue forKey: ITEM_KEY];
            [rowDict setObject: priceValue forKey: PRICE_KEY];
            [rowDict setObject: groupValue forKey: GROUP_ID_KEY];
            [rowDict setObject: dateValue forKey: DATE_ADDED_KEY];
            [shoppingListItems addObject: rowDict];
            // release our interest in all the value items
            [dateValueS release];
            [primaryKeyValue release];
            [itemValue release];
            [priceValue release];
            [groupValue release];
            [rowDict release];
}
```

One thing worth noting is the handling of dates. You might have noticed in the previous code that there's no C function to retrieve a date. You might think that you could get a numeric value, say, one that represents the number of milliseconds since a certain "epoch" date, but it doesn't work that way (instead, you just get back the year of the date value). What works is to get a string representation of the date and to then parse that with NSDateFormatter. You just have to keep in mind that SQLite date values don't maintain a time zone, and if you use the database's own date commands—like we did in make-table-script by using the DATETIME('NOW') statement—then those times are assumed to be in UTC. Actually, that makes for a pretty good system for consistent timekeeping: always convert times to UTC before you insert, assume all times retrieved from the database are in UTC, and convert to the local time zone as necessary.

So, in the previous code, we use an NSDateFormatter to convert the database's string representation into an NSDate. This formatter is initialized in initWithNibName:bundle:.

`SQLite/DatabaseShoppingListComplete/Classes/ListByPriceViewController.m`

```objc
dateFormatter = [[NSDateFormatter alloc] init];
[dateFormatter setTimeZone: [NSTimeZone timeZoneWithAbbreviation:@"UTC"]];
[dateFormatter setDateFormat: @"yyyy-MM-dd HH:mm:ss"];
```

We then create a formatter, set its time zone to UTC, and provide the date string format returned by the database, in Locale Data Markup Language (LDML) format (aka *Unicode Technical Standard 35* [Uni04]).

Assuming that loadDataFromDb (called from viewWillAppear:) has created the instance variable shoppingListItems array, the tableView:cellForRowAtIndexPath: method can grab an index from this table by its index, cast it to an NSDictionary, and use the name/value pairs to populate a table cell. With multiple items of interest in each cell, this is an ideal time to use custom IB-designed table cells, as introduced in Section 5.7, *Custom Table View Cells*, on page 94. For the example we've included in the downloadable code, we've provided fields for the item name, price, and group; you can of course lay out your cells differently and use some or all of the fields retrieved from the database.

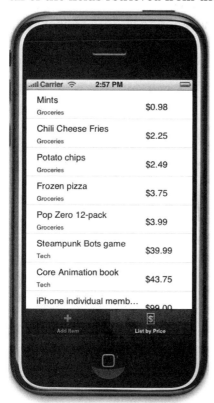

Once a cell has either been dequeued from the table or been loaded from the nib, setting its values is just a matter of reading from the array of dictionaries created earlier by loadDataFromDb.

SQLite/DatabaseShoppingListComplete/Classes/ListByPriceViewController.m
```
UILabel *itemLabel = (UILabel*) [myCell viewWithTag:1];
UILabel *groupLabel = (UILabel*) [myCell viewWithTag:2];
UILabel *priceLabel = (UILabel*) [myCell viewWithTag:3];
NSDictionary *rowVals =
        (NSDictionary*) [shoppingListItems objectAtIndex: indexPath.row];
NSString *itemName = (NSString*) [rowVals objectForKey: ITEM_KEY];
```

```
itemLabel.text = itemName;
int groupid = [(NSNumber*) [rowVals objectForKey: GROUP_ID_KEY] intValue];
groupLabel.text =  GROUP_NAMES [groupid];
NSNumber *price = (NSNumber*) [rowVals objectForKey: PRICE_KEY];
priceLabel.text = [priceFormatter stringFromNumber: price];
```

This is a fairly simple example of what you can do with the iPhone's SQLite 3 database. SQLite 3 is a powerful database, offering triggers, JOINs (including some OUTER JOINs), and much of the SQL-92 standard. In fact, SQLite's web page says it's easier to list the SQL-92 features it doesn't implement (check out *SQL Features That SQLite Does Not Implement* [Hip09] for details) than to list all those that it does. So, if your persistence needs include a database, the iPhone's built-in support will probably be all you need.

Chapter 11
Core Data

We hope you have been convinced by Chapter 10, *The SQLite Database*, on page 185 that you want to use a database. However, if you are one of the majority developers for whom SQL is just not that exciting, then you are in luck. Core Data provides a wonderfully easy-to-use wrapper around SQL that lets us Objective-C developers spend our time thinking in objects instead of queries.

Don't get us wrong—SQL is a great language and has a ton of power. But it has a totally different conceptual model than objects, and it can be a pain in the neck to manually switch your thinking back and fourth. That is where Core Data shines; it provides all the power of the database and none of the hassles of the conceptual switching. You can use Core Data and never even know or care that the data is going into a relational database. That's the beauty of the framework.[1]

In this chapter, we'll look at how to use Core Data to create and use the persistent data in an application. Please grab the CoreData/Conference project from the download code bundle. The chapter leaves out some of the detail of things that we have covered extensively in the previous chapters, and it might help to have the completed code to compare your code to as you work through the example. Much of the code you will need to make this app work is intentionally left out of the discussion because we have covered the concepts in previous chapters. Topics like creating new view controllers and their associated nib files are not covered at all. Let's get started with a look at the application we are going to build.

---

1. Of course, understanding SQL and relational databases can help when debugging, but that knowledge is not required to use and understand Core Data.

11.1 The Conference Application

To help us explore the Core Data APIs, we'll build an application called Conference. This application will help users manage which sessions they want to attend at a conference. For each track in the conference, we will display a list of sessions in that track. Users then select the sessions they want to attend.

The Conference application has four screens. Let's look at each one in turn. The opening screen displays a list of tracks. The user can add and delete tracks, and when the Edit button is pushed, selecting a track takes you to the editing screen for the track information. Here is the app in edit mode with three tracks added:

The list of tracks displayed here comes from Core Data. All the Track objects are fetched from persistent storage, and that list becomes the underlying data model for this table. We will talk a lot more about the classes that Core Data provides to make this job easier in Section 11.5,

Track Table View, on page 219 where we discuss the RootViewController that backs this table view.

Tracks can be added to the list by tapping on the + button in the top-right side. When that button is clicked, we will call on Core Data to make the new object and place it into the *managed object context* so it can be persisted.[2] Clicking any of the red – buttons on one of the rows will allow the user to delete the object. After the changes are made, as you'd expect, we save them by asking Core Data to persist the changes.

The next screen allows the user to edit the track's name and abstract. This screenshot shows the IT track selected:

As the text of the name or abstract for a track is edited, the track's properties are updated. These changes too are registered, but we don't have to code that. Core Data is watching our objects and makes note

---

2. We are going to go over the managed object context in detail in the next section.

of any changes. When the user finishes here, the changes are persisted again.

When the track list is not in editing mode, selecting a track takes the user to the list of sessions for that track. Here is a screenshot of the list of sessions for the iPhone track:

Pressing the + button, as you'd expect, creates a new Session. We use Core Data to create the new object and persist it to the database.

And finally, selecting a particular session takes you to the editing screen for that session. A screenshot of that interface is shown on the next page.

When the user is finished making changes to the session information, the text is placed back on the session object, and as you'd expect, Core Data makes note of the changes since it's watching the session object. In fact, Core Data is watching all objects that are managed by it. Changes we make to any object that we get from Core Data are catalogued so they can be persisted.

As you can see from the application's description, we are going to be making heavy use of Core Data to provide the functionality of our Conference application. We have not talked in much detail yet about how Core Data provides all its functionality or how we get it set up to do so, so the next section is going to explain how to get Core Data set up and a bit of how it works.

11.2 The Core Data Stack

Let's get started by creating the new project. Our data model is list oriented and hierarchical, so we want to start with the navigation-based application template. Call it Conference, and don't forget to select the box that says "Use Core Data for storage."

A lot of what is created by the template is old hat by now. The familiar app delegate and root view controller classes and the associated nib files are familiar. We have seen most of this stuff several times now, and we know what it is. However, when we start to dig into what we've made by instantiating this template with "Use Core Data for storage" enabled, we start to see some subtle differences. Before we dig into implementing our application, let's take a look at the application delegate and see what has been added to our familiar template.

For starters, the app delegate class (ConferenceAppDelegate) has quite a bit more to it than we have seen in the past. While we have typically seen a window and viewController of one sort or another in this class, several new properties are related to Core Data now. Here is the header file:

CoreData/Conference01/Classes/ConferenceAppDelegate.h

```
Line 1  @interface ConferenceAppDelegate : NSObject <UIApplicationDelegate> {
    -       NSManagedObjectModel *managedObjectModel;
    -       NSManagedObjectContext *managedObjectContext;
    -       NSPersistentStoreCoordinator *persistentStoreCoordinator;
    5
    -       UIWindow *window;
    -       UINavigationController *navigationController;
    -   }
    -
   10   @property (nonatomic, retain, readonly)
    -           NSManagedObjectModel *managedObjectModel;
    -   @property (nonatomic, retain, readonly)
    -           NSManagedObjectContext *managedObjectContext;
    -   @property (nonatomic, retain, readonly)
   15           NSPersistentStoreCoordinator *persistentStoreCoordinator;
    -   @property (nonatomic, readonly) NSString *applicationDocumentsDirectory;
    -
    -   @property (nonatomic, retain) IBOutlet UIWindow *window;
    -   @property (nonatomic, retain) IBOutlet
   20           UINavigationController *navigationController;
    -
    -   - (IBAction)saveAction:sender;
    -
    -   @end
```

This set of four new properties work together in what is called the Core Data *stack*. This stack of objects is the basis of how Core Data works. These are the objects that give Core Data its really cool feature set. The four objects that make up this stack are the persistent object store, the persistent store coordinator, the managed object context, and the managed object model. Here is a diagram depicting the four objects and their relationships. A discussion of each object follows.

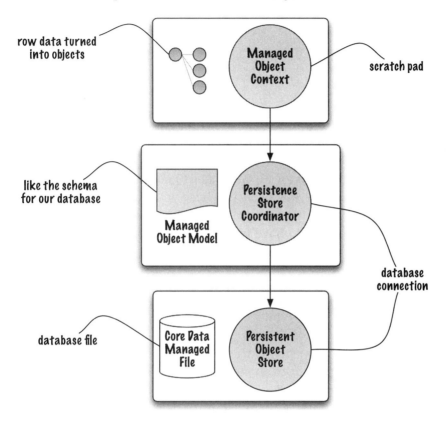

The *persistent object store* (or POS for short) performs all the lowest-level translation of object-speak to data as well as managing the opening and closing the underlying file. It is created, managed, and heavily used by the persistent store coordinator. Consider the SQLite POS as an example; objects go into the POS, SQL comes out and is pushed through the SQLite API, and a change to the database file results. When objects are needed, a request comes in from the persistent store coordinator and is translated by the POS to SQL queries. The SQL is then sent through the SQLite API, and the returned records are used by the persistent store coordinator to produce objects.

Core Data ships with three POS implementations: SQLite, binary, and memory. I recommend always using the SQLite implementation because it's the easiest to debug and see what is actually persisting. The binary implementation is good for putting really small data sets into if you don't want the performance overhead of SQL and your data is very simple. The memory implementation is good for temporary data that does not need to persist beyond the POS.

The *persistent store coordinator* (or PSC for short) provides a generalized cover over the persistent object store. In the most general case, you can have multiple object stores, but on the iPhone you rarely will. Where the persistent object store is specific to a particular store, the persistent store coordinator is more general. The persistent store coordinator uses the managed object model to help it understand the form and layout of the objects that are being persisted to the stores.

The *managed object model* (or MOM for short) contains the description, or the metadata, of your model. It is where you describe the *entities*[3] and their properties that make up the model of your application. In Conference, we are going to have two entities, Session and Track.

The last object in the stack, and the one we will be interacting with the most, is the *managed object context* (or MOC for short). The MOC is used as a scratch pad. Objects are pulled through the stack into the MOC and then kept there while we change them. All inserts, deletes, and updates to the set of objects in the MOC are held until we tell the MOC to save. At that point, the MOC's list of changes is pushed down through the stack, at each step translated closer to the eventual language of the POS where it eventually becomes native (that is, SQL statements for the SQLite POS) and sent to the persistent storage.

This stack of objects is created for us by the app delegate. All the code to build out the stack is generated for us when we check that little "Use Core Data for storage" checkbox. Let's go look at how these objects are created for us.

11.3 Building the Core Data Stack

You probably noticed that each of the Core Data–related properties is read-only in the ConferenceAppDelegate header file. That is because each of them is actually *derived* (or built when asked for). Instead of relying on some outside object to create and set these objects, the app

---

3. An entity is something that is modeled, like the Session in our example. An entity captures features of the modeled thing, such as its attributes and relationships to other entities.

delegate creates them. Let's look at the implementation provided by the template and see how the stack is created.

Let's start on the bottom of the stack with the persistent store coordinator and persistent object store. Here is the code:

CoreData/Conference01/Classes/ConferenceAppDelegate.m

```
Line 1  - (NSPersistentStoreCoordinator *)persistentStoreCoordinator {
     -      if (persistentStoreCoordinator != nil) {
     -          return persistentStoreCoordinator;
     -      }
     5
     -      NSURL *storeUrl = [NSURL fileURLWithPath:
     -                          [[self applicationDocumentsDirectory]
     -                              stringByAppendingPathComponent: @"Conference.sqlite"]];
     -
    10          NSError *error = nil;
     -      persistentStoreCoordinator =
     -      [[NSPersistentStoreCoordinator alloc]
     -          initWithManagedObjectModel: [self managedObjectModel]];
     -      if (![persistentStoreCoordinator
    15          addPersistentStoreWithType:NSSQLiteStoreType
     -          configuration:nil
     -          URL:storeUrl
     -          options:nil
     -          error:&error]) {
    20          // Handle error
     -      }
     -
     -      return persistentStoreCoordinator;
     -  }
```

Three things are happening in this method. First we grab the documents directory and add Conference.sqlite to the end; we use that path to create a URL for our database file. Next we create an instance of NSPersistentStoreCoordinator with the managed object model on line 11. And finally we configure a new persistent object store starting on line 14. The persistent object store uses the SQLite type and stores the database in the file specified by storeURL. The last argument is a pointer to an error; if something goes wrong, then error will be set to an instance of NSError that contains detailed information about what went wrong.

The applicationDocumentsDirectory property is used to find the application's Documents directory. It is calculated every time the get method is called. For more information, see Section 10.3, *Copying to the Documents Directory*, on page 192.

With these few lines of code, we have the Core Data stack connected to a SQLite database file. Let's look at how we load the managed object model next. Here is the code:

```
CoreData/Conference01/Classes/ConferenceAppDelegate.m
```
```
- (NSManagedObjectModel *)managedObjectModel {
  if (managedObjectModel != nil) {
    return managedObjectModel;
  }
  managedObjectModel = [[NSManagedObjectModel mergedModelFromBundles:nil]
                        retain];
  return managedObjectModel;
}
```

In this code, we use the mergedModelFromBundles: to grab all the model files that are in our application's bundle and merge them into one MOM. This class method looks through the whole application bundle looking for model files, and each one is loaded and merged into the overall MOM. In practice, though, most iPhone applications will have only one model.

The last piece of the stack is the managed object context. Here is the code to create the MOC:

```
CoreData/Conference01/Classes/ConferenceAppDelegate.m
```
```
- (NSManagedObjectContext *) managedObjectContext {
  if (managedObjectContext != nil) {
    return managedObjectContext;
  }

  NSPersistentStoreCoordinator *coordinator =
          [self persistentStoreCoordinator];
  if (coordinator != nil) {
    managedObjectContext = [[NSManagedObjectContext alloc] init];
    [managedObjectContext setPersistentStoreCoordinator: coordinator];
  }
  return managedObjectContext;
}
```

This code does two things: it grabs the persistent store coordinator and then uses that to configure the managed object context.

Now that we have seen how the Core Data stack is created and we have talked about how it is used, we can start to dig into how we are going to use Core Data in Conference. We will start by modeling the classes that make up our model.

11.4 Modeling

Now that we have seen the way the template has set up Core Data for us, let's start fleshing out our application by building our managed object model.

The managed object model is a central part of how Core Data interacts with your model. Core Data uses the entity when it translates between the world of Objective-C and the world of SQLite. The model is similar to the schema of your database with a little bit of extra information about how to translate from Objective-C to SQL and back. Roughly speaking, each entity corresponds to a table in the database, and each attribute corresponds to a column in that table. On the Objective-C side, each entity corresponds to a class, and each attribute corresponds to a property on that class. There is, of course, a ton of detail behind how a MOM is used to map objects into rows, and vice versa, but you don't have to fully understand all that thanks to the great tools available in Xcode. Let's take a quick look at the UI. Double-click Conference.xcdatamodel in Xcode. You should see something that looks more or less like this screenshot:

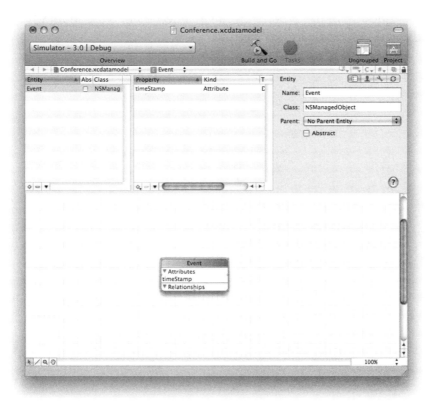

We are going to modify this template data model so it captures our Track and Session entities. Roughly speaking, we need to accomplish these tasks:

- Change the template-provided Event entity to the Track entity.
- Add the name and trackAbstract attributes and the sessions relationship to the Track entity.
- Create the Session entity.
- Add the name, sessionAbstract, and sessionID attributes and the track relationship to the Session entity.
- Generate the Track and Session classes and add them to the project.

With these two entities and the relationship between them modeled, Core Data will have the information it needs to make them persistent.

Start by renaming the Event entity to Track. Select the entity; then in the Entity Attributes inspector, change the name from Event to Track. Also, change the class name from NSManagedObject to Track. Track has two attributes, name and trackAbstract. Both of them are strings with no constraints (length, regex, and so on). Add them by selecting the + button under the properties list (the list to the right of the entity list) and choosing Add Attribute. When you have added them, you should have something that looks roughly like this:

Make sure you turn off Optional for each of the attributes. Although this is not necessary, it will help make sure that no bogus data makes it into our database. As you have probably noticed, there are lots of other ways we can customize the constraints that Core Data will place on our attributes. For example, if we were to provide a regular expression (also known as a *regex*) in the Attributes inspector, Core Data would ensure that whatever value was placed in that attribute matches the regex before it saved the value. If the value fails any of the constraints, a validation error is raised. We can use the error to create a user visible/understandable message to show so the user knows what to fix. We are not going to take the time to go into detail on that, but the *Core Data documentation* [App09b] has more information on presenting errors to your users.

Now that we have the Track entity, we need to create the Session entity. Under the list of entities (where Track currently shows up on the top-left side), click the + button. Rename the new entity to Session, change its class name to Session, and add three string attributes: sessionID, name, and sessionAbstract. Again, all the Session attributes should have no constraints, and the Optional switch should be turned off. When you are done, you should have something that looks like this:

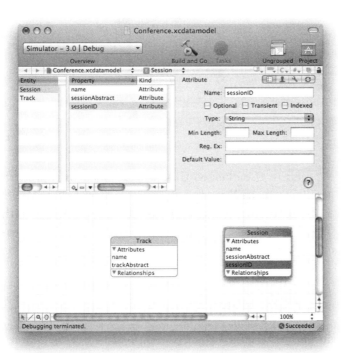

Now that we have the two entities set up, we need to create the relationship between them. Select the Track entity, and click the + button under the property list (top-left side). Choose Add Relationship, and name the new relationship sessions. Make the relationship's destination the Session entity, choose the to-many checkbox, and make sure the delete rule is Cascade. Next we need to add the inverse relationship to the Session entity. Select the Session entity, add a relationship, name it track, set its destination to Track, and choose its inverse to be sessions. Set the delete rule on the new relationship to Nullify, and turn Optional off. When you are done, the model should look something like this:

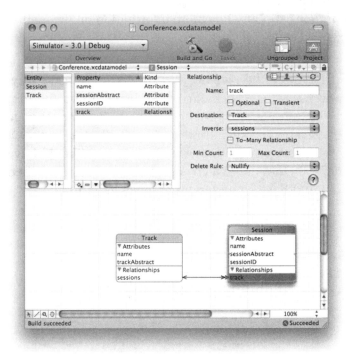

Our next step is to generate the classes that will be used to represent this model in our application. In Xcode, add a new group called Model Classes, and select it. Go back to the model, select all the entities with ⌘-a, and then choose File > New File from the menu. Select the Managed Object Class from the dialog box that pops up, as in the next screenshot.[4]

---

4. If you don't have the model editor as the front window when you select the File > New File menu item, you won't see the Managed Object Class option in the New File Wizard.

Click Next on the next page of the wizard. You don't need to change anything here, but make sure the Conference target is selected. On the final page of the wizard, make sure the checkbox next to Session and Track is selected so that the generator uses both our entities when it makes the classes. The page should look like this:

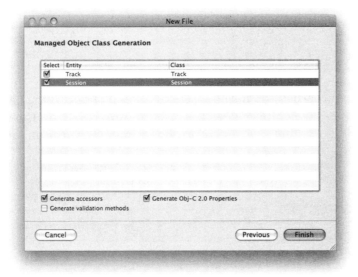

After you click the Finish button, you should have a header and implementation file for both the Track and Session classes.

Let's take a look at the code the generator made for us. Here is the header file for the Session class:

`CoreData/Conference02/ModelClasses/Session.h`

```objc
@class Track;

@interface Session : NSManagedObject {
}

@property (nonatomic, retain) NSString * sessionID;
@property (nonatomic, retain) NSString * name;
@property (nonatomic, retain) NSString * sessionAbstract;
@property (nonatomic, retain) Track * track;

@end
```

Not much that is unexpected here. The generator created a property for each of the attributes and relationships in the model, all of them set to retain their object values. Next let's look at the implementation. Here is the code from the .m file:

`CoreData/Conference02/ModelClasses/Session.m`

```objc
@implementation Session

@dynamic sessionID;
@dynamic name;
@dynamic sessionAbstract;
@dynamic track;

@end
```

This file has very little substance too. The new and interesting piece is the use of @dynamic for the properties. This declaration says to the compiler that the properties will have get/set method pairs provided for them at runtime, so the compiler does not need to provide them. Core Data provides these methods for us, so we don't have to worry about them.

Let's look at the Track class next. Here is the header file:

`CoreData/Conference02/ModelClasses/Track.h`

```objc
Line 1  @class Session;
     -
     -  @interface Track : NSManagedObject {
     -  }
     5
     -  @property (nonatomic, retain) NSString * trackAbstract;
     -  @property (nonatomic, retain) NSString * name;
```

```objc
    @property (nonatomic, retain) NSSet* sessions;

@end

@interface Track (CoreDataGeneratedAccessors)
- (void)addSessionsObject:(Session *)value;
- (void)removeSessionsObject:(Session *)value;
- (void)addSessions:(NSSet *)value;
- (void)removeSessions:(NSSet *)value;
@end
```

The top half is probably what you expected, which is a property for every attribute and relationship in the entity. The CoreDataGeneratedAccessors category might be new, however, so let's look at it in a bit more detail. An Objective-C category is a way for us to add methods to a class. On line 12, the CoreDataGeneratedAccessors category is declared for the Track class. All the methods declared in this category (until the @end) become part of the interface. Now let's look at the implementation:

`CoreData/Conference02/ModelClasses/Track.m`

```objc
@implementation Track

@dynamic trackAbstract;
@dynamic name;
@dynamic sessions;

@end
```

Not much here... the methods that are declared in the CoreDataGeneratedAccessors category are generated by Core Data at runtime (thus the name of the category). We need only the category in the header file so that we can call the methods without a compiler warning. This little bit of Objective-C magic is brought to you by Key-Value Coding (KVC). The Apple *Key-Value Coding* [App08e] documentation has a bunch of the detail. For now, though, you can ignore the detail and just know that you can call addSessionsObject: on any instance of Track.

Now that we have our model in place, Core Data will be able to persist it. Let's look at the code needed to make the Track table view work.

11.5 Track Table View

The good thing is you already know how table views work. They ask their data source for the number of sections, for the number of rows in each section, and then for the cell in a particular location. That part of what we are about to look at is well understood now. The new piece of this is the adaptor that sits between Core Data and the table view data source API. The class that provides that adaptation is NSFetchedResultsController (or FRC for short). Let's look at how it's used in the RootViewController. After we look at how we use it, we will look at the code we used to create it.

Let's start with the basic data source methods. The first is numberOfSectionsInTableView:, and as you know, this method is supposed to return to the table view the number of sections in the displayed data. To do that, we simply ask the fetched results controller like this:

CoreData/Conference02/Classes/RootViewController.m

```
- (NSInteger)numberOfSectionsInTableView:(UITableView *)tableView {
    return [[fetchedResultsController sections] count];
}
```

The fetched results controller knows how many sections are in the data that it's managing, so it is easy for it to return that. The next important method in the data source is the tableView:numberOfRowsInSection: method. This method returns the row count for the particular section. Again, the fetched results controller knows this information, so we can simply ask it. The code looks like this:

CoreData/Conference02/Classes/RootViewController.m

```
- (NSInteger)tableView:(UITableView *)tableView
  numberOfRowsInSection:(NSInteger)section {
    return [[[fetchedResultsController sections] objectAtIndex:section]
            numberOfObjects];
}
```

The tableView:cellForRowAtIndexPath: method is called to get the cell configured and ready to display. And again, the fetched results controller comes to our aid with the objectAtIndexPath: method. Here is the code:

CoreData/Conference02/Classes/RootViewController.m

```
Line 1  - (void)configureCell:(UITableViewCell *)cell withTrack:(Track *)track {
    -       cell.textLabel.text = track.name;
    -       cell.detailTextLabel.text = track.trackAbstract;
    - }
    5
    - - (UITableViewCell *)tableView:(UITableView *)tableView
    -         cellForRowAtIndexPath:(NSIndexPath *)indexPath {
    -     static NSString *CellIdentifier = @"Cell";
```

```
10      UITableViewCell *cell = 
                [tableView dequeueReusableCellWithIdentifier:CellIdentifier];
        if (cell == nil) {
          cell = [[[UITableViewCell alloc] initWithStyle:UITableViewCellStyleSubtitle
                                         reuseIdentifier:CellIdentifier] autorelease];
15      }

        Track *track = [fetchedResultsController objectAtIndexPath:indexPath];
        [self configureCell:cell withTrack:track];

20      return cell;
    }
```

The first method in this code block does the cell configuration by setting the textLabel's text to the name of the track and setting the detailTextLabel's text to the trackAbstract. The cell configuration was pulled out into a separate method because we need to call it from two places. The first one is the tableView:cellForRowAtIndexPath:, and the second we will look at shortly.

Since the FRC knows the data model, it can easily tell us the object we need. The code for tableView:cellForRowAtIndexPath: is boilerplate until we get to line 17. Here is where the FRC comes through for us again. We simply ask for the object at the index path, and it returns it. As you can see, the FRC makes populating a table view with Core Data a breeze. Now let's look at how we created the FRC and talk about some of the additional functionality that we can use from the FRC.

The fetched results controller, as we said earlier, is an adaptor to make fetching and displaying data from Core Data in a UITableView easier. To do that, the FRC uses a bunch of other objects to make it able to perform its duties. Here is a diagram showing the pieces:

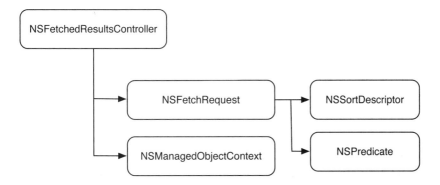

The fetched results controller uses the fetch request and the managed object context to get the data. The fetch request can optionally contain a sort descriptor or predicate. The sort order will, as you probably expected, cause the results of the fetch to be sorted. The predicate will limit the number of objects that come back from the fetch. We don't have space to go into detail, but you can see *Core Data: Apple's API for Persisting Data Under Mac OS X* [Zar09].

In addition to making it really easy to populate the table view, the FRC also manages memory very aggressively. It will ensure that only the objects that are actually needed are kept in memory and will flush out the others whenever a memory warning happens.

Here is the code used to create the FRC for the track table view:

CoreData/Conference02/Classes/RootViewController.m
```
- (NSFetchedResultsController *)fetchedResultsController {
  if (fetchedResultsController != nil) {
    return fetchedResultsController;
  }

      NSFetchRequest *fetchRequest = [[NSFetchRequest alloc] init];
      NSEntityDescription *entity =
  [NSEntityDescription entityForName:@"Track"
              inManagedObjectContext:managedObjectContext];
      [fetchRequest setEntity:entity];

      NSSortDescriptor *sortDescriptor = [[NSSortDescriptor alloc]
                              initWithKey:@"name" ascending:YES];
      [fetchRequest setSortDescriptors:[NSArray arrayWithObject:sortDescriptor]];

      NSFetchedResultsController *aFetchedResultsController =
  [[NSFetchedResultsController alloc] initWithFetchRequest:fetchRequest
                                      managedObjectContext:managedObjectContext
                                        sectionNameKeyPath:nil
                                                 cacheName:@"Root"];
  aFetchedResultsController.delegate = self;
      self.fetchedResultsController = aFetchedResultsController;

      [aFetchedResultsController release];
      [fetchRequest release];
      [sortDescriptor release];

      return fetchedResultsController;
}
```

The code does three things: create the fetch request, add a sort descriptor to that fetch request, and then create the fetched results controller object. The fetch request has a lot of features that allow us to optimize the way the fetch happens. For this simple example, it won't matter, but for big data sets, it's a good idea to dig into the detail (*Core Data* [Zar09] or *Core Data Documentation* [App09b] are both good references).

Next is creating the sort order; here we are saying that we want the track objects to be ordered ascending on the name property. Sorting is important for our users. If we don't impose a sort ordering in the FRC's fetch request, then each time we go back to the database, the order might be different.[5]

We create the FRC using the fetch request object we just built and the managed object context that we got from the app delegate. (We did not show that code, but it's in the applicationDidFinishLaunching: method on the ConferenceAppDelegate from the downloaded code.) We use **nil** for the section key because we don't want any sections. In applications that need a section, you'd specify the name of a property here that the FRC would use to determine which section the object belongs in.

The FRC manages sections for our table view; all we need do is specify a key path. This is yet another "magic" use of Key-Value Coding. Again, the Apple *Key-Value Coding* [App08e] documentation is the place to go for the details. The fetched results controller uses the key path to get the set of values that make up the section names. For each fetched object, the FRC calls valueForKeyPath:. These values are then made unique and become the name for each section. When using this feature, we find it easier to debug if we've placed the table view in the Grouped style or implemented the tableView:titleForHeaderInSection: method[6] so the sections are obvious.

We also set the cache name for our fetched results controller to Root. The FRC caches a bunch of information about the sections under this name. At startup, if there is a cache with this name, the FRC looks to make sure it can use it and, if so, loads that instead of recomputing the section information.

We set our RootViewController to be the delegate of the fetched results controller on line 21. Being the delegate becomes important in the next section where we look at adding new tracks.

---

5. Databases store their data in sets, which are unordered.
6. Get the section information from the FRC by calling [frc.sections objectAtIndex:section], and then get the title from the section information's name property.

Apart from formatting and changing the name of the entity, this code is what we got from the template.

11.6 Fetching the Tracks

The fetched results controller has done a great job making it really easy for us to display the data in our table view. Now let's look at how the FRC gets the data in the first place.

Since we want the tracks to show up at launch time, we place the code to fetch them in the viewDidLoad method. Here is the code:

CoreData/Conference02/Classes/RootViewController.m
```
- (void)viewDidLoad {
  [super viewDidLoad];

      // Set up the edit and add buttons.
  self.navigationItem.leftBarButtonItem = self.editButtonItem;

  UIBarButtonItem *addButton = [[UIBarButtonItem alloc]
          initWithBarButtonSystemItem:UIBarButtonSystemItemAdd
                              target:self
                              action:@selector(insertTrack)];
  self.navigationItem.rightBarButtonItem = addButton;
  [addButton release];

      NSError *error = nil;
      if (![self.fetchedResultsController performFetch:&error]) {
  // handle the error...
      }

  self.title = @"Tracks";
}
```

After setting up the buttons, we send performFetch: to the fetched results controller. The FRC then fetches the objects from the database according to the fetch request. Behind the scenes it also sets up all the data structures it needs to provide the section count and other stuff that makes the FRC good at providing the data to a table view.

11.7 Change the Tracks

Now that we have seen how the fetched results controller helps us get the data into the table view, let's look at what happens as we change that data. The FRC helps us keep the table view in sync with the changes through delegation. Before we dive into how that works, let's look at the insertTrack method to see how we add new objects. Here is the code:

CoreData/Conference02/Classes/RootViewController.m
```
- (void)insertTrack {
  self.firstInsert = [self.fetchedResultsController.sections count] == 0;

  NSManagedObjectContext *context =
        [self.fetchedResultsController managedObjectContext];
  NSEntityDescription *entity =
        [self.fetchedResultsController.fetchRequest entity];
  Track *track =
        [NSEntityDescription insertNewObjectForEntityForName:[entity name]
                                      inManagedObjectContext:context];

  [track setValue:@"Track" forKey:@"name"];
  [track setValue:@"This is a great track" forKey:@"trackAbstract"];

  NSError *error = nil;
  if (![context save:&error]) {
    // Handle the error...
  }
}
```

The first thing we do in this method is check to see whether this is the first insert. We use that property later in the FRC delegate methods. Next we create a new track object. Since we are managing our objects with Core Data, we need to create them via Core Data methods. The implementation of insertNewObjectForEntityForName:inManagedObjectContext: ensures that the proper Core Data–related initialization happens. Next we use the KVC method setValue:forKey: to set the name and trackAbstract properties. Although we could have used the property set methods (that is, track.name = @"Track";), it is very common to see Core Data code use the KVC approach.

Lastly, we ask the context to save the changes by calling the save: method. We ignore the error, which as you'd expect is bad practice in a production application. NSError is a general class that can contain lots of data; it is difficult to describe a general approach to handling them. However, some errors can be recovered from and some cannot. If there is something wrong with the database file and it can't be opened, then there is little you or your user can do to recover. If, however, the error is simple, like the object has been changed in another managed object context, you can present a UIAlertView to the user asking whether they want to revert or overwrite. During the early phase of development, we typically log the errors to the console with NSLog. Then as part of "finishing up" the application, we take a long look at where we're logging errors and decide which (if any) types of errors we want to present to the user. The bottom line is that error handling is very application specific, and you should think about how you want to handle the errors in your application.

Now that we've seen how we add tracks, let's look at deleting them. Being a pro at table views by now, you know that we implement the tableView:commitEditingStyle:forRowAtIndexPath: method to commit the edits that users indicate (via the red Delete button, for example). Here is the code for the method:

CoreData/Conference02/Classes/RootViewController.m
```
- (void)tableView:(UITableView *)tableView
commitEditingStyle:(UITableViewCellEditingStyle)editingStyle
forRowAtIndexPath:(NSIndexPath *)indexPath {

  if (editingStyle == UITableViewCellEditingStyleDelete) {
    NSManagedObjectContext *context =
            [fetchedResultsController managedObjectContext];
    Track *track = [fetchedResultsController objectAtIndexPath:indexPath];
    [context deleteObject:track];

            // Save the context.
    NSError *error;
    if (![context save:&error]) {
      // Handle the error...
    }
  }
}
```

In this method, we are grabbing the selected track and then telling the context to delete the object by calling the deleteObject: method. Then we ask the context to save the changes with the save: method.

You might have noticed in both the insert and delete case here that we are not updating the table view. That is because the FRC tells us when a change has been processed by calling one of its delegate methods. Let's look at those next.

Being the delegate of the fetched results controller ensures that we get notification as changes are made to the persistent state of our data. There are four delegate methods in the NSFetchedResultsControllerDelegate protocol. The controllerWillChangeContent: method lets the delegate know that the FRC is about to change the data. This is a great place to let the table view know that you are about to start tweaking its set of rows and sections by calling beginUpdates. When the changes are finished being processed, the FRC sends the controllerDidChangeContent: to its delegate. This is the spot you want to tell the table view that you are done making changes by calling endUpdates.

During the process of persisting the changes, the FRC sends the controller:didChangeObject:atIndexPath:forChangeType:newIndexPath: for every changed object. Here is our implementation:

CoreData/Conference02/Classes/RootViewController.m

```
- (void)controller:(NSFetchedResultsController *)controller
  didChangeObject:(id)anObject
      atIndexPath:(NSIndexPath *)indexPath
    forChangeType:(NSFetchedResultsChangeType)type
     newIndexPath:(NSIndexPath *)newIndexPath {

  if(NSFetchedResultsChangeUpdate == type) {
    [self configureCell:[self.tableView cellForRowAtIndexPath:indexPath]
            withTrack:anObject];
  } else if(NSFetchedResultsChangeMove == type) {
    [self.tableView reloadSections:[NSIndexSet indexSetWithIndex:0]
              withRowAnimation:UITableViewRowAnimationFade];
  } else if(NSFetchedResultsChangeInsert == type) {
    if(!self.firstInsert) {
      [self.tableView insertRowsAtIndexPaths:[NSArray arrayWithObject:newIndexPath]
                    withRowAnimation:UITableViewRowAnimationRight];
    } else {
      [self.tableView insertSections:[[NSIndexSet alloc] initWithIndex:0]
                withRowAnimation:UITableViewRowAnimationRight];
    }
  } else if(NSFetchedResultsChangeDelete == type) {
    NSInteger sectionCount = [[fetchedResultsController sections] count];
    if(0 == sectionCount) {
      NSIndexSet *indexes = [NSIndexSet indexSetWithIndex:indexPath.section];
      [self.tableView deleteSections:indexes
                withRowAnimation:UITableViewRowAnimationFade];
    } else {
      [self.tableView deleteRowsAtIndexPaths:[NSArray arrayWithObject:indexPath]
                    withRowAnimation:UITableViewRowAnimationFade];
    }
  }
}
```

In this method, we update the table view depending on what type of change happened. For updates, we call the configureCell:withTrack: method that we discussed earlier. For moves, we update the whole section. We update the whole section because the sort order of the items in the track list is based on an editable field (the name property). When that value is changed, it is more efficient to just update the whole section than it is to update the individual cells. For inserts, we have to distinguish between the first insert and subsequent inserts. Since the application starts with no data at all, the table view is told there are zero sections when it first asks numberOfSectionsInTableView:. When we insert the first row, we are inserting a section as well. Telling the table view to insert a row into a nonexistent section causes the table view to become confused. Finally, on a delete, we tell the table view which rows are being deleted unless we are on the final object. When deleting the final object, we need to tell the table view to delete the section, or it gets confused similarly to the insert issue.

For the typical use case of our users making changes in an "editing view," this is the perfect place to update the table view. However, keep in mind that you almost certainly don't want to update the table view in this method if you are importing large numbers of objects in a background thread. If you have a "batch update" use case, just reload the table view in the controllerDidChangeContent: instead.

We are almost done with our overview of Core Data. We have seen the following:

- How Core Data's stack of objects work together to provide us with data and to push our changes back into the persistent store

- How to use Core Data objects to set up an NSFetchedResultsController

- How to use an NSFetchedResultsController to populate and keep a table view up-to-date

The last remaining piece of dealing with table views that we have not looked at yet is how Core Data can help us with navigating into the detail of our objects.

11.8 Navigation

Navigation from a table view is intended to take us from a general set of data to a more specific set of data. In our case, there are two forms of specific data. One is the list of sessions that make up our track; the other is the data about the track itself. Recall that we want to navigate to the list of sessions when a track is selected, but if the table view is in editing mode, we want to navigate to a screen that lets the user manipulate the name and trackAbstract properties.

Here is the code to accomplish that:

CoreData/Conference02/Classes/RootViewController.m

```
- (void)tableView:(UITableView *)tableView
didSelectRowAtIndexPath:(NSIndexPath *)indexPath {
  Track *track = [[self fetchedResultsController] objectAtIndexPath:indexPath];
  if(YES == self.editing) {
    self.trackEditingVC.selectedTrack = track;
    [self.navigationController pushViewController:self.trackEditingVC animated:YES];
  } else {
    self.sessionsVC.selectedTrack = track;
    self.sessionsVC.title = track.name;
    [self.navigationController pushViewController:self.sessionsVC animated:YES];
  }
}
```

As you'd expect, we have a view controller for each of the possible destinations. When we are in editing mode, we push the trackEditingVC onto the navigation stack. While the trackEditingVC is active, it will need access to the currently selected track, so we set its selectedTrack property before pushing it onto the stack.

When we are not in editing mode, we push the sessionsVC. The sessionsVC also needs to know the selected track, so we set its selectedTrack property as well. We also set the title to the track's name.

Since the mechanics of navigation were covered in detail previously, we are not going to go into any of the detail here. Instead, we are going to cherry-pick some interesting methods from these other view controllers to highlight how we use the managed objects (instances of Track and Session) and Core Data to implement the functionality.

First let's look at the TrackEditingViewController. When the user is done editing the trackAbstract's text view, we get the textViewDidEndEditing: delegate method called (because the view controller is the text view's delegate). Here is the code:

CoreData/Conference02/Classes/TrackEditingViewController.m

```
- (void)textViewDidEndEditing:(UITextView *)textView {
  self.selectedTrack.trackAbstract = textView.text;
}
```

Notice the simplicity of what we are doing here. All we do is change the property value. Core Data takes care of the rest. From this one line of code, Core Data knows that the track object has changed, and it knows that a save needs to happen and can take care of making that save happen with a single simple method invocation. Not only does this work for really simple stuff like string values, but it also works for relationships, as we will see in a minute with the SessionsViewController.

When the user is done making changes, she hits the Done button on the top-left side. That button invokes the done method. The code is here:

CoreData/Conference02/Classes/TrackEditingViewController.m
```
- (void)done {
  [self.abstractText resignFirstResponder];
  [self.nameField resignFirstResponder];
  [self.navigationController popViewControllerAnimated:YES];
}
```

Telling the two text input objects to resignFirstResponder makes sure that the values typed into them are pushed into the selectedTrack object. Then we tell the navigationController to pop the top view controller (which is this TrackEditingViewController object). As the view controller is relinquishing control and removing its view from the stack of views, it will eventually get the viewWillDisappear:. We choose to put the save in this notification method. Here is the code:

CoreData/Conference02/Classes/TrackEditingViewController.m
```
- (void)viewWillDisappear:(BOOL)animated {
  [super viewWillDisappear:animated];
  NSError *error = nil;
  if (![self.selectedTrack.managedObjectContext save:&error]) {
            // Handle the error...
  }
}
```

It's starting to look familiar by now, right? Call the save: method, and all the changes get pushed down to the database. Notice that we get the managedObjectContext from the selectedTrack. All managed objects know which managed object context they belong to, which makes saving them super easy.

Now that we have seen the editing view controller, let's look at the SessionsViewController for some more interesting code that will help you understand how to use Core Data to edit the detailed objects in your model. Recall that as the sessionsVC is pushed onto the navigation stack in the RootViewController's implementation of tableView:didSelectRowAtIndexPath:, its selectedTrack property is set. Here is the set method:

CoreData/Conference02/Classes/SessionsViewController.m

```objc
- (void)setSelectedTrack:(Track *)track {
  if(track != _selectedTrack) {
    [_selectedTrack release];
    _selectedTrack = [track retain];
    self.fetchedResultsController = nil;
    [self.tableView reloadData];
  }
}
```

This is more or less a run-of-the-mill set method until we get to setting the fetchedResultsController property to **nil**. To understand why that is necessary, let's look at the implementation of the get method for the fetchedResultsController property:

CoreData/Conference02/Classes/SessionsViewController.m

```objc
- (NSFetchedResultsController *)fetchedResultsController {
  if (nil == _fetchedResultsController) {
    NSFetchRequest *fetchRequest = [[NSFetchRequest alloc] init];
    NSManagedObjectContext *context = self.selectedTrack.managedObjectContext;
    NSEntityDescription *entity =
    [NSEntityDescription entityForName:@"Session"
               inManagedObjectContext:context];
    [fetchRequest setEntity:entity];

    NSPredicate *pred = [NSPredicate predicateWithFormat:@"track = %@",
                         self.selectedTrack];
    [fetchRequest setPredicate:pred];

    NSSortDescriptor *sortDescriptor = [[NSSortDescriptor alloc]
                                        initWithKey:@"name" ascending:YES];
    [fetchRequest setSortDescriptors:[NSArray arrayWithObject:sortDescriptor]];

    NSFetchedResultsController *aFetchedResultsController =
    [[NSFetchedResultsController alloc] initWithFetchRequest:fetchRequest
                                        managedObjectContext:context
                                          sectionNameKeyPath:nil
                                                   cacheName:@"Sessions"];
    aFetchedResultsController.delegate = self;
    self.fetchedResultsController = aFetchedResultsController;
    NSError *error = nil;
    if (![self.fetchedResultsController performFetch:&error]) {
      // handle the error...
    }
    [aFetchedResultsController release];
    [fetchRequest release];
    [sortDescriptor release];
  }
      return _fetchedResultsController;
}
```

Most of this code looks the same as the method we reviewed in detail from the RootViewController with the exception of the entity name and the predicate its the same code. On line 11 is where we do something that might be a bit unexpected. Here we are creating an NSPredicate that restricts the list of Session objects to being just the ones associated with the selectedTrack. In this way, we are able to use all the convenience of the fetched results controller and have its content restricted to the proper list of sessions. We can also manage updating the table view in the same way we updated the RootViewController's table view, namely, via the NSFetchedResultsControllerDelegate methods.

Just as creating the fetched results controller is largely the same except for small tweaks, adding sessions and removing sessions are very similar to adding and removing tracks. Let's first look at the code for deleting. Here is the implementation of the tableView:commitEditingStyle:forRowAtIndexPath: from the SessionsViewController:

CoreData/Conference02/Classes/SessionsViewController.m

```
- (void)tableView:(UITableView *)tableView
commitEditingStyle:(UITableViewCellEditingStyle)editingStyle
forRowAtIndexPath:(NSIndexPath *)indexPath {

    if (editingStyle == UITableViewCellEditingStyleDelete) {
        Session *session = [self.fetchedResultsController objectAtIndexPath:indexPath];
        [self.selectedTrack removeSessionsObject:session];
        NSManagedObjectContext *context = self.selectedTrack.managedObjectContext;
        [context deleteObject:session];
        // Save the context.
        NSError *error;
        if (![context save:&error]) {
            // Handle the error...
        }
    }
}
```

Notice that except for the difference in type (track vs. session), this code is the same as the code from RootViewController. The only real difference is on line 7. In addition to removing the session from the managed object context, we also remove it from the track's session list.

Now let's look at the inserting of new sessions. Here is the code:

CoreData/Conference02/Classes/SessionsViewController.m

```
- (void)insertSession {
    self.firstInsert = [self.fetchedResultsController.sections count] == 0;
    NSString *nextId = [self nextSessionIdentifier];
        // Create a new instance of the entity managed by the
    // fetched results controller.
```

```objc
    NSManagedObjectContext *context = self.selectedTrack.managedObjectContext;
        Session *session =
    [NSEntityDescription insertNewObjectForEntityForName:@"Session"
                                  inManagedObjectContext:context];

        // If appropriate, configure the new managed object.
        [session setValue:@"Session" forKey:@"name"];
        [session setValue:@"This is a great session" forKey:@"sessionAbstract"];
    [session setValue:nextId forKey:@"sessionID"];

    [self.selectedTrack addSessionsObject:session];

        // Save the context.
    NSError *error = nil;
    if (![context save:&error]) {
                // Handle the error...
    }
}
```

There are two pieces of this code that differ from the code to insert a track that we looked at earlier. The first is on line 16 where we add the newly created session to the selectedTrack. Notice the use of the addSessionsObject: method. Recall that this method is added automatically by Core Data to our Track class. So, even though we don't provide an implementation, we can call the method without error or warning.

The next interesting line in this method is on line 3. The Session has a unique session identifier, and we use this method to calculate the unique ID by incrementing the largest value in the database by one. Let's look at that code:

CoreData/Conference02/Classes/SessionsViewController.m

```objc
- (NSString *)nextSessionIdentifier {
  NSString *nextId = @"Session01";

  NSManagedObjectContext *ctx = self.selectedTrack.managedObjectContext;
  NSEntityDescription *entity = [NSEntityDescription entityForName:@"Session"
                                    inManagedObjectContext:ctx];
        NSFetchRequest *fetchRequest = [[NSFetchRequest alloc] init];
        [fetchRequest setEntity:entity];
  NSString *predFormat = @"sessionID = max(sessionID)";
  NSPredicate *pred = [NSPredicate predicateWithFormat:predFormat];
  [fetchRequest setPredicate:pred];

  NSError *error = nil;
  NSArray *values = [ctx executeFetchRequest:fetchRequest error:&error];
  if(0 != [values count]) {
    Session *session = [values objectAtIndex:0];
    NSString *maxId = [session valueForKey:@"sessionID"];
```

```
        NSString *number = [maxId stringByTrimmingCharactersInSet:
                           [NSCharacterSet letterCharacterSet]];
        NSString *name = [maxId stringByTrimmingCharactersInSet:
                         [NSCharacterSet decimalDigitCharacterSet]];
        NSNumberFormatter *formatter =
        [[[NSNumberFormatter alloc] init] autorelease];
        NSNumber *value = [formatter numberFromString:number];
        nextId = [NSString stringWithFormat:@"%@%02d", name, [value intValue] + 1];
    }
    return nextId;
}
```

In this method we do three things: first create a fetch request, second use the fetch request to get the session with the maximum sessionID, and third parse the sessionID to get the integer value from the end so we can increment it. The interesting part of this method is the fact that we can use aggregate functions in our predicates. Predicates are a very powerful tool that we have to limit what we get back from the database to exactly what we want. The Apple *Predicate Programming Guide* [App08g] documentation has a lot of detail on predicates.

As you can probably see now, Core Data is a powerful framework that allows us to easily get at data in a database by doing some simple modeling and then using that model. We have only just begun to scratch the surface of what is possible.

Chapter 12
Connecting to the Internet

One of the most revolutionary aspects of the iPhone is the effect it has had on mobile Internet use. Before the iPhone, browsing the Web or performing other network-based tasks was so unappealing that few users bothered to try it, and fewer still stuck with it. So, it was a shock in February 2008 when Google's head of mobile operations told a conference that Google was seeing 50 times more searches from iPhone OS devices than from any other mobile browser.[1] In March, measurement firm M:Metrics said that its studies showed that 84.8 percent of iPhone users said they accessed news and information with their phones, as compared to 13.1 percent of the overall mobile phone market and 58.2 percent of smart-phone owners.[2]

So, you don't need us to tell you that using the network is critical to your users. For many iPhone applications, the network provides the value of the application. In this chapter, you'll learn how to use the most important network features of the iPhone: loading web pages, reading and writing network streams, and parsing the data provided by web services.

12.1 Building a Browser in Ten Minutes with UIWebView

Perhaps the most typical network task performed by many network applications is to load a web page. You may not plan to compete with the excellent Safari web browser that comes with the iPhone—in fact, you probably shouldn't—but you may have a network component to your application that's easiest to just bring up in a web view.

---

1. *Google Homes in on Revenues from Phones* [PT08]
2. *IPhone Users Love That Mobile Web* [Sto08]

If this is what you need, you're in luck, because the functionality of Safari's WebKit engine is available to you in the form of the UIWebView.

This particular component is so easy to use and so capable that to start this chapter, we'll write a basic browser, capable of handling real-world web pages, in ten minutes. In fact, we've even called the sample project TenMinuteBrowser, which you'll find in the sample code's NetworkIO directory.

So, set your clock, and here we go!

Setting Up the Project

Start a new Xcode project, using the View-based Application template. Call it TenMinuteBrowser. This will create header and implementation files for TenMinuteBrowserAppDelegate and TenMinuteBrowserViewController. You won't need to touch the app delegate at all.

Our browser will just have a URL field and a large view of the page content. So, go into the TenMinuteBrowserViewController, and add IBOutlets for urlField (a UITextField) and webView (a UIWebView). Since we're going to be managing a text field and will want to dismiss the keyboard, add the UITextFieldDelegate protocol declaration too. Finally, add an instance method to handle the clicking of the Go button, say, handleGoTapped. Your @interface should look like the following:

NetworkIO/TenMinuteBrowser/Classes/TenMinuteBrowserViewController.h
```
@interface TenMinuteBrowserViewController : UIViewController
                                <UITextFieldDelegate> {
        IBOutlet UITextField *urlField;
        IBOutlet UIWebView *webView;
}

-(IBAction) handleGoTapped;

@end
```

Building the Browser GUI in IB

OK, one minute down, and we're ready to go to Interface Builder. Double-click TenMinuteBrowserViewController.xib to launch IB. The nib already has its File's Owner set to your view controller class, with a single View wired up and ready to configure. Double-click to open the view, and add three objects to it:

- A UITextField across most of the top, for the URL
- A UIButton with the text Go

Figure 12.1: VIEW FOR A SIMPLE BROWSER IN INTERFACE BUILDER

- And a UIWebView filling most of the view below the text field and the Go button

If you're cruising along, you can customize the text field with your favorite website's URL or set the view's background color. At any rate, the result should look more or less like Figure 12.1.

How are you doing? Four minutes in maybe? Not to worry, we'll make it. The next thing we need to do is to wire up the GUI components to the outlets and actions. Ctrl+click the File's Owner to open its list of outlets and received actions.

- Connect the urlField outlet to the text field.
- Connect the webView outlet to the UIWebView.
- Connect the handleGoTapped action to the Go button, selecting the Touch Up Inside event.
- Finally, Ctrl+click the text field to expose its outlets, and connect its Delegate to File's Owner.

Save up and quit Interface Builder.

Implementing the Browser

Check the clock. Are we going to make it in less than ten minutes? Don't worry, we don't have far to go. Back in Xcode, open TenMinuteBrowserViewController.m. We'll need to write a method to get the URL from the text field and have the web view load that site; this method will be called when the user clicks the Go button or when they hit Return on the pop-up keyboard.

If you care to check the UIWebView documentation—later, since we're on the clock—you'll see that it has a simple loadRequest: method that takes one argument, an NSURLRequest. So, all you need to load the web page is to get an object of that class...which you can get from an NSURL...which you can create from an NSString...which is what the text property of the text field is. So, loading the web page is just going to take a few lines:

NetworkIO/TenMinuteBrowser/Classes/TenMinuteBrowserViewController.m
```
-(void) loadURL {
        NSURL *url = [[NSURL alloc] initWithString: urlField.text];
        NSURLRequest *request = [[NSURLRequest alloc] initWithURL: url];
        [webView loadRequest: request];
        [request release];
        [url release];
}
```

Now all you have to do is call this method when the Go button is tapped, remembering to also dismiss the keyboard:

NetworkIO/TenMinuteBrowser/Classes/TenMinuteBrowserViewController.m
```
-(IBAction) handleGoTapped {
        [urlField resignFirstResponder];
        [self loadURL];
}
```

And you can use the text field's textFieldShouldReturn: method to handle the user hitting the keyboard's Return button instead of the Go button:

NetworkIO/TenMinuteBrowser/Classes/TenMinuteBrowserViewController.m
```
- (BOOL)textFieldShouldReturn:(UITextField *)textField {
        if (textField == urlField) {
                [self handleGoTapped];
        }
        return YES;
}
```

Still under ten minutes? Good thing, because you're done. Click Build and Go to run your app in the simulator. Type in a fully formed URL (in

Figure 12.2: RENDERING A WEBSITE WITH UIWEBVIEW

other words, with the http://), click Go, and the results should look like Figure 12.2.

Got Five More Minutes?

As you can see, all the substantial work in this application is done by the UIWebView. Once you've loaded the page, this view—backed by the WebKit engine for rendering HTML, interpreting JavaScript, and handling the network communication—does all the work for handling your web interactions, including submitting forms, navigating to new pages, running client-side browser apps, and so on.

If you can spare a few more minutes, it's pretty trivial to build this out into a full-featured browser. The most obvious thing lacking from the example is the usual forward and back buttons. These are trivial to implement with the UIWebView's goForward and goBack methods. You could also add a "busy" indicator to indicate when the view is loading a page by providing a delegate that implements the UIWebViewDelegate protocol, which provides the callbacks webViewDidStartLoad:, webViewDidFinishLoad:, and webView:didFailLoadWithError:.

Another thing you might want to do is support rotation so you can use the browser in landscape mode. The UIWebView works like any other UIView in this regard: first, override shouldAutorotateToInterfaceOrientation: to accept landscape orientations. Then open the view with Interface Builder, select the web view, and turn on the horizontal and vertical springs with the Size inspector.

Even if you're not planning on developing a browser, the UIWebView has other compelling uses. Although UIKit doesn't provide a styled text component for iPhone apps, you can style HTML to your heart's content with CSS and put that styled HTML into a UIWebView. In fact, this is an excellent way to provide an About screen for your application, because you can provide links to your application's home page, emails for tech support, or even dialable phone number links, all by just authoring HTML.

To do this, instead of loading a page from the Web, you can include your HTML, CSS, and images in the application bundle and then find them inside the bundle. In the example project, we've implemented this by pulling up a *aboutbook.html* page from inside the bundle if the URL starts with the string *about:*. Making a URL from a path in the bundle is just a matter of converting the path string to an NSURL.

NetworkIO/TenMinuteBrowser/Classes/TenMinuteBrowserViewController.m
```
NSString *aboutPath =
        [[NSBundle mainBundle] pathForResource:@"aboutbook"
                                        ofType:@"html"];
url = [[NSURL alloc] initFileURLWithPath: aboutPath];
```

Create an NSURLRequest from this, just like before, and load it in the UIWebView, and you'll see your local file rendered in the web view, as shown in Figure 12.3, on the facing page.

12.2 Reading Data from the Network

Of course, not everything on the Internet is a web page. Heck, not everything on port 80 is a web page. You might want to use it for exchanging data with a web service or tuning in to web radio. Many applications will have their own reasons for connecting to the network and using the data for their own application-specific purposes. Fortunately, if you

Figure 12.3: SHOWING A LOCAL ABOUT PAGE WITH UIWEBVIEW

want to open a URL and retrieve its data, Cocoa's URL Loading System makes the process exceptionally simple.

The URL Loading System is a set of classes that let you work with four essential URL protocols: http://, https://, ftp://, and file://. By way of example, let's explore what it takes to read in the source of a web page (or, technically, the content of any supported URL) and display it to the user. An example project, SimpleCocoaURLReader, is provided in the sample code in code/NetworkIO/. It offers the user a layout much like the simple browser in Section 12.1, *Building a Browser in Ten Minutes with UIWebView*, on page 235, except that instead of a UIWebView, the raw source is displayed in a simple UITextView. You can see the completed example running and showing the source of the Pragmatic Programmer home page, http://www.pragprog.com/, in Figure 12.4, on the following page.

Figure 12.4: DISPLAYING THE SOURCE OF A WEBSITE LOADED VIA URL LOADING SYSTEM

As you would expect, the UI particulars of this app are identical to the WebKit demo. Clicking the Load button or pressing Return at the end of the text field calls a loadURL method that's responsible for retrieving the content from the URL specified in the text field. However, in the browser example, calling loadRequest: was the end of the story. Here, handling the request is just the beginning.

In the URL Loading System, you take the NSURLRequest and create an NSURLConnection from it. The request specifies what you want and how you want it; its methods return the URL, whether the HTTP method is GET or POST, the data you want to submit via a POST, and so on.[3] A corresponding NSURLResponse, representing the server's reply, is

---

3. These properties are read-only. Use NSMutableURLRequest if you need to set them to nondefault values.

automatically generated later. By contrast, with these two objects, the NSURLConnection represents the *action* of connecting to the server with the request and getting the response. And, by design, you are meant to interact with this connection in a highly asynchronous way, as you'll soon see.

NetworkIO/SimpleCocoaURLReader/Classes/URLLoaderViewController.m

```
NSURLRequest *request = [[NSURLRequest alloc] initWithURL: url];
NSURLConnection *connection = [[NSURLConnection alloc]
                                initWithRequest:request
                                delegate:self];
[connection release];
[request release];
```

Yes, that's right—you create an NSURLConnection and immediately release it, because you'll never need it again. What the heck is going on here?

The connection is set up with a delegate, in this case self, so your object will get a wake-up call when there's something you need to handle. In the URL Loading System, you're notified of everything interesting— the opening or closing of the connection, the receipt of more content, errors, authentication challenges—in the form of callbacks to a delegate object's various methods. These methods are described in NSURLConnection's documentation; the most basic callbacks are as follows:

- connection:didReceiveResponse: indicates that enough data has been received to compose an NSURLResponse object.

- connection:didReceiveData: provides an NSData wrapper around the most recent block of bytes retrieved from the connection.

- connectionDidFinishLoading: signals that the download has completed normally.

- connection:didFailWithError: indicates that the download failed, supplying an NSError to explain the failure.

To provide a display of the URL's source, the implementation of these callbacks is pretty simple. When you create the connection, you clear the UITextView and start spinning a UIActivityIndicatorView with the startAnimating method. When you receive a response, you do...nothing, though you're welcome to inspect the NSURLResponse object if it has data that interests you (for example, instead of the example's spinning activity indicator, you could use a UIProgressView, set its maximum value from

> **Joe Asks...**
>
> **What If I Need to Use Something Other Than HTTP or FTP?**
>
> We've given the URL Loading System a lot of space because it's one of those 80-20 (or 90-10) situations. The overwhelming majority of application network traffic is going to be HTTP on port 80. Since port 80 is almost never blocked by proxies or firewalls, engineers have spent the last decade or two putting all sorts of functionality on it, from remote procedure calls to streaming audio. It makes you wonder whether the port-blocking network admins would have been better off leaving more ports open so that different services would operate on distinct ports, instead of overloading the meaning of port 80. But we digress....
>
> If you need to implement a non-HTTP, non-FTP protocol, you're basically proposing classic socket programming: open input and/or output streams to some host on some port, and start reading and writing data in some defined or de facto protocol. Cocoa on the iPhone doesn't help for this. On the Mac, there's an NSHost class, which you pass to NSStream's getStreamsToHost:port:inputStream:outputStream: method, but those aren't part of the public iPhone API and therefore should not be used.
>
> Instead, you need to use Core Foundation and the CFNetwork framework, which provides network abstractions that you call with procedural C functions. Although it has some higher-level support for HTTP and FTP, if you want to get really down and dirty, you'll use socket APIs directly. To wit, the functionality of that missing Cocoa method is provided by the CFNetwork function CFStreamCreatePairWithSocketToHost().

the response's expectedContentLength, count bytes as you receive data, and update the progress bar each time).

The event you probably care most about is receiving data. The delegate's connection:didReceiveData: method will be called repeatedly with new data, until the end of the stream is reached. Whatever you care to do with the data as it downloads, you do it here.

NetworkIO/SimpleCocoaURLReader/Classes/URLLoaderViewController.m

```
- (void)connection:(NSURLConnection *)connection
                          didReceiveData:(NSData *)data {
    NSLog (@"connectionDidReceiveData");
    NSString *newText = [[NSString alloc]
                        initWithData:data
                        encoding:NSUTF8StringEncoding];
    if (newText != NULL) {
        [self appendTextToView:newText];
        [newText release];
    }
}
```

The example code's implementation is pretty trivial; it makes an NSString from the NSData and appends it to the UITextView. Obviously, your needs might be different. For example, if you needed to keep the entire contents of the URL in memory, you could create an NSMutableData object to which you would then append each received NSData object. You could also implement file downloading by opening an output stream to a file, as shown in Section 8.7, *Writing Data to Files*, on page 158, and writing each received block of data to it.[4]

The other two methods you'll need to handle in pretty much any case are the "end" and "error" events. You don't have to explicitly close or release any connection-related resources with the URL Loading System, so you may have only UI-related tasks to attend to, such as disabling an activity indicator or providing an error message if the connection reports an error.

NetworkIO/SimpleCocoaURLReader/Classes/URLLoaderViewController.m

```
- (void) connectionDidFinishLoading: (NSURLConnection*) connection {
    [activityIndicator stopAnimating];
}
```

---

4. Cocoa for Mac OS X has an NSURLDownload class for this purpose, but it is not present on iPhone OS.

```
-(void) connection:(NSURLConnection *)connection
                                didFailWithError: (NSError *)error {
    UIAlertView *errorAlert = [[UIAlertView alloc]
                    initWithTitle: [error localizedDescription]
                    message: [error localizedFailureReason]
                    delegate:nil
                    cancelButtonTitle:@"OK"
                    otherButtonTitles:nil];
    [errorAlert show];
    [errorAlert release];
    [activityIndicator stopAnimating];
}
```

And that, remarkably enough, is it. No baby-sitting of a connection, no resources to carefully release in every end and error state, none of the nagging details that you might have dealt with in other network streaming APIs. Plus, the URL Loading System is asynchronous by design: while other APIs make it natural to block the UI in a tight "read bytes until done" loop, this API encourages you to work in a more resilient event-driven mind-set.[5]

12.3 HTTP Authentication

One thing our asynchronous reader can't do for us yet is to read a URL that's password-protected. Requests for content in an authentication realm are typically going to get back an HTTP response status of 401 (Unauthorized) and user-readable boilerplate text explaining the rejection of the request. Fortunately, the URL Loading System makes it very easy to deal with HTTP authentication challenges.

Setting Up a Password-Protected Website

First, though, you'll want to set up an authentication realm of your own to test against. There's no way to know whether your authentication code is working unless you test it against a real password-protected website. Fortunately, your Mac already has the industry-standard Apache web server, so you can use that for your testing.

Go to the System Preferences on your Mac, select Sharing, and turn on Web Sharing if it's not already enabled, as shown in Figure 12.5, on the facing page.

---

5. Actually, you *can* force NSURLConnection to operate in a blocking mode by calling sendSynchronousRequest:returningResponse:error:, if you really think you need to do so. Given the high latency and the failure potential of wireless connections, we'd rather you didn't.

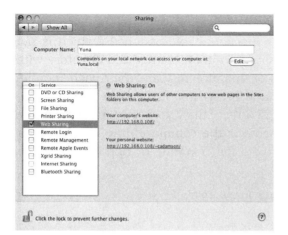

Figure 12.5: ENABLING WEB SHARING IN SYSTEM PREFERENCES

This starts up your Apache web server. Make note of your computer's address, since that's what you'll be using later from your iPhone or the simulator.

Next, you'll need to edit some of the Apache configuration files manually. On Mac OS X, the Apache-related files are stored in different parts of the filesystem. We'll start by creating a directory in the web root—the folder that is presented as the top level of your website—which will then be password-protected by setting up a configuration file.

By default, your Apache web root is /Library/WebServer/Documents. So, in the Finder, go to the top level of your boot drive or partition, and navigate to that directory. Create a folder, and give it a simple name like iphone.[6] Now, with your text editor of choice, create a simple web page in this directory. Ours has just a bit of text and a screenshot from an earlier chapter:

```
<html><head><title>OK!</title></head>
<body>
<h1>You're In!</h1>
<img src="FilesystemExplorer-app-dir.png"/>
</body>
</html>
```

---

6. You may need to be an administrator on your Mac to create and edit files in this directory.

The next step is to lock down this directory with Apache authentication. To do this, you need to edit files in /etc/apache2/, which is not visible from the Finder. So, using Terminal, change the directory with cd /etc/apache2. The contents of this directory are the Apache configuration, specifically httpd.conf and everything it imports. To edit these files, you'll need root access and a text editor you can launch from the command line, such as vi, emacs, or the command-line launchers for popular Mac text editors like mate for TextMate or bbedit for BBEdit. First, edit the httpd.conf file:[7]

⇒ ```
Yuna:apache2 cadamson$ sudo emacs httpd.conf
Password:
```

You could set up the authentication directory right here in the master httpd.conf file, but that's somewhat bad form, because it creates an administration hassle once you can't tell what parts of the file came with Mac OS X and what parts you added. Instead, jump to the end of the file, and just insert a directive to include all configuration files in the users directory.

```
# Add a vhost to authenticate the iphone directory for testing
Include /etc/apache2/users/*.conf
```

Now you're ready to set up your password-protected directory. Change to the users directory and edit a new file, with any name you like, provided it ends in .conf so the <Include> from the master configuration file will pick it up. iphone-test-dir.conf would be a good choice. In this file, you'll use Apache's <Directory> directive to declare that your iphone directory needs special handling; specifically, it needs password protection. The directive is detailed in Apache's *Authentication, Authorization and Access Control* [Apa09] documentation, but a simple example shouldn't be hard to follow:

```
#
# Basic user/pass authentication for iphone directory
#
<Directory "/Library/WebServer/Documents/iphone">
        AuthType Basic
        AuthName "Authenticate or die"
        AuthUserFile /etc/apache2/iphonedirpasswd
        Require user chris
</Directory>
```

---

7. For vi and emacs, you'll want to launch the editor with sudo so you have write access. With BBEdit and TextMate, you don't need to; you'll be challenged for an administrator password the first time you try to save the file.
```

This directive does the following:

- Indicates that the directory in question is /Library/WebServer/Documents/iphone.
- Chooses Apache's "basic" username/password authentication module.
- Provides a user-readable name for the authentication realm: "Authenticate or die."
- Defines the password file that will store usernames and passwords for this realm.
- Lists the usernames that will be allowed to authenticate, in this case chris. You can use multiple users or create groups, but for your own testing, one user should be plenty.

The authentication directive tells Apache to look in the file /etc/apache2/iphonedirpasswd for usernames and passwords, so you'll obviously need to set that up. You create Apache user with the htpasswd command, which you'll need to run as root.

⇒
```
Yuna:apache2 cadamson$ sudo htpasswd -c /etc/apache2/iphonedirpasswd chris
Password:
New password:
Re-type new password:
Adding password for user chris
Yuna:apache2 cadamson$
```

The command shown here instructs htpasswd to create the specified file (via the -c option) and provides the username we want to create. Somewhat confusingly, the first "Password:" prompt is actually the shell challenging you for your administrator password in order to sudo the command, that is, to run as root. The "New password" that you're prompted for twice is the Apache user password you want to create. Since you'll be typing this on an iPhone keyboard, you may want to keep it simple: perhaps test instead of ph33rth3l337hax0rd00d.

With your password file created and Apache configured to use it, it's now time to test with your desktop browser. Restart Apache with sudo apachectl restart, and then navigate to the password-protected directory by appending its name to your web address, as shown in Sharing Preferences. In our case, that was http://192.168.0.108/iphone/, though your IP address will probably be different. If you've configured everything correctly, your web browser should pop up a username and password challenge dialog box. If it doesn't seem to work, check the Apache logs in /var/log/apache2/ to see whether any errors were reported when you restarted Apache.

Handling HTTP Authentication

Now that you have a password-protected site to test against, you can add HTTP authentication support to your application.

In the URL Loading System, HTTP authentication is signaled by the delegate callback method connection:didReceiveAuthenticationChallenge:, which provides an NSURLAuthenticationChallenge object. This object represents the state of the challenge-response dialogue between your client and the server. For example, you can use the protectionSpace method to get the "realm" on the server that requires authentication, or you can use the previousFailureCount method to find out whether the response is from a failed attempt to supply a username and password.

Perhaps most importantly, the sender method returns an object implementing the NSURLAuthenticationChallengeSender protocol. Since this object represents the sender of the challenge, it's what you'll interact with to answer the authentication challenge. Your options are to either answer the challenge by providing credentials like a username and password (useCredential:forAuthenticationChallenge:), try to proceed without providing credentials (continueWithoutCredentialForAuthenticationChallenge:), or give up and abort the request (cancelAuthenticationChallenge:).

Let's try it in practice. In the example code, we implement connection:didReceiveAuthenticationChallenge: by looking at the previous failure count:

NetworkIO/SimpleCocoaURLReader/Classes/URLLoaderViewController.m

```
- (void)connection:(NSURLConnection *)connection
      didReceiveAuthenticationChallenge:
      (NSURLAuthenticationChallenge *)challenge {
    if ([challenge previousFailureCount] != 0) {
        // if previous failure count > 0, then user/pass was rejected
        NSString *alertMessage = @"Invalid username or password";
        UIAlertView *authenticationAlert =
        [[UIAlertView alloc] initWithTitle:@"Authentication failed"
                           message:alertMessage
                           delegate:nil
                           cancelButtonTitle:@"OK"
                           otherButtonTitles:nil];
        [authenticationAlert show];
        [authenticationAlert release];
        [alertMessage release];
        [activityIndicator stopAnimating];
    } else {
        // show and block for authentication challenge
```

```
                    AuthenticationChallengeViewController *challengeController =
                        [[AuthenticationChallengeViewController alloc]
                                initWithNibName:@"AuthenticationChallengeView"
                                bundle:[NSBundle mainBundle]
                                loader: self
                                challenge: challenge];
                    [self presentModalViewController:challengeController
                                animated:YES];
                    [challengeController release];
        }
}
```

If the failure count is nonzero (line 4), then we got here as a result of a failed username/password response, so we show the user an error alert (lines 6–15). However, if this is a new challenge, we'll show a password challenge view in a modal (that is, blocking) pop-up (lines 19–26).

This AuthenticationChallengeViewController offers a simple view, shown in Figure 12.6, on the next page with username and password fields. Its two actions send control back to the view controller that spawned it, either calling a "cancel" method or providing the username and password to an "OK" method.

Both methods dismiss the modal username/password view controller and retrieve the NSURLAuthenticationChallengeSender from the challenge. The difference is that the cancel method sends it the cancelAuthenticationChallenge: message, while the "OK" method creates an NSURLCredential object from the username and password and sends it as a response to the challenge:

NetworkIO/SimpleCocoaURLReader/Classes/URLLoaderViewController.m

```
- (void) handleAuthenticationOKForChallenge:
                    (NSURLAuthenticationChallenge *) aChallenge
                    withUser: (NSString*) username
                    password: (NSString*) password {
        // try to reply to challenge
        NSURLCredential *credential = [[NSURLCredential alloc]
                    initWithUser:username
                    password:password
                    persistence:NSURLCredentialPersistenceForSession];
        [[aChallenge sender] useCredential:credential
                    forAuthenticationChallenge:aChallenge];
        [credential release];
        [self dismissModalViewControllerAnimated:YES];
}
```

Figure 12.6: GUI to provide username and password

Notice that the NSURLCredential initializer takes a persistence argument. This can be one of three NSURLCredentialPersistence constants: NSURLCredentialPersistenceNone for credentials that are immediately forgotten, NSURLCredentialPersistenceForSession for credentials that are stored for the duration of the current session, and NSURLCredentialPersistencePermanent for credentials stored permanently in the user's keychain. The actual behavior seems to vary between the simulator and the device; trying out each persistence mode on the device, "permanent" credentials survive application restarts (and even the removal and reinstallation of the application), while the others don't.

At this point, you're done, whether or not the username and password are accepted. If they are, you'll start getting calls to connection:didReceiveData: with the contents of the URL you've now gained access to. If they were wrong, you'll get another call to connection:didReceiveAuthenticationChallenge:, and this time the NSURLAuthenticationChallenge's previousFailureCount will be nonzero, indicating that the challenge has been issued anew in response to an incorrect username/password combination.

12.4 Parsing XML from Web Services

Now let's stop for a moment and think about what you're going to do with the data you get from the network. Our example of just showing downloaded HTML in a text view is easy to program but not particularly practical. Eventually, you're probably going to want to *parse* the downloaded data.

In many cases, a service is exposed via some kind of network API, which specifies what a client sends (and how) and what the service sends in response. Broadly speaking, web APIs that use HTTP can be referred to as *web services*.[8] And, more often than not, the data returned by the service is XML.

Let's take a concrete example. Twitter has a very open, easy-to-use public API detailed on its developer wiki at http://apiwiki.twitter.com/. Its REST API can even be used from a browser. For example, the "public timeline" that shows the most recent updates from all users is available as http://twitter.com/statuses/public_timeline.format, where format is one of xml, json, rss, and atom. Thus, to get a plain XML view of the timeline, you just need to load the URL http://twitter.com/statuses/public_timeline.xml. In fact, you can use this URL in the SimpleCocoaURLLoader example and see the raw XML reply in the text view. Here's a fragment of the returned XML, edited to show the structure and some of the most interesting fields:

```
<?xml version="1.0" encoding="UTF-8"?>
<statuses type="array">
<status>
  <created_at>Sun Apr 19 17:02:43 +0000 2009</created_at>
  <id>1559022041</id>
  <text>Wife and kids shopping. Must write and finish webservices
        section this afternoon.</text>
  <source>web</source>
  [...]
```

---

8. Although some definitions limit web services to only those that adhere to specific standards, like SOAP, the term is increasingly used for any HTTP-based machine-to-machine service.

```xml
<user>
  <id>12604932</id>
  <name>Chris Adamson</name>
  <screen_name>invalidname</screen_name>
  <location>Grand Rapids, MI</location>
  <description>I write, edit, and code stuff. I also raise
               children and sometimes clean things</description>
  [...]
  </user>
</status>
[...]
</statuses>
```

Now let's take the next step and parse out the useful information. We'll just retrieve the names of the users and the text of their statuses, commonly known as *tweets*.

Start a new view-based project, SimpleTwitterClient, and use IB to edit SimpleTwitterClientViewController.xib to offer a large text view, an "Update tweets" button, and an activity indicator, as shown in Figure 12.7, on the facing page. Then in the header file, declare IBOutlet properties for activityIndicator and tweetsView, and declare an event-handling IBAction method for the button, updateTweets. Connect the two outlets and the button's Touch Up Inside event as appropriate.

Loading the Twitter Data

To implement the updateTweets method, we need to do two things: get the XML data from the Twitter web service and parse out the parts of it that interest us. You already know how to do the first part: create an NSURLRequest, create an NSURLConnection with it, and write the delegate methods to handle the life-cycle events of receiving the data and handling errors. Many of the Twitter APIs require HTTP authentication, so if you wanted to support those features, then you'd also implement callbacks like connection:didReceiveAuthenticationChallenge:.

But what do we want to do with the data? Let's look ahead to what parsing it will require. The NSXMLParser class has two initializer methods: initWithContentsOfURL: and initWithData:. There are two good reasons not to use the former: it may block the GUI while downloading all the data, and it can't handle the HTTP authentication challenge issued by some of the Twitter APIs. So, we'll plan on using initWithData:, which means we can just collect all the downloaded XML into an NSMutableData and hand that to the parser when we're done downloading.

Figure 12.7: EDITING SIMPLETWITTERCLIENTVIEWCONTROLLER.XIB VIEW IN IB

So, define NSMutableData *tweetsData in the header file. Now let's implement the updateTweets method to allocate tweetsData and start loading from the URL.

NetworkIO/SimpleTwitterClient/Classes/SimpleTwitterClientViewController.m

```
-(IBAction) updateTweets {
        tweetsView.text = @"";
        [tweetsData release];
        tweetsData = [[NSMutableData alloc] init];
        NSURL *url = [NSURL URLWithString:
                @"http://twitter.com/statuses/public_timeline.xml"];
        NSURLRequest *request = [[NSURLRequest alloc] initWithURL: url];
        NSURLConnection *connection = [[NSURLConnection alloc]
                        initWithRequest:request
                        delegate:self];
        [connection release];
        [request release];
        [activityIndicator startAnimating];
}
```

Once the request is initialized, the URL Loading System will start calling our delegate callback methods, likely starting with connection:didReceiveResponse:, followed by a series of connection:didReceiveData: delegate callbacks that provide the XML data. All we need to do is to stuff each new block of data into tweetsData.

NetworkIO/SimpleTwitterClient/Classes/SimpleTwitterClientViewController.m

```
- (void)connection:(NSURLConnection *)connection didReceiveData:(NSData *)data {
    [tweetsData appendData: data];
}
```

Finally, when the URL Loading System reaches the end of the stream, it calls back to connection:didFinishLoading:. We can stop spinning the activityIndicator and call a yet-to-be-written parseTweets method.

NetworkIO/SimpleTwitterClient/Classes/SimpleTwitterClientViewController.m

```
- (void) connectionDidFinishLoading: (NSURLConnection*) connection {
    [activityIndicator stopAnimating];
    [self startParsingTweets];
}
```

Parsing the Twitter Data

Now we have all the XML in an NSData object. To parse it, we'll use an NSXMLParser. There are two broad classes of XML parsing approaches. A *DOM parser* creates a tree-structured model in memory, in which each node of the tree maps to an element of the XML markup. When you need the deep structure of the tree, this can be helpful. But often, you don't. Moreover, DOM parsers often make you wait until the whole tree is parsed before you can do anything with the data. The alternative approach is the *event-driven parser*, in which the parser notifies interested parties as it parses each XML tag and lets the listeners decide what, if anything, to do. This is potentially lighter and allows your code to do useful work during the parse, perhaps displaying incomplete but useful data before the parse is completed.

The NSXMLParser follows this event-driven approach. If you look at its documentation, you'll see a very familiar pattern. After creating the parser, you set a delegate,[9] and then you receive various callback methods as the XML is parsed: parserDidStartDocument:, parser:foundCharacters:, parser:didStartElement:namespaceURI:qualifiedName:attributes:, and so on.

---

9. NSXMLParser uses an *informal* delegate protocol, meaning there's just a list of callback methods in the documentation and not a defined protocol that your class needs to declare that it implements.

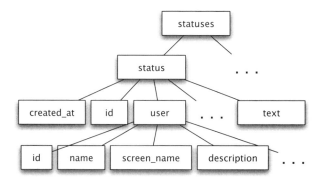

Figure 12.8: DIAGRAM OF XML RESPONSE RECEIVED FROM TWITTER WEB SERVICE API

This is the same asynchronous pattern you saw in Section 12.2, *Reading Data from the Network*, on page 240, and earlier still in Section 8.5, *Asynchronous File Reading*, on page 148.

Now that you know how you're going to be called, it's time to think about the data that will be coming back to you in those callbacks. Let's take another look at the XML response we get from the Twitter API. An overall view of the response as a tree structure is shown in Figure 12.8.

The Twitter timeline APIs send back XML with a single <statuses> element, which contains a number of status children. Each of these is one tweet, with elements providing a timestamp, unique ID, and other metadata for the tweet, as well as a <text> element with the tweet text itself. The <status> also contains a <user> element that describes the user who posted the tweet. This element has child elements that describe the user: their <name>, <screen_name>, <location>, and so on.

There's a nested tree structure here, but we don't necessarily have to concern ourselves with it. Within each <status>, our simple name-and-message-only example can limit itself to processing the <text> and <name> tags and ignore the rest. For now, let's limit ourselves to that approach.

We'll begin by creating an NSXMLParser with the downloaded tweetData and have it immediately start parsing.

```
NetworkIO/SimpleTwitterClient/Classes/SimpleTwitterClientViewController.m
```
```objc
- (void) startParsingTweets {
    NSXMLParser *tweetParser = [[NSXMLParser alloc] initWithData:tweetsData];
    tweetParser.delegate = self;
    [tweetParser parse];
    [tweetParser release];
}
```

This will immediately start calling the parsing callback methods, if they exist. So, we need to figure out what each of these methods should do. Ultimately, we're building a big string of name-message pairs to put in the text view, so we'll need an NSMutableString that we can repeatedly append to as we get each tweet parsed. To build each tweet, we need to watch for status elements, and once one of those has begun, watch for name and text elements, which can be stored as name-value pairs in an NSMutableDictionary until the status element ends and we add the tweet to the tweetsString. Within each of those elements, we just need the text, but there's a catch: parser:foundCharacters: isn't guaranteed to provide all a tag's text in one callback. In fact, it rarely does. So, we want an NSMutableString to hold an element's text as it's built up with repeated callbacks.

Given all that, you'll need to define the following instance variables in the header file:

```
NetworkIO/SimpleTwitterClient/Classes/SimpleTwitterClientViewController.h
```
```objc
NSMutableString *tweetsString;
NSMutableDictionary *currentTweetDict;
NSString *currentElementName;
NSMutableString *currentText;
```

Now we're ready. First, we handle the parser:didStartDocument: message. When we get this callback, we can allocate the tweetsString:

```
NetworkIO/SimpleTwitterClient/Classes/SimpleTwitterClientViewController.m
```
```objc
- (void)parserDidStartDocument:(NSXMLParser *)parser {
    [tweetsString release];
    tweetsString = [[NSMutableString alloc]
                    initWithCapacity: (20 * (140 + 20)) ];
    currentElementName = nil;
    currentText = nil;
}
```

Since tweets can be only 140 characters long and the Twitter web service API states that only 20 tweets are provided per call, we can actually make a decent guess as to how big a string we'll need. Add another 20 characters for a typical username and whitespace, and we can allocate our NSMutableString with enough room for twenty 160-character items before it has to resize itself.

Once the document starts, we need to start handling element parsing callbacks. We'll get parser:didStartElement: callbacks for every single element, in a classic tree-parsing order: first statuses, then its first status child, then status's first child (perhaps created_at), and so on, until we finally get an element with no children. We'll get one or more parser:foundCharacters: callbacks for its text and then parser:didEndElement:. The parser then returns to its parent and either finds another child element to begin or the parent's text, and so on.

We don't want to handle every element. In fact, there are only three we care about: status, name, and text. The first of these is special, because it tells us to start watching for its children. The others are the ones we'll put in the currentTweetDict. To watch for these, let's add NSSet *interestingTags to the header file. Here's how we'll set that up. First, **#define** the elements you're interested in:

NetworkIO/SimpleTwitterClient/Classes/SimpleTwitterClientViewController.m

```
#define INTERESTING_TAG_NAMES @"text", @"name", nil
```

Then, the first chance you get (viewDidLoad, for example), initialize the NSSet with these values. Notice that initWithObjects: takes a variable-length list of objects, terminated by **nil**. We could put that list here, but using the **#define** at the top of the file is more findable and readable:

NetworkIO/SimpleTwitterClient/Classes/SimpleTwitterClientViewController.m

```
- (void) viewDidLoad {
    [super viewDidLoad];
    interestingTags = [[NSSet alloc] initWithObjects: INTERESTING_TAG_NAMES];
}
```

Now we're ready to handle the elements we're interested in, as the parser discovers them. We begin an element with the callback parser:didStartElement:namespaceURI:qualifiedName:attributes::

NetworkIO/SimpleTwitterClient/Classes/SimpleTwitterClientViewController.m

```
Line 1  - (void)parser:(NSXMLParser *)parser didStartElement:(NSString *)elementName
    -              namespaceURI:(NSString *)namespaceURI
    -              qualifiedName:(NSString *)qualifiedName
    -              attributes:(NSDictionary *)attributeDict {
    5      if ([elementName isEqualToString:@"status"]) {
    -          [currentTweetDict release];
    -          currentTweetDict = [[NSMutableDictionary alloc]
    -                  initWithCapacity: [interestingTags count]];
    -      }
```

```
10                else if ([interestingTags containsObject: elementName]) {
                      currentElementName = elementName;
                      currentText = [[NSMutableString alloc] init];
                  }
    }
```

This implementation does two things. If the element is *status*, then it represents the beginning of a new tweet and therefore sets up an NSMutableDictionary to hold name-value pairs of any interesting child elements. Otherwise, if the element is one of the "interesting tags," meaning *name* or *text*, then we hold on to the name of the current element and set up an NSMutableString to hold its value. Notice in the **else if** on line 10 how NSSet's containsObject: makes it easy to look through the set of interesting tag names. That's why we set it up earlier.

The next message we expect from the parser is one or more callbacks with text from an element:

NetworkIO/SimpleTwitterClient/Classes/SimpleTwitterClientViewController.m

```
- (void)parser:(NSXMLParser *)parser foundCharacters:(NSString *)string {
    [currentText appendString:string];
}
```

This takes the characters and appends them to the currentText string, where we build up the value of a name-value pair. The callback doesn't tell you which element the text is part of, because you should already know that from a previous "start element" call. Notice, though, that we don't check to see that the current element is "interesting" before appending the text. This is a sneaky little optimization: if we allocate a currentText only for interesting elements and **nil** it when we end *any* element, then this append message will go nowhere when it's called inside noninteresting elements.

So, we have a plan to handle the beginning of an element and the text within a pair of tags we care about. Now let's look at what we do when the parser finds the end of the element:

NetworkIO/SimpleTwitterClient/Classes/SimpleTwitterClientViewController.m

```
- (void)parser:(NSXMLParser *)parser didEndElement:(NSString *)elementName
        namespaceURI:(NSString *)namespaceURI
        qualifiedName:(NSString *)qName {
    if ([elementName isEqualToString:currentElementName]) {
        [currentTweetDict setValue: currentText forKey: currentElementName];
    } else if ([elementName isEqualToString:@"status"]) {
```

```
            [tweetsString appendFormat:@"%@: %@\n\n",
                    [currentTweetDict valueForKey:@"name"],
                    [currentTweetDict valueForKey:@"text"]];
        }
        [currentText release];
        currentText = nil;
}
```

We want to do two things when an element ends. If it's one of the "interesting" tags, then we know that there is no more text for this element, and we can store the name-value pair using the currentElementName and the accumulated currentText. On the other hand, if it's a status element that has ended, then we've parsed all the metadata for one tweet, and we can use its values. For this simple example, all we do is to append the tweet's username and message to the mutable tweetsString. We also need to release and **nil** out any currentText object that we may have allocated in parser:didStartElement:.

When all the parsing wraps up, we'll get the parserDidEndDocument: callback. When this happens, all we need to do is to replace the text view's contents with the accumulated tweetsString:

NetworkIO/SimpleTwitterClient/Classes/SimpleTwitterClientViewController.m
```
- (void)parserDidEndDocument:(NSXMLParser *)parser {
        tweetsView.text = tweetsString;
}
```

And that's all the work we need to do to parse Twitter's public timeline. Build and Go to run the app, and click the update button. The result is shown in Figure 12.9, on page 263.

More XML Parsing Options

This is intentionally a pretty simple example, and a few extensions should be obvious. First, if you want to use most of the Twitter web service API—such as the services for getting a user's "friends timeline" or posting their own tweets—then you'll need to use HTTP authentication, as described in Section 12.3, *HTTP Authentication*, on page 246.

It's also possible you'll want to pursue a deeper and more sophisticated parse. For example, it might be useful to do more of a DOM-style parse and maintain the parent-child relationships of the source XML. This is made surprisingly easy by Objective-C delegation. To build out the tree structure, you create an element class that *does its own parsing*, meaning that it is this element class that implements the NSXMLParser's

Arranging Code with #pragma mark

If you have a single class that implements multiple protocols, as well as overriding inherited methods from its superclasses, it can get disorganized quickly. One tool for keeping your sets of methods straight is the **#pragma mark** directive. Any text after the **#pragma mark** becomes a menu item in the method/function list menu in Xcode's editor. Here's a sample usage:

#pragma mark NSURLConnection callbacks

We've put a few of these in SimpleTwitterClientViewController.m to help organize the methods, and here's the resulting menu:

```
  INTERESTING_TAG_NAMES
✓ @implementation SimpleTwitterClientViewController
    -updateTweets
    -startParsingTweets
  UIViewController methods
    -viewDidLoad
  NSURLConnection callbacks
    -connection:didReceiveData:
    -connection:didFailWithError:
    -connectionDidFinishLoading:
  NSXMLParser callbacks
    -parserDidStartDocument:
    -parser:didStartElement:namespaceURI:qualifiedName:attributes:
    -parser:didEndElement:namespaceURI:qualifiedName:
    -parser:foundCharacters:
    -parserDidEndDocument:
  TODO: add http authentication for friends timeline

  ViewController boilerplate
    -didReceiveMemoryWarning
    -dealloc
```

Notice how we also got a "TODO" menu item. Any one-line comments of the form // TODO: text also get added to this menu. For clarity, you can also use a single hyphen character as the **#pragma mark** text to create a simple separator line in the menu.

Figure 12.9: VIEWING TWITTER'S PUBLIC TIMELINE WITH NSXMLPARSER

informal delegate protocol. You create a root element and make it the parser's delegate. The element then handles the various callbacks, and when it gets the callback for a new element in the XML, it creates a new element object and makes this new element the parser's delegate. Conversely, when the parser finds the end of an XML element, the element object returns the delegate to its parent. The root element ultimately gets the parserDidEndDocument: callback, and at that point, it's the parent to a tree of objects representing the contents of the source XML. For full details of this recursive approach, see Apple's *Event-Driven XML Programming Guide for Cocoa* [App08b].

One other option to consider is a pair of projects at Google Code. TouchXML is a lightweight replacement for Mac OS X's XML support, which contains a number of classes not available on iPhone OS. Based on the open source libxml2 library, it supports parsing from an NSData or NSString, offers basic XPath support and a set of navigation-style methods to retrieve siblings, children, and so on. Another project, TouchJSON, is a parser and generator for the JSON format used by some web services. Both projects are part of Google's collection of iPhone OS open source projects, Touch Code, at http://code.google.com/p/touchcode/.

12.5 Sending Mail from Your Application

Thus far, we've covered some of the most typical uses of the network: browsing the Web, downloading data from web servers, and processing web service data. One final very common activity we'll consider in this chapter is using email. Every iPhone and iPod touch already has a capable email application, and in iPhone OS 3.0, developers gain access to its standard UI for sending email.

The optional MessageUI framework contains only one class and one delegate protocol. Our example will be correspondingly simple. Create a view-based application called InAppMailer. Add the MessageUI.framework to the project, add #import <MessageUI/MessageUI.h> in InAppMailerViewController.h, and then add the protocol <MFMailComposeViewControllerDelegate> to the **@interface** declaration. While in this header file, declare an IBOutlet to a UITextView called mailHistoryView, make it a **@property**, and declare an IBAction called composeMailTapped.

The GUI will be trivial: open InAppMailerViewController.xib, and add a text view (connected to the mailHistoryView outlet of File's Owner) and a button titled "Compose mail" whose Touch Up Inside event is connected to File's Owner's composeMailTapped action. You now have a button to bring up the mail composer view and a text view to output the results.

To create an email message, you create an instance of the MFMailComposeViewController, which manages a view that looks like the "new mail" view of the Mail application (see Figure 12.10, on the facing page). This view controller handles all the events when the mail-composer view is showing, such as displaying the keyboard when the user is in an editable field or bringing up a picker when the user clicks the From line to select one of his or her accounts.

Figure 12.10: COMPOSING MAIL

Typically, you create the MFMailComposeViewController in code and display it modally. So, that's what we'll do when a user taps the button:

NetworkIO/InAppMailer/Classes/InAppMailerViewController.m
```
-(IBAction) composeMailTapped {
        if (! [MFMailComposeViewController canSendMail]) {
                UIAlertView *cantMailAlert = [[UIAlertView alloc]
                        initWithTitle:@"Can't mail"
                        message:@"This device is not able to send e-mail"
                        delegate:NULL
                        cancelButtonTitle:@"OK"
                        otherButtonTitles:NULL];
                [cantMailAlert show];
                [cantMailAlert release];
                return;
        }

        MFMailComposeViewController *mailController =
                [[[MFMailComposeViewController alloc] init] autorelease];
        [mailController setMessageBody:@"My app can send mail" isHTML:NO];
        mailController.mailComposeDelegate = self;
        [self presentModalViewController:mailController animated:YES];
}
```

The first thing you do here is to call the class method canSendMail so you can see whether the device is even configured for sending mail; this example shows an error alert and performs an early return if it is not.

If the device can send mail, you create an instance of MFMailCompose-ViewController and, optionally, prepopulate various fields of the message. You can set the subject, recipients, CC and BCC recipients, and message body (as plain-text or HTML), as well as add attachments. In this example, we use setMessageBody:isHTML: to prepopulate the body with the string My app can send mail. Note that since you will be displaying the mail composition view, the user will be able to change or delete any field you populate—the class does not allow you to force the user to send anything without their awareness and intent. The controller also has a mailComposeDelegate property that you should set in order to be notified when the user finishes working with the mail composer. With any fields prepopulated and the delegate set, you can slide in the mail composing view with presentModalViewController.

The MFMailComposeViewControllerDelegate protocol defines a single callback method: mailComposeController:didFinishWithResult:error:. This indicates that the user has finished with the mail composer. The second parameter provides one of four MFMailComposeResult enumerated values: MFMailComposeResultSent if the user composed and sent a message, MFMailComposeResultSaved if he or she tapped Cancel and opted to save the message to the Drafts folder in the Mail app, MFMailComposeResultCancelled if the user canceled and didn't save, and MFMailComposeResultFailed. You can get more detail about the failure case by retrieving the error parameter.

In our example app, let's just log the result and time to the text view:

NetworkIO/InAppMailer/Classes/InAppMailerViewController.m

```
- (void)mailComposeController:(MFMailComposeViewController*)controller
          didFinishWithResult:(MFMailComposeResult)result
                        error:(NSError*)error {
    if (error) {
        UIAlertView *cantMailAlert = [[UIAlertView alloc]
                initWithTitle:@"Mail error"
                message: [error localizedDescription]
                delegate:NULL
                cancelButtonTitle:@"OK"
                otherButtonTitles:NULL];
        [cantMailAlert show];
        [cantMailAlert release];
    }
```

```
        NSString *resultString;
        switch (result) {
                case MFMailComposeResultSent:
                        resultString = @"Sent mail"; break;
                case MFMailComposeResultSaved:
                        resultString = @"Saved mail"; break;
                case MFMailComposeResultCancelled:
                        resultString = @"Cancelled mail"; break;
                case MFMailComposeResultFailed:
                        resultString = @"Mail failed"; break;
        }
        mailHistoryView.text = [NSString stringWithFormat:
                @"%@%@ at %@\n", mailHistoryView.text,
                resultString, [NSDate date]];
        [controller dismissModalViewControllerAnimated:YES];
}
```

When we get this callback, the MFMailComposeViewController is done with its work, so we call dismissModalViewControllerAnimated to slide it off-screen.

There are a couple of things you might want to do with the completed email but discover you can't. The fields of the email are exposed only via the setter methods we saw earlier, so you cannot programmatically look at the finished controller and pick out the addressee, email body, or other contents of the composed message. You'll also notice that you can't programmatically send or save the message. The MessageUI framework is designed to leave the user in control, and letting apps send unseen email on the user's behalf or access the contents of a message is presumably a privacy concern.

Chapter 13
Peer-to-Peer Networking

iPhones and iPod touches are everywhere, each with several wireless connectivity options. In this chapter, we'll look at how applications on iPhones can discover the services they need and connect to other nearby iPhones or laptops using wifi or Bluetooth. First we'll look at the service discovery protocol described by Bonjour, and then we'll look at the new Game Kit framework added in iPhone 3.0. Game Kit offers two important new communications technologies for developers: Bluetooth wireless communication and voice chat.

13.1 Using Ad Hoc Network Services with Bonjour

In Chapter 12, *Connecting to the Internet*, on page 235, we connected to URLs that either are implicit in the nature of an application (for example, one that uses a public web service) or are entered manually by the user into the device. But there are cases in which the user really shouldn't need to know or care about the network details about what they're connecting to. Particularly in LAN-based applications such as finding peers for a chat application or printing documents to a printer, the user is interested only in the functionality of something on the network; they care about *what it does*, not *where it is*.

With the iPhone OS, the case for Bonjour-enabled networking may be even stronger. iPhones and iPod touches will always be entering and exiting local networks, and rather than expecting users to enter new hostnames or addresses on these networks, it's far more appropriate to let the devices dynamically discover the services they'll need.

This service-oriented mind-set is the philosophy of *Bonjour*. In Bonjour, the user is often unaware of hostnames, ports, or paths. Instead, Bonjour applications expose network services: printers, chat peers, iTunes music collections, and so on.

To explore how this works, we'll rely upon the fact that your Mac can Bonjour-enable its Apache web server. We'll use this ability to let the phone browse sites on the LAN without ever typing in an address.

Bonjour-Enabling Apache

First, though, we'll need to explicitly enable Bonjour-browsing of your Apache websites. In Mac OS X 10.5 (Leopard), Bonjour is disabled by default in your Apache configurations, so you may need to turn it on, unless you upgraded from previous versions of Mac OS X where it was enabled by default. Look at the file /etc/apache2/other/bonjour.conf. Leopard's default file, which does not allow Bonjour discovery of your Mac's websites, looks like this:

```
<IfModule bonjour_module>
    RegisterUserSite customized-users
</IfModule>
```

We can edit this file to expose the computer's "main" web directory, individual user directories, or both. To expose the main website (that is, http://localhost/), add the directive *<RegisterDefaultSite>*. You can expose user directories (for example, http://localhost/~cadamson) with *<RegisterUserSite>* directives, providing either a username or the value all-users. To expose both the main site and all user websites, your bonjour.conf will look like this:

```
<IfModule bonjour_module>
        # RegisterUserSite customized-users
        RegisterDefaultSite
        RegisterUserSite all-users
</IfModule>
```

Restart Apache either by deselecting and reselecting the Web Sharing entry in System Preferences' Sharing pane or by performing a sudo apachectl restart on the command line.

To ensure your sites are visible via Bonjour, launch Safari, and click the Bookmarks button. You'll notice that among the collections, there's an entry for Bonjour. Click this, and you should see your computer name and all the users for whom you've enabled Bonjour web sharing, as shown in Figure 13.1, on the facing page.

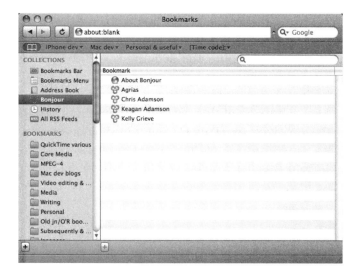

Figure 13.1: BROWSING BONJOUR-ENABLED APACHE WEB SHARES WITH SAFARI

Assuming you can see your main site (your computer's name as defined in the Sharing preferences, Agrias in this example) and whatever users you've enabled, you're ready to browse Bonjour with the iPhone.

13.2 Bonjour Service Discovery

To find services of interest with Bonjour, the iPhone OS provides Cocoa and Core Foundation APIs, NSNetServices, and CFNetServices. The Cocoa one, which we'll focus on here, is highly dynamic and fairly easy to use, isolating you from many of the low-level networking details of Bonjour. In fact, to begin browsing for a service, all you need to do is create an NSNetServiceBrowser (which we assign to an instance variable called bonjourBrowser) and provide it with the domain you're searching in (a blank string if you're just browsing the LAN), a service "type," and a delegate to handle callbacks from the service discovery process.

NetworkIO/BonjourWebBrowser/Classes/RootViewController.m

```
-(void) startSearchingForWebServers {
    bonjourBrowser = [[NSNetServiceBrowser alloc] init];
    [bonjourBrowser setDelegate: self];
    [bonjourBrowser searchForServicesOfType:@"_http._tcp" inDomain:@""];
}
```

The only line here that isn't self-explanatory is line 4, using the arbitrary string "_http._tcp" to define the service to search for. This string consists of two parts: a string to define the semantics of the service (here, _http) and a transport mechanism (_tcp), separated by periods. Well-known service types are documented on the website at http://www.dns-sd.org/ServiceTypes.html, which as of this writing contains more than 400 Bonjour types, including daap for iTunes' Digital Audio Access Protocol, ipp for Internet Printing Protocol, nfs for the old Network File System, and many more. Within this list, you'll find http, with the following description:

```
World Wide Web HTML-over-HTTP
Tim Berners-Lee <timbl at W3.org>
Protocol description: RFC 2616
Defined TXT keys: u=<username> p=<password> path=<path to document>
(see txtrecords.html#http)
NOTE: The meaning of this service type, though called just "http", actually
denotes something more precise than just "any data transported using HTTP".
The DNS-SD service type "http" should only be used to advertise content that:
 * is served over HTTP,
 * can be displayed by "typical" web browser client software, and
 * is intented primarily to be viewed by a human user.
...
```

In other words, http is the service type that you would expect a typical web server to advertise via Bonjour to web browsers and what we'll search for to find the websites exposed by Apache.

Once you call searchForServicesOfType:inDomain: on the NSNetServiceBrowser, your delegate will start getting callbacks informing it of the state of the search. If your search fails, perhaps because you've used an invalid service type (notice how you have to precede the type name and protocol with underscore characters—and don't forget the . that separates them), you'll immediately receive the failure callback netServiceBrowser:didNotSearch:. However, if the search begins successfully, you'll get the callback netServiceBrowserWillSearch:, which you might use to indicate to the user that a long-running search is underway (for example, by spinning an activity indicator).

Each time a matching service is discovered, you'll get the callback netServiceBrowser:didFindService:moreComing:, whose arguments are an NSNetService describing the matching service and a **BOOL** indicating whether the net service browser is waiting for additional services. Typically, a delegate will use this callback to assemble a list of available services. In the example application we've provided, NetworkIO/BonjourWeb-

Figure 13.2: BROWSING BONJOUR-ENABLED APACHE WEB SHARES WITH NSNETSERVICEBROWSER

Browser, matching services are used to build a UITableView list of available services, as shown in Figure 13.2. Along with users' home pages on your LAN, you might discover other devices; Sciezka in this figure is a Bonjour-discoverable laser printer with a web-based configuration.

But what if a service goes away? In this case, you'll receive the callback netServiceBrowser:didRemoveService:moreComing:, which works just like the didFindService version, except that it indicates the disappearance of a service. In the example code, we handle this case by removing the given service from the list; you can try it by running the example, going to Sharing in the System Preferences, and turning off Web Sharing. Once you do, all the discovered web services will disappear from the list in the simulator or on the device.

So, once you have a service, what do you do with it? Just finding the *service record*, as we've done here, isn't necessarily enough to use the service. There's a separate *resolution* step that provides complete details about the service, filling in the NSNetService's host, port, and other fields that are uninitialized when you first discover the service.

The example code takes a look at the service in viewDidLoad: to see whether it has already been resolved from an earlier call. You can tell by looking to see whether the service's hostName or port return meaningful values. If they don't, then you need to kick off resolution.

NetworkIO/BonjourWebBrowser/Classes/WebPageViewController.m
```
- (void)viewDidLoad {
       [super viewDidLoad];
       // start resolution if necessary, otherwise just get the path and show
       // page. see http://developer.apple.com/qa/qa2004/qa1389.html for more
       // on why this is necessary
       if ([netService hostName] != nil) {
              [self loadPageFromService];
       } else {
              [resolutionActivityIndicator startAnimating];
              [netService setDelegate: self];
              [netService resolveWithTimeout: RESOLUTION_TIMEOUT];
       }
}
```

As with service discovery, the Cocoa API makes service resolution a highly dynamic process. You begin resolution by setting a delegate on the service and calling resolveWithTimeout:, which takes as its argument an NSTimeInterval in seconds. After this, your code just handles callbacks as the resolution proceeds.

If resolution fails, you'll get the netService:didNotResolve: callback. However, if it succeeds, you'll get netServiceDidResolveAddress:, at which point the NSNetService object will have all the details you need to communicate with the service: a host, a port, and a collection of service-specific name-value pairs in the TXT record.

Once you've resolved the service, you can get its host, port, and other metadata. The example code uses the resolved service to build a URL, create an NSURLRequest from that, and load it into a UIWebView, as introduced in Section 12.1, *Building a Browser in Ten Minutes with UIWebView*, on page 235.

NetworkIO/BonjourWebBrowser/Classes/WebPageViewController.m

```
Line 1   -(void) loadPageFromService {
             // get path from the TXT record
             NSDictionary *txtRecordDictionary =
                     [NSNetService dictionaryFromTXTRecordData:
      5                  [netService TXTRecordData]];
             NSData *pathData =
             (NSData*) [txtRecordDictionary objectForKey: @"path"];
             NSString *path = [[NSString alloc] initWithData: pathData
                         encoding:NSUTF8StringEncoding];
     10
             // see http://www.dns-sd.org/txtrecords.html#http for the rules
             // on getting url from service data

             // build URL from host, port, and path
     15      NSString *urlString = [[NSString alloc]
                     initWithFormat: @"http://%@:%d/%@",
                     [netService hostName],
                     [netService port],
                     path];
     20      NSURL *url = [[NSURL alloc] initWithString: urlString];
             urlLabel.text = urlString;
             self.title = [netService name];
             NSURLRequest *request = [[NSURLRequest alloc] initWithURL:url];
             [webView loadRequest: request];
     25      // stop activity indictator; could also do this when web view either
             // completes or errors, by providing UIWebView delegate
             [resolutionActivityIndicator stopAnimating];

             [request release];
     30      [url release];
             [urlString release];
             [path release];
             // pathData and txtRecordDictionary don't get released, because
             // they were merely "gotten" and not retained
     35  }
```

It may be helpful to think back to what we will need to get from the NSNetService in order to load the web page. To create a URL, we need a host, port, and path. The host and port are easy, because they can be retrieved with the methods hostName and port. The path is trickier. For the machine's main web page, the path may be an empty string, whereas for the user pages, it'll be something like ~cadamson. As defined by the http service type, the TXT record stores the path in a name-value pair, keyed with the string path. So, our example code can put together a suitable URL by pulling these items out of the resolved NSNetService. Lines 3 through 9 get the path as a block of bytes and

Figure 13.3: VIEWING A WEB PAGE DISCOVERED VIA BONJOUR

convert them to an NSString. With this done, lines 15 through 20 create a URL, from which you can create an NSURLRequest and hand that off to a UIWebView. The discovered and rendered web page is shown in Figure 13.3.[1]

Probably the most important thing to note in this example is that once Bonjour has found and resolved the service, its job is done. In this example, the actual *use* of the service (loading the web page) is entirely done by the UIWebView. In some other application, you would set up your own connection to the service. If the service is defined as using HTTP, you could use the URL Loading System as described in Section 12.2, *Reading Data from the Network*, on page 240. If it's some cus-

---

1. We added the "(cadamson)" to /Users/cadamson/Sites/index.html to make sure we had found the right page. Your personal site may look different.

tom protocol, then you would need to open a socket connection, given the host and port in the resolved service, and perhaps some protocol-specific metadata provided by the TXT record. But at any rate, once Bonjour resolves a service, you're back in the realm of plain old network I/O.

This example has considered only the client side of Bonjour, under the presumption that small mobile devices are more likely to be clients than servers. However, the entire Bonjour stack is present in iPhone OS, so you can use your iPhone as a server, announcing it to clients with Bonjour's publication APIs. Having said that, if what you want to connect to is another iPhone OS device, then you could use the Bonjour-based Game Kit framework, introduced in the next section, to operate as a server or peer.

Your Cocoa code that uses the NSNetServices APIs can also be compiled and run on Mac OS X, so you can share code between the two platforms. Apple also provides Bonjour implementations for Windows and Linux and even offers a Java API, so you have a lot of choices if you want to use those platforms to provide services that your iPhone apps can use. To learn more about Bonjour, the creator of Zeroconf and the editor of the book you're reading have published the final word on the subject, *Zero Configuration Networking, The Definitive Guide* (O'Reilly), which covers Bonjour from the highest to the lowest level of the standard and its APIs.

13.3 Game Kit Overview

Many applications could benefit from the iPhone's ad hoc Bluetooth networking; just imagine being able to exchange business e-cards or documents even when you're not on a local wifi network. Moreover, the voice chat features introduced in iPhone OS 3.0 work on any kind of network, not just the short-range Bluetooth, and can therefore be useful in a wide range of communication apps. Despite the name, Game Kit's features are useful to many kinds of applications, not just games.

Game Kit is provided by the optional framework GameKit.framework, which consists of just three classes (GKPeerPickerController, GKSession, and GKVoiceChatService), along with three formal protocols (GKPeerPickerControllerDelegate, GKSessionDelegate, and GKVoiceChatClient).

It's important to realize that the Bluetooth networking and voice chat features provided by Game Kit are *entirely independent*. The voice chat

> **You May Need Two Devices!**
>
> As of this writing, the iPhone family has a mixed level of support for Game Kit networking. The original iPhone's Bluetooth does not work with Game Kit, and the first-generation iPod touch doesn't even have Bluetooth. That leaves the iPhone 3G, the 3GS, and the second-generation iPod touch as the only devices that can use this framework's networking classes.
>
> Unfortunately, the simulator in the iPhone 3.0 SDK does not support Game Kit networking over Bluetooth, even on Macs with Bluetooth. Instead, it runs the Game Kit protocol over the computer's internet connection, either wifi or Ethernet. This means that for the time being you either need *two* supported devices to test and run Game Kit networking code or need *two* Macs running the simulator, since the nature of networking is to have multiple participants.*
>
> ---
> *. In fact, this chapter's sample code was developed in the iPhone Lab at WWDC 2009, because it was the only way we could wrangle two Game Kit–capable devices.

can be run over any network connection, perhaps the Bluetooth connection provided by Game Kit, or over a wireless connection written with one of the lower-level networking APIs, such as CFNetwork or even BSD sockets. You could even support both kinds of networking in a single application.

Also, although Game Kit's networking feature uses Bonjour to advertise, discover, and resolve services over Bluetooth, it does not directly use the Bonjour APIs, as covered in Section 13.1, *Using Ad Hoc Network Services with Bonjour*, on page 269. Instead, a GKSession class wraps both the service discovery and the sending and receiving of data via Bluetooth. Still, a little bit of Bonjour remains evident: when advertising your service, you still use the same kind of service ID string that regular Bonjour uses.

13.4 Setting Up a Bluetooth-Networked Game

Let's begin by focusing on these Bluetooth networking classes. In the next few sections, we'll build a peer-to-peer game that connects and communicates via Bluetooth. The game is pathologically simple, but

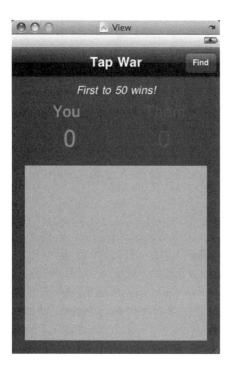

Figure 13.4: VIEW FOR P2PTAPWARVIEWCONTROLLER IN INTERFACE BUILDER

even at this level, we'll start to see some of the concerns that come up when developing a networked game.

Our game is called P2PTapWar, and the concept is really simple: once two players are connected, each needs to tap a box on the screen as quickly as possible. Each tap counts as one point and will be communicated to the opponent over the air. The first player to 50 taps wins.

This example uses the single-view application template. The only class we'll be working with is P2PTapWarViewController. The view for this class, as defined in P2PTapWarViewController.xib, is shown in Figure 13.4. It contains a navigation bar that's used only for presenting a title and a Find Bar Button Item (we don't actually use navigation in this application), a few labels for the scores, and a big gray box that our users will be frantically tapping.

To get started, you'll need to go to P2PTapWarViewController.h and set up IBOutlet properties for the UILabels playerTapCountLabel and opponentTapCountLabel, as well as the UIBarButtonItemstartQuitButton. You'll also need to define two IBActions: handleStartQuitTapped and handleTapViewTapped. With these declarations made, you can open P2PTapWarViewController.xib in Interface Builder and connect the outlets. For the actions, you need to show the Connections inspector (⌘ 2) for the Bar Button Item and connect its selector to the File's Owner's handleStartQuitTapped method. For the tap view, there are no actions at first, because UIViews don't accept user input and therefore don't generate actions. Use the Identity inspector (⌘ 4) to change its class to UIControl, and then connect the Touch Down event to the File's Owner object's handleTapViewTapped.

13.5 Setting Up a Peer Picker

Game Kit's networking classes provide two things: an abstraction of a Bluetooth wireless *session* and a *peer picker* user interface for finding and connecting to other devices over Bluetooth. Since the picker will help us find a peer and set up a session, let's start there and reveal the session-related details as we go along.

The peer picker is controlled with the class GKPeerPickerController and is quite minimal: it has instance methods to show and dismiss the picker, a property to define which kinds of connections to offer (Bluetooth or otherwise), and a delegate. To show the picker when the Find button is tapped, here's all we need to do:

NetworkIO/P2PTapWar/Classes/P2PTapWarViewController.m
```
-(IBAction) handleStartQuitTapped {
        if (! opponentID) {
                actingAsHost = YES;
                GKPeerPickerController *peerPickerController =
                        [[GKPeerPickerController alloc] init];
                peerPickerController.delegate = self;
                peerPickerController.connectionTypesMask =
                        GKPeerPickerConnectionTypeNearby;
                [peerPickerController show];
        }
}
```

This method shows the picker only if opponentID, an NSString instance variable you'll need to define in the header file, is non-**nil**. Peers in Game Kit are identified by peerID strings, and since we need only one for this game, we can use the presence or absence of this ivar as a flag for whether we've already found an opponent.

Assuming we don't have an opponent, we set an actingAsHost flag variable (to be explained shortly) to a temporary value and then allocate the GKPeerPickerController. The picker needs a delegate to call back with connection updates, which can be this view controller itself, provided we declare in the header file that we implement the GKPeerPickerControllerDelegate protocol. We'll be implementing its various methods in a moment.

The next step is to indicate which kinds of connections the peer picker should offer, via the connectionsTypeMask property, which takes a bit field of possible values. Game Kit defines two: GKPeerPickerConnectionTypeNearby represents a Bluetooth connection, while GKPeerPickerConnectionTypeOnline is any other kind of wireless connection. If you combine these values with the | operator, the first step of the picker will be to give the user a choice between the two types of connections. However, Game Kit offers only a Bluetooth API, so if the user chooses Online, the picker will dismiss, and your delegate will get a callback to peerPickerController:didSelectConnectionType:. This would be your signal to start setting up your own network connection, presumably with one of the low-level networking APIs like CFNetwork or BSD sockets. For this example, we offer only the Nearby connection, so the user will go directly to a GUI that shows Bluetooth peers. Finally, with the object all set up, we call the show method to display the picker, as shown in Figure 13.5, on the next page, with a single discovered peer.

13.6 Providing a Peer Picker Delegate

Most of our application's interaction with a peer picker is in your implementation of the GKPeerPickerControllerDelegate protocol. The delegate is responsible for providing a session object to the picker and responding when a peer connects. It also can opt to react when the picker is canceled without choosing a peer. If you use GKPeerPickerConnectionTypeOnline to make the picker offer non-Bluetooth connections, you would be notified of that choice in a callback to peerPickerController:didSelectConnectionType:, at which point you would need to set up that connection.

For our Bluetooth-only application, the most important thing for our delegate to do is to implement the peerPickerController:sessionForConnectionType: method. This is called when the picker starts up and needs to fetch a GKSession object. The session object is what implements the Bluetooth connection between multiple devices. We need one session to connect to an opponent, so we declare the gkSession as an instance variable in P2PTapWarViewController.h and lazily instantiate it the first time it's needed, that is, when the picker asks for it.

Figure 13.5: GKPeerPickerController showing Bluetooth peers

NetworkIO/P2PTapWar/Classes/P2PTapWarViewController.m
```
-(GKSession*) peerPickerController: (GKPeerPickerController*) controller
           sessionForConnectionType: (GKPeerPickerConnectionType) type {
    if (!gkSession) {
        gkSession = [[GKSession alloc]
                initWithSessionID:AMIPHD_P2P_SESSION_ID
                displayName:nil
                sessionMode:GKSessionModePeer];
        gkSession.delegate = self;
    }
    return gkSession;
}
```

peerPickerController:sessionForConnectionType: passes in the picker that is requesting the session, along with one of the type constants (which we can assume to be GKPeerPickerConnectionTypeNearby, the only type we configured the picker to use). The GKSession object has a single designated initializer taking three parameters, each of which merits careful consideration:

- initWithSessionID: The sessionID is an NSString that uniquely identifies the protocol that your application uses to communicate. By using unique IDs, an application can avoid connecting to a device that might be using Game Kit to provide a completely different (and incompatible) communication protocol. Apple recommends that you use a Bonjour service ID, the same kind of 14-character ID we saw introduced in Section 13.2, *Bonjour Service Discovery*, on page 271. If you're going to be defining your own communication protocols, as we will here, you should choose a Bonjour service ID and register it with the Bonjour service registry at http://dns-sd.org/. For this example, we have registered the service ID amiphd-p2p[2] and **#define**'d it in the header file as AMIPHD_P2P_SESSION_ID. On the other hand, if you provide **nil** for this argument, a service ID is generated from your application's bundle identifier.

- displayName: This is the name presented to peers. It is meant as a displayable, human-readable string, so if you were writing a game that let the user choose a name, you would use that here. If you send **nil** for this argument, as we do here, the device's name is used.

- sessionMode: There are three session modes you can declare. A server uses GKSessionModeServer to indicate that it advertises itself on the network but does not search for other services. Clients use GKSessionModeClient to indicate that their session will search for services but not provide them. And a peer, set up with GKSession-ModePeer, acts as both a server and a client, advertising itself as a service and searching for others. Peering is the most typical use of Game Kit's networking API and is what we use here.

The other thing we do when we set up the GKSession is to provide a delegate, which will be asynchronously updated with events from the session, such as the connection and disconnection of peers. Since we want the view controller to serve as the delegate, we need to add GKSessionDelegate to the list of implemented protocols in the header file's **@interface** declaration.

Once we provide a session, presumably one with a useful delegate, there's almost nothing else the picker delegate needs to do. The picker is really responsible only for setting up a connection, whereas all the networking happens in the GKSession and its delegate. In fact, the only thing left to do for the picker delegate is to call [picker dismiss]; to get rid of the picker's UI, in peerPickerController:didConnectPeer:toSession:. The

---

2. This string is based on this book's internal code name and is not otherwise meaningful.

picker will dismiss itself if canceled, although you can implement peerPickerControllerDidCancel: if you want to be notified of that event.

13.7 Network Game Logic

The next step is to start working with the GKSession to send data to a peer and to use the GKSessionDelegate to handle the connection, disconnection, and receipt of data from the peer. But before we do that, we need to ask ourselves an important question: just how are we going to use the connection to communicate game data?

Well, part of the answer is easy enough: the GKSession class has two methods for sending data, sendData:toPeers:withDataMode:error: and sendDataToAllPeers:withDataMode:error:. But just *what* are we going to send?

There are actually interesting problems involved in maintaining the state of a game across a network connection. If each peer has a model of the game's state, how do you keep them in sync? How much time and bandwidth will that require? What happens if you don't?

Our game seems to have a simple requirement: keep track of how many times each player has tapped their box. But there's more to it than that. We also have to coordinate the start of the game; you couldn't have one player choose an opponent and start tapping while the peer is still deciding whether to even accept the connection. And you need to coordinate the end of the game; if one player wins, the other needs to be told to stop.

All of this has implications for our network protocol. We are sending different kinds of messages, and the protocol should be robust enough to handle this. It might be helpful to design a protocol that can be extended with new kinds of messages in the future.

We've chosen a flexible, if inefficient, approach for our simple game. Every message will be a set of key-value pairs, packed into an NSData object by an NSKeyedArchiver, as introduced in Section 8.8, *Property Lists and NSCoding*, on page 161. We'll define a set of message keys in the header file, and peers that receive messages will look for these keys.

NetworkIO/P2PTapWar/Classes/P2PTapWarViewController.h
```
#define START_GAME_KEY @"startgame"
#define END_GAME_KEY   @"endgame"
#define TAP_COUNT_KEY  @"taps"
```

In this simple version, one side will use START_GAME_KEY in a message to start the game. Whenever either peer gets a tap event, it will send TAP_COUNT_KEY with its current tap count as the value, and if it has reached the number of taps needed to win, it can include END_GAME_KEY to notify the other side that the game is over.

If you decided you wanted to use a timer rather than a maximum number of taps, you could add another message key for sending time updates. For such a case, you would want to designate one peer whose time value is considered canonical by any other peers and send out updates of the "official" game time; if each peer had its own timer, there's a chance that their respective clocks could be just enough off to cause a problem with the consistency of the game's state.

One other consideration with sending data over the wire has to do with what Game Kit calls the data *mode*. Messages can be sent in a "reliable" or "unreliable" fashion. Unreliable messages are sent once and neither checked for success nor retried; reliable messages are repeatedly sent until they get through. Experienced network developers will recognize this as the difference between UDP and TCP.

Since we want to maintain state across the network, you might think we should send a reliable message for each tap. But the nature of this game is that we can afford lossiness, in the interest of performance. Rather than sending a message that just says "opponent tapped," we can send a message with the total tap count. If this doesn't get through, it probably doesn't matter, because we'll be sending another message a fraction of a second later, when the player taps again, which has all the data the other side needs (that is, the tap count).

In more complex games, this strategy of tolerating network lossiness sometimes implies the use of *dead reckoning*, the estimation of game state given the current state. If your app is modeling moving objects and misses an expected update from a network peer, you can continue moving those objects according to your game physics and then make a correction when you do get updated network data.

As a general strategy, see what you can send unreliably, and then use reliable transmissions for things that really count. For example, although we can tolerate a missed tap count update, we don't want to risk missing messages that start or end the game, so these should be sent reliably.

13.8 Communicating via the GKSession

Having mapped out a strategy for sending game data across Bluetooth, we can now implement our protocol with Game Kit's communication methods. We'll want to be able to handle state changes from peers (i.e., when the opponent connects or disconnects), send data to the opponent, and receive data from the opponent.

Sending Data

We need to send data to a peer every time the tap view is tapped, so let's go ahead and implement our handleTapViewTapped event handler:

NetworkIO/P2PTapWar/Classes/P2PTapWarViewController.m

```
-(IBAction) handleTapViewTapped {
        playerTapCount++;
        [self updateTapCountLabels];
        // did we just win?
        BOOL playerWins = playerTapCount >= WINNING_TAP_COUNT;
        // send tap count to peer
        NSMutableData *message = [[NSMutableData alloc] init];
        NSKeyedArchiver *archiver =
                [[NSKeyedArchiver alloc] initForWritingWithMutableData:message];
        [archiver encodeInt:playerTapCount forKey: TAP_COUNT_KEY];
        if (playerWins)
                [archiver encodeBool:YES forKey:END_GAME_KEY];
        [archiver finishEncoding];
        GKSendDataMode sendMode =
                playerWins ? GKSendDataReliable : GKSendDataUnreliable;
        [gkSession sendDataToAllPeers: message withDataMode:sendMode error:NULL];
        [archiver release];
        [message release];
        // also end game locally
        if (playerWins)
                [self endGame];
}
```

This obviously calls a few internal game methods that we haven't written yet, starting with the call to update the score locally with update-TapCountLabels. The critical part of the method is after this, however: an NSKeyedArchiver is created to pack an NSMutableData with key-value pairs for our message. The updated tap count is added to the message, and if it equals the tap count needed to win the game, the END_GAME_KEY is added as well. We then call GKSession's sendDataToAllPeers:withDataMode:error: method in reliable mode if it includes the END_GAME_KEY, unreliably otherwise. Finally, there's a little more local logic to end the game locally if necessary, with yet-to-be-written endHostedGame and endJoinedGame methods.

That takes care of the sending, but there's clearly quite a bit we haven't accounted for, including the receipt of messages and the game startup. These tasks aren't initiated by our application but are instead performed by the delegate methods, which handle asynchronous events from the session.

Handling State Changes

Let's start with session:didReceiveConnectionRequestFromPeer:, which is called when one party receives a request from another to connect. When the GKSession is connected via the peer picker, this callback is received only by the player who was asked to join the game, not by the one who chose the opponent in the picker. This gives us a chance to make the requesting player the *host*, a designation we use so that only one party actually starts the game.

NetworkIO/P2PTapWar/Classes/P2PTapWarViewController.m

```
- (void)session:(GKSession *)session
            didReceiveConnectionRequestFromPeer:(NSString *)peerID {
    actingAsHost = NO;
}
```

Assuming that this player accepts the request, each side's delegates will get a callback to session:peer:didChangeState:, with the state GKPeerStateConnected. A number of other states can be reported this way, but for now, let's just implement some logic to set up the game when a peer connects:

NetworkIO/P2PTapWar/Classes/P2PTapWarViewController.m

```
- (void)session:(GKSession *)session peer:(NSString *)peerID
       didChangeState:(GKPeerConnectionState)state {
    switch (state)
    {
        case GKPeerStateConnected:
                    [session setDataReceiveHandler: self withContext: nil];
                    opponentID = peerID;
                    actingAsHost ? [self hostGame] : [self joinGame];
                    break;
    }
}
```

When a connection is received, the first thing this method does is to call setDataReceiveHandler:withContext: on the GKSession. This is critical, because it gives the session an object that is capable of receiving data over the network. The handler object is not specified with a formal protocol, but it has to implement a callback method with the following signature:

- (**void**) receiveData:(NSData *)data fromPeer:(NSString *)peer
            inSession: (GKSession *)session context:(**void** *)context;

setDataReceiveHandler:context: also takes a context that is passed back to the receiveData:fromPeer:inSession:context method. As a **void***, this context reference can be any kind of pointer, including all Objective-C objects. We don't need a context object for this game, so we set it to **nil**.

Next, our state-change handler remembers the peer ID of the opponent as the instance variable opponentID and either starts or joins the game based on whether this player is the host. Both of these methods need to update the local state and GUIs, but only the host needs to send a "start game" message over the connection. Here are the hostGame and joinGame methods, along with the initGame and updateTapCountLabels convenience methods they both call:

NetworkIO/P2PTapWar/Classes/P2PTapWarViewController.m

```
-(void) updateTapCountLabels {
        playerTapCountLabel.text =
                [NSString stringWithFormat:@"%d", playerTapCount];
        opponentTapCountLabel.text =
                [NSString stringWithFormat:@"%d", opponentTapCount];
}

-(void) initGame {
        playerTapCount = 0;
        opponentTapCount = 0;
}

-(void) hostGame {
        [self initGame];
        NSMutableData *message = [[NSMutableData alloc] init];
        NSKeyedArchiver *archiver = [[NSKeyedArchiver alloc]
                initForWritingWithMutableData:message];
        [archiver encodeBool:YES forKey:START_GAME_KEY];
        [archiver finishEncoding];
        NSError *sendErr = nil;
        [gkSession sendDataToAllPeers: message
                        withDataMode:GKSendDataReliable error:&sendErr];
        if (sendErr)
                NSLog (@"send greeting failed: %@", sendErr);
        // change state of startQuitButton
        startQuitButton.title = @"Quit";
        [message release];
```

```
        [archiver release];
        [self updateTapCountLabels];
}

-(void) joinGame {
        [self initGame];
        startQuitButton.title = @"Quit";
        [self updateTapCountLabels];
}
```

In startGame, you can again see how we use an NSKeyedArchiver to build a message in an NSMutableData, which as a subclass of NSData is appropriate for use with the GKSession's sendDataToAllPeers:withDataMode:error: method.

Receiving Data

Now that we've handled state changes from opponents,[3] the last remaining task is to deal with the data we receive from a peer. We created the outgoing data with an NSKeyedArchiver, so to unpack it on the receiving end, we'll use an NSKeyedUnarchiver.

NetworkIO/P2PTapWar/Classes/P2PTapWarViewController.m

```
- (void) receiveData: (NSData*) data fromPeer: (NSString*) peerID
             inSession: (GKSession*) session context: (void*) context {
        NSKeyedUnarchiver *unarchiver =
             [[NSKeyedUnarchiver alloc] initForReadingWithData:data];
        if ([unarchiver containsValueForKey:TAP_COUNT_KEY]) {
                opponentTapCount = [unarchiver decodeIntForKey:TAP_COUNT_KEY];
                [self updateTapCountLabels];
        }
        if ([unarchiver containsValueForKey:END_GAME_KEY]) {
                [self endGame];
        }
        if ([unarchiver containsValueForKey:START_GAME_KEY]) {
                [self joinGame];
        }
        [unarchiver release];
}
```

As you can see, the unarchiver gets the data received by the GKSession and looks for some of the known keys. If it sees TAP_COUNT_KEY, it unpacks the value and updates the score display, whereas if END_GAME _KEY appears, it calls a method to end the game, cleans up the local

---

3. Actually, a fully robust app would want to handle some of the other state changes, such as gracefully dealing with a peer that has disconnected.

state, disconnects all peers from the GKSession, and calls a convenience method to show a victory or defeat alert, both of which are shown in Figure 13.6, on the facing page.

NetworkIO/P2PTapWar/Classes/P2PTapWarViewController.m

```
-(void) showEndGameAlert {
        BOOL playerWins = playerTapCount > opponentTapCount;
        UIAlertView *endGameAlert = [[UIAlertView alloc]
                initWithTitle: playerWins ? @"Victory!" : @"Defeat!"
                message: playerWins ? @"Your thumbs have emerged supreme!":
                        @"Your thumbs have been laid low"
                delegate:nil
                cancelButtonTitle:@"OK"
                otherButtonTitles:nil];
        [endGameAlert show];
        [endGameAlert release];
}

-(void) endGame {
        opponentID = nil;
        startQuitButton.title = @"Find";
        [gkSession disconnectFromAllPeers];
        [self showEndGameAlert];
}
```

That's everything you need to build and deploy this peer-to-peer Bluetooth game. To review, we used a GKPeerPickerController to present the user with a GUI to select an opponent. We provided the picker with a GKSession to handle the local Bluetooth networking and added delegate methods so this session could pass along asynchronous events like peers connecting. On the GKPeerStateConnected event, we set up the game, using the session to send data to the peer and providing the session with a "data receive handler" that could process incoming messages from the peer.

13.9 Voice Chat

Along with Bluetooth local networking, the other feature provided by Game Kit is peer-to-peer chat. As mentioned earlier, these two features are completely independent: you can use the voice chat with the Bluetooth network we set up in the previous sections or over a wifi connection that you've set up. Let's look in general terms at how voice chat works.

Voice chat uses just two classes. The GKVoiceChatService represents a single, shared access point to voice chat functionality. You get a refer-

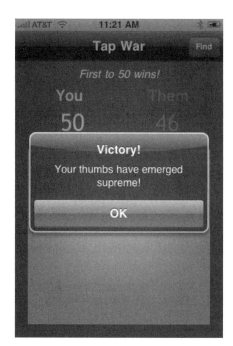 

Figure 13.6: END-OF-GAME ALERTS FOR PEERS IN A GAME KIT-NETWORKED GAME

ence to this singleton with the class method defaultVoiceChatService. With it, you can initiate a voice chat via startVoiceChatWithParticipantID:error, which takes a participantID string uniquely identifying the party to chat with.

You might wonder who is going to map this string to a actual remote peer to chat with. The answer is, *you do*. You need to provide the service with an object implementing the GKVoiceChatClient protocol. This object provides the network connection used by the GKVoiceChatService and determines how a message to a given participantID is actually routed and delivered. Similarly, when it receives voice data over its network connection, it calls the service's receivedData:fromParticipantID: or receivedRealTimeData:fromParticipantID: to deliver the voice data to the voice chat service so that the received data can be decoded and played out to the speaker or headphones.

If you wanted to overlay voice chat atop P2PTapWar, you could do so in a fairly straightforward way. By treating the voice chat participantID as equivalent to the Bluetooth network's peerID, it would be straightforward for your GKVoiceChatClient to implement voiceChatService:sendData:toParticipantID: by calling the GKSession's sendData:toPeers:withDataMode:error: method. The only potential tricky part is that your protocol has to be able to distinguish between voice data and other game data, but this would be easy enough to handle in our app by adding a new key (say, VOICE_CHAT_DATA) to our set of possible message contents. On the receiving end, the NSData containing the voice data would be unpacked by the data handler and forwarded to that device's GKVoiceChatService.

A few details regarding the voice chat service have more to do with the processing of audio than with networking, which we'll note briefly here. The GKVoiceChatService has a level metering API, so you can provide visual feedback to your user of the respective power levels of both the local and remote speakers. The methods for level-metering are highly analogous to those provided by the AV Foundation framework (covered in Chapter 16, *Playing and Recording Audio*, on page 327) and the Music Player APIs in the Media Player framework (covered in Chapter 15, *iPod Library Access*, on page 303), so once you've worked with those, adding level metering to your voice chat will be straightforward.

One other important consideration for voice chat is that all media apps on iPhone OS are expected to communicate their needs and intents to the audio system in advance. The APIs to do this are covered in Section 16.7, *Interacting with Audio Sessions*, on page 353, and in the context of supporting voice chat, there are two primary concerns. First, you need to declare that your application will play and capture audio simultaneously so the system will silence background audio (for example, from the iPod application) and reserve the audio input for your application. You do this by telling the audio session that your application wants to use the category AVAudioSessionCategoryPlayAndRecord. You also will want to use the audio session to verify that audio input is even available: while iPhones always have a microphone available, later iPod touches can capture only if a suitable microphone is attached. You'll see how to perform both of these tasks in the later media chapters.

Chapter 14

Video Playback

The iPhone is a great media device. Its wide, bright, high-resolution screen is surprisingly comfortable for watching your own in-flight movies, and its support for audio is surprisingly deep. Not only does the iPhone play the most popular audio formats, it also includes exceptional support for working with audio at a number of levels, from simple playback to low-level processing of raw audio samples.

We'll begin our survey of iPhone OS media with video. This is actually a little atypical, because audio is usually simpler than video, so you'll often learn about a platform's audio capabilities before you dig into video. But in the case of the iPhone, the video playback API is extremely simple, and it turns out that playing a video is one of the easiest things you can do with the SDK.

14.1 Video Playback with MPMoviePlayerController

The iPhone SDK's video API is very simple, but unfortunately, it's also simplistic. You really don't get to do anything with video *but* play it back. On an iPhone 3GS, you can also record video into the picture library (see Section 20.4, *Capturing Video*, on page 406). But beyond that, you really can't do much else with video.

The video playback API is largely contained in a single class, MPMoviePlayerController, an object whose capabilities let you do the following:

- Load a video from a URL
- Play and pause the video programmatically
- Maintain properties to represent the movie's background color, user control behavior, and scaling mode (whether the movie is stretched or cropped to fill some or all of the screen)

- Produce notifications that allow interested code to be alerted to state changes in the playback

Let's see what is available by way of a short example. In Xcode, create a new view-based application called SimpleVideoPlayer.

The MediaPlayer framework is not included by default by the Xcode templates. You'll need to add it to your project. In Xcode's Groups & Files list, expand Targets, and choose the SimpleVideoPlayer application icon. Bring up its Inspector with the Info toolbar button or ⌘I, and select the General tab. The bottom of this pane shows the currently linked frameworks, with a + button below it to add new frameworks, like MediaPlayer.framework.[1]

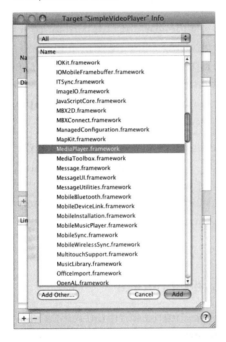

You'll also need to #import <MediaPlayer/MediaPlayer.h> in your SimpleVideoPlayerViewController.h header file.

Building the GUI

Our GUI is going to be surprisingly simple. We'll provide a Play button to kick off playback and a text field to log events.

---

1. You can also add frameworks by right-clicking the Frameworks folder in Groups & Files and choosing Add > Existing Frameworks.

Open SimpleVideoPlayerViewController.xib in IB, and build a GUI that looks like the following:

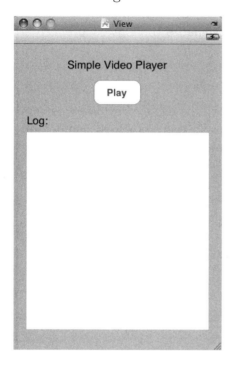

Where are we going to display our movie? We don't need to worry about providing a view to contain the movie. Once the user starts playing a movie with the MPMoviePlayerController, the controller takes over your screen, autorotating to a landscape orientation if necessary, and plays the video. There's no view object to embed in a view and wire up in IB, because video playback is an all-or-nothing proposition on the iPhone.

So, the GUI we provide will need only to handle the tap on the Play button and write messages out to the text field. Therefore, we need one outlet and one action. While you're in SimpleVideoPlayerViewController.h, go ahead and add an instance variable for the MPVideoController. The header should look like this:

MediaPlayback/SimpleVideoPlayer/Classes/SimpleVideoPlayerViewController.h

```
@interface SimpleVideoPlayerViewController : UIViewController {
    MPMoviePlayerController *moviePlayer;
    UITextView *logView;
}
- (IBAction) playVideo;
@property (retain, nonatomic) IBOutlet UITextView *logView;
@end
```

With the outlet and action defined, you should be able to use IB to connect the text field to the logView outlet on File's Owner and connect the button's Touch Up Inside event to the playVideo: action. Do so, save the nib, and quit IB. Don't forget to **@synthesize** the logView property in the implementation file too.

Selecting the Movie

Now that we've set up the GUI, all we need to do is to provide a video and create the controller. MPMoviePlayerController has a single designated initializer, initWitContenthURL:. This can take either a **file://**-style URL referring to a file somewhere in the application's filesystem sandbox or a network URL (presumably an **http://** URL, like http://www.subfurther.com/video/running-start-iphone.m4v, a copy of the movie used for this chapter's screenshots). The downloadable sample assumes you'll drop a single MPEG-4 file[2] called movie.m4v into your project file to add it to the application bundle by dragging it to the Resources folder in your project's Groups & Files. Given that, creating the MPMoviePlayerController in the viewDidLoad method is pretty much trivial:

MediaPlayback/SimpleVideoPlayer/Classes/SimpleVideoPlayerViewController.m
```
NSString *videoPath = [[NSBundle mainBundle]
        pathForResource:@"movie" ofType:@"m4v"];
if (videoPath == NULL)
        return;
NSURL *videoURL = [NSURL fileURLWithPath: videoPath];
moviePlayer = [[MPMoviePlayerController alloc] initWithContentURL:videoURL];
```

With the controller created and assigned to the property moviePlayer, the implementation of playVideo is a one-liner: [moviePlayer play]; Build and Go, and then tap the Play button to see your video in the simulator.[3]

---

2. See Section 14.3, *Supported Media Formats*, on page 299 for information on supported formats.
3. There is a bug in iPhone SDK 2.2 where playing video in the simulator will fail with an error if you have paired a Bluetooth device to your Mac. Remove your Bluetooth devices

> **Streaming Support in iPhone OS 3.0**
>
> In iPhone OS 3.0, your URL can also point to an *HTTP Live Streaming* URL, a new format created by Apple and submitted to the Internet Engineering Task Force as a proposed standard.* In this format, the server splits up a large media file or live source into smaller segments that can be sent to clients over HTTP. See Apple's *HTTP Live Streaming Overview* (App09a) for a description of the format and some sample URLs.
>
> Most of the work in HTTP live streaming is done on the server; an iPhone client simply needs to provide the URL to the MPMoviePlayerController. It turns out this also supports Shoutcast-style MP3-over-HTTP audio streams, but since the MPMoviePlayerController takes over your whole screen, the audio-only Shoutcast stream turns your display into a big gray QuickTime logo... not an ideal user experience!
>
> ---
> *. *HTTP Live Streaming (Internet-Draft)* (Pan09)

The MPMoviePlayerController has a handful of properties you can set or get. The scalingMode allows you to determine whether video that doesn't match the iPhone screen size should be scaled to fill one dimension (MPMovieScalingModeAspectFit), both dimensions with possible cropping (MPMovieScalingModeAspectFill), both dimensions with a possible modification of the aspect ratio (MPMovieScalingModeFill), or not scaled at all (MPMovieScalingModeNone). The other interesting property is movieControlMode, which governs the controls that will be presented when you tap on the screen during playback: a full set of timeline and volume controls (MPMovieControlModeDefault), volume controls only (MPMovieControlModeVolumeOnly), or no controls at all (MPMovieControlModeHidden).

14.2 Receiving Notifications from the Movie Player

The MPMoviePlayerController provides notifications of state changes and other events during playback. By handing these notifications, a player app could present a "loading" indicator until the movie was ready to play and then bring up a Play button. Or you could have a noninteractive "cut scene" movie in a game that would then return to the game when the movie finished playing.

---

in System Preferences, or build and run on the device, to get this example working.

Currently, three notifications are defined:

- MPMoviePlayerContentPreloadDidFinishNotification indicates that preloading has finished and the controller is ready to play the movie, or an error has occurred.

- MPMoviePlayerPlaybackDidFinishNotification notifies listeners that playback has finished.

- MPMoviePlayerScalingModeDidChangeNotification indicates that the user has used the fill control to change the movie's scaling mode.

To get one of these notifications, you use the NSNotification API, providing the NSNotificationCenter's default instance with the name of the notification you want to receive, and a selector (that is, a method name) to receive callbacks when the notification is sent out. So, in viewDidLoad, you can register for a "playback finished" notification with a one-line call:

MediaPlayback/SimpleVideoPlayer/Classes/SimpleVideoPlayerViewController.m

```
[[NSNotificationCenter defaultCenter] addObserver:self
        selector: @selector (playbackFinished:)
        name:@"MPMoviePlayerPlaybackDidFinishNotification"
        object:nil];
```

To log messages to the text view, we can create a simple method to reset the text view to its current text, the date, and a log message:

MediaPlayback/SimpleVideoPlayer/Classes/SimpleVideoPlayerViewController.m

```
- (void) appendTextToLogView: (NSString*) text {
        logView.text = [NSString stringWithFormat: @"%@%@: %@\n",
                logView.text, [NSDate date], text];
}
```

Then, when the notification calls back to the playbackFinished method, it just logs the string Finished to this method:

MediaPlayback/SimpleVideoPlayer/Classes/SimpleVideoPlayerViewController.m

```
- (void) playbackFinished: (NSNotification*) notification {
        [self appendTextToLogView:@"Finished"];
}
```

With a similar log message added to playVideo, the main view looks like this after you've finished playing a video:

14.3 Supported Media Formats

When you play back audio or video, you need to know the media formats the iPhone can and can't play and how to produce media that will work on the device.[4]

There's a big difference between audio/video *container formats* and the contents inside them. Some formats, like QuickTime .mov files, allow almost any kind of audio or video *codec*: an H.264 video track with MP3 audio inside a .mov is perfectly legal (and will play on the iPhone). Other formats are locked to their contents, like .mp3s, which contain *only* MP3 audio data.

---

4. The available formats are described on the consumer-facing iPhone support pages such as http://www.apple.com/iphone/specs.html, but they're worth summarizing briefly.

Here's a summary of the most significant media containers and contents that are supported on the iPhone:

Audio Format	Contents
.aif	PCM (big-endian integer samples). AIFC format (also .aif) can also handle IMA4, μ-law, A-law, and others.
.wav	PCM (little-endian integer samples), μ-law, A-law.
.aac, .m4a, .mp4, .3gp, .3g2	AAC.
.mp3	MPEG-1 Layer 3 ("MP3").
.caf	All of the previous, plus Internet Low-Bitrate Codec (iLBC), Apple Lossless (ALAC), IMA4, Adaptive Multi-Rate (AMR), and others.

A/V Format	Contents
.mp4, .m4v, .3gp, .3g2	MPEG-4 Part 2 Simple video, H.264 Baseline Profile video. AAC audio.
.mov	Any of the previous video codecs. Any of the previous audio codecs.

Of the containers, the Core Audio File (.caf) format may be the least well known but is the most flexible. Like a QuickTime movie (.mov), it is content-agnostic and therefore can contain *any* audio codec. It also maintains time-to-sample tables that make skipping around a file faster than in purely stream-based formats like .mp3. In fact, Apple generally recommends using CAF for packaging your application sounds.

Some of the supported codecs have specific limitations on bitrate, options, size, or other parameters, so we encourage you to take a look at the *iPhone Application Programming Guide* [App09c] for specific guidance. For the audio formats, you can convert between the many formats and codecs supported by Core Audio with the command-line utility /usr/bin/afconvert. In fact, its help message displays the supported formats and contents; just type the following command into Terminal.app or xterm:

```
afconvert --help
Usage:
afconvert [option...] input_file [output_file]
```

```
Options: (may appear before or after arguments)
    { -f | --file } file_format:
        '3gpp' = 3GP Audio (.3gp)
                 data_formats: 'aac ' 'samr'
        '3gp2' = 3GPP2 Audio (.3g2)
                 data_formats: 'aac ' 'samr'
. . .
```

If you're already a seasoned compressionist, you can tune your output to these specs, but for the average developer, there are a lot of details to get wrong. One option to consider is to use QuickTime's Movie to iPhone exporter, which creates an iPhone-compatible movie with no compression options whatsoever. Many users think you have to purchase QuickTime Pro to access QuickTime's export functionality, but it turns out that QuickTime Pro unlocks export features only in the QuickTime browser plug-in and the QuickTime Player application. You can use its exporter in other applications, such as Final Cut Express and Final Cut Pro (via the menu item Export > Using QuickTime Conversion) or iMovie (add your video to a new project, and then choose Share > Export Using QuickTime), without buying QuickTime Pro. In these applications, the QuickTime export dialog box will offer a pop-up menu with a choice of export formats, one of which will be iPhone:[5]

If you're a QuickTime or QTKit developer, you can also do this export in a Mac application of your own with the iPhone Export Component. See Apple's *Technical Note TN2188: Exporting Movies for iPod, Apple TV and iPhone* [App09d] for all the specifics.

---

5. Actually, there will be two. The Movie to iPhone (Cellular) exporter creates a smaller, lower-bandwidth file for distribution over the slower EDGE network.

Chapter 15
iPod Library Access

Chances are, by the time a user downloads and installs your application, they've already used iTunes to put their favorite music, podcasts, movies, and other media on their iPhone or iPod touch. In earlier versions of the iPhone OS, this media was completely off-limits to your application. Now, starting with iPhone SDK 3.0, your application has some access to the user's music library. This means your application can play the user's favorite music, examine the contents of their library, and even control the background iPod player.

In this chapter, we'll look at the new features provided by the Media Player framework. We'll interact with the iPod application while it plays, and then we'll search through the user's music library to find and play their songs, podcasts, and audiobooks.

15.1 Monitoring iPod Playback

Let's say you want to interact with the iPod player by monitoring what the user is playing. That's where we'll start our look at the Media Player framework. The native media player is special: it can run in the background, unlike applications written with the public SDK. This is why you can start playing audio within the iPod application and then quit the application or even lock the screen while the audio continues playing.

Creating the Music GUI

For a sample, let's create a flippable utility application whose front side shows the current item from the iPod application and provides a few simple controls. Later in this chapter, we'll use the flip side to create a

> **The iPod Application**
>
> The iPhone and iPod touch use different names for their media player applications. On the iPhone, there is a single iPod application, which allows you to browse and play both audio and video. On the iPod touch, there are separate Music and Videos applications. Despite this seeming difference, the functionality from the developer's point of view is the same: the object returned by [MPMusicPlayerController iPodMusicPlayer] is functionally the same on both platforms. So, when we refer to "the iPod application," we mean the native media player, regardless of how it's presented to the end user.

user interface to pick items from the library to play. Note that the simulator doesn't have a music player application, so to run this example, you will have to build for and install to an actual device.

Create a utility application with Xcode, called MusicLibraryClient. This will set up view controller classes for the main view and flip side, along with custom view classes that we won't need to customize further.

Open the MainView.xib file in IB (not MainWindow.xib), and lay out the GUI elements shown in Figure 15.1, on the facing page. These components are as follows:

- A label for the current playback time, in a very large font size (we used 48-point Trebuchet).

- A Play/Pause button. Text is OK; in the downloadable example, we've provided a button that uses a PlayButton.png for the normal-state image and a PauseButton.png for the selected state.

- A slider for moving back and forward through the song. This action is commonly called *scrubbing*. The slider's value will represent the current time within the audio, in seconds, since that's the format the Media Player library uses for its current time and duration properties. Add small labels at both ends of the slider for the start time (always 0:00) and the song's duration.

- A UIImageView for the item's artwork. Since some iPod library items are likely not to have cover art, it helps to have an "empty art" image to fall back on, shown here as the initial image for the view.

Figure 15.1: LAYOUT FOR MUSICLIBRARYCLIENT'S MAIN VIEW

- Labels for the song title, artist, and album. These will have slightly different meanings for podcasts and audiobooks: episode/artist/podcast and title/author/book, respectively.

- The "info" button to switch to the flip-side view is provided for you and is already wired up.

You'll then need to declare IBOutlets to the components whose values will change at runtime. Declare the following instance variables, make them properties with **@property** and **@synthesize**, and wire them up as usual:

MediaLibrary/MusicLibraryClient/Classes/MainViewController.h
```
UILabel *currentTimeLabel;
UISlider *currentTimeSlider;
UIButton *playPauseButton;
UILabel *currentItemDurationLabel;
UILabel *currentItemTitleLabel;
UILabel *currentItemArtistLabel;
UILabel *currentItemAlbumLabel;
UIImageView *currentItemArtworkView;
```

Since we'll be using a nonstandard framework, we need to add it to our project. In the project view, add MediaPlayer.framework to the Frameworks folder, and add #import <MediaPlayer/MediaPlayer.h> to MainViewController.h.

Getting the Current Playback Status

To populate these various fields, we obtain a controller object that manages the background application. This class is called MPMusicPlayerController and can create two kinds of instances. The iPod Music Player is an object that interacts with the system's media player, allowing you to get information about what's currently playing; to start, pause, and stop playback; and to provide a new queue of items to play. The other kind of player is the *application music player*. This has the same functionality as the iPod player but is completely independent of it: using this object doesn't affect the system media player state. You'd use this player if you wanted to get audio from the user's iPod library and play it in your own application, such as for background music in a game. If you use the application player, the iPod player may coexist with it; this probably isn't desirable, since you'd potentially have two songs playing at once. You could get the iPod player object and stop it, but it's better to use the techniques described in Section 16.7, *Interacting with Audio Sessions*, on page 353 to tell the system how you want to mix with system audio of all types.

You get the player of your choice with a class method, either iPodMusicPlayer or applicationMusicPlayer. Once you have the player, you can use a number of properties to manage the playback mode or state; the most-useful ones are the self-explanatory currentPlaybackTime, volume, repeatMode, and shuffleMode. There's also a read-only playbackState whose various MPMusicPlaybackState enumerated values can be used for setting the state of our Play/Pause button:

MediaLibrary/MusicLibraryClient/Classes/MainViewController.m
```
- (void) updatePlayState {
    MPMusicPlayerController *iPodController =
        [MPMusicPlayerController iPodMusicPlayer];
    playPauseButton.selected =
        (iPodController.playbackState == MPMusicPlaybackStatePlaying);
}
```

We also want information about the item that's playing. The nowPlayingItem provides an MPMediaItem that describes what's currently playing, paused, or queued to play. The MPMediaItem represents one playable item: a song, a podcast episode, a section of an audiobook, and so on. It has only two public methods, but one of them, valueForProperty:, provides us with everything we need to know about the currently playing item (and later, with items found in the media library). There is an extensive set of defined keys for retrieving metadata properties, in three general groups:[1]

- *General media item property keys*: Traits common to any media item: MPMediaItemPropertyTitle, MPMediaItemPropertyArtist, MPMediaItemPropertyPlaybackDuration, and far more than are practical to list here.

- *Podcast item property keys*: Metadata items unique to podcasts. The only such key currently defined is MPMediaItemPropertyPodcastTitle.

- *User-defined property keys*: Properties that relate to the use of the media more than its intrinsic traits: MPMediaItemPropertyPlayCount, MPMediaItemPropertyLastPlayedDate, and so on.

Retrieving the properties of the currently playing item will allow us to populate most of the labels we set up and to set the cover art for the UIImageView. For the latter, we should have a fallback in case the user hasn't set the album art in iTunes. Assuming you've put an empty album PNG in your project's resources, declare an instance variable UIImage *emptyAlbumImage; in the header, **#define** a EMPTY_ALBUM_FILE_NAME, and set up the UIImage early in your viewDidLoad.

MediaLibrary/MusicLibraryClient/Classes/MainViewController.m

```
emptyAlbumImage = [[UIImage alloc] initWithContentsOfFile:
        [[NSBundle mainBundle]
            pathForResource:EMPTY_ALBUM_FILE_NAME ofType:@"png"]];
```

With that set up, we can write a updateCurrentiPodItemMetadata method to get the now-playing item and populate most of the UI components with its various properties, or we can blank them out if no item is currently playing.

MediaLibrary/MusicLibraryClient/Classes/MainViewController.m

Line 1
```
- (void) updateCurrentiPodItemMetadata {
    MPMusicPlayerController *iPodController =
        [MPMusicPlayerController iPodMusicPlayer];
```

---

1. See the MPMediaItem documentation for a complete list and the data types returned for each property.

```
                MPMediaItem *nowPlayingItem= [iPodController nowPlayingItem];
                if (nowPlayingItem == nil) {
                        currentTimeSlider.maximumValue = 1.0;
                        currentTimeSlider.value = 0;
                        currentItemDurationLabel.text = @"00:00";
                        currentItemTitleLabel.text = @"(nothing playing)";
                        currentItemArtistLabel.text = nil;
                        currentItemAlbumLabel.text = nil;
                        currentItemArtworkView.image = emptyAlbumImage;
                } else {
                        NSNumber *durationNumber = [nowPlayingItem
                                valueForProperty:MPMediaItemPropertyPlaybackDuration];
                        currentTimeSlider.maximumValue = [durationNumber floatValue];
                        currentItemTitleLabel.text = [nowPlayingItem
                                valueForProperty:MPMediaItemPropertyTitle];
                        currentItemArtistLabel.text = [nowPlayingItem
                                valueForProperty:MPMediaItemPropertyArtist];
                        currentItemAlbumLabel.text = [nowPlayingItem
                                valueForProperty:MPMediaItemPropertyAlbumTitle];
                        MPMediaItemArtwork *coverArt = [nowPlayingItem
                                valueForProperty:MPMediaItemPropertyArtwork];
                        if (coverArt)
                                currentItemArtworkView.image = [coverArt
                                        imageWithSize: currentItemArtworkView.frame.size];
                        else
                                currentItemArtworkView.image = emptyAlbumImage;
                }
        }
```

Many of these property values are NSStrings, which makes it easy to set them as the text of the labels in our GUI. One exception is the duration property (line 14), which is an NSNumber representing the playback duration in seconds as an NSTimeInterval (that is, a **double**). Also notice the retrieval of the "item art," typically album cover art or a podcast logo,[2] on line 23. This property value is an MPMediaItemArtwork object, from which you can get a scaled UIImage via imageWithSize:. For this example, we'll request an image exactly the size of the UIImageView. If the property is **nil**, we'll just use our emptyAlbumArt instead.

Working with Time Properties

One label we haven't updated is the current time display. It's easy enough to get the current time from the MPMusicPlayerController's cur-

---

2. This property does not support "enhanced podcasts," which update the artwork as the episode plays; you'll always get the same episode-specific artwork regardless of the current playback position.

rentPlaybackTime method. This returns the number of seconds elapsed since the beginning of the track, as an NSTimeInterval (that is, a **double**). The trick is how we present it to the user, who wouldn't be well served by a big floating-point number.

With a little bit of string-formatting voodoo, we can get an NSString to show the time, padded to a specific number of digits, so that seconds less than 10 will get a leading 0. Provide a formatted value for the duration label by adding the following line to the updateCurrentiPodItemMetadata method, in the **else** block:

```
currentItemDurationLabel.text = [NSString stringWithFormat: @"%02d:%02d",
            [durationNumber intValue] / 60,
            [durationNumber intValue] % 60];
```

More important, you can now update the current time display by getting the currentPlaybackTime property from the MPMusicPlayerController:

MediaLibrary/MusicLibraryClient/Classes/MainViewController.m

```
- (void) updateCurrentiPodItemTime {
    MPMusicPlayerController *iPodController =
            [MPMusicPlayerController iPodMusicPlayer];
    MPMediaItem *nowPlayingItem= [iPodController nowPlayingItem];
    if (nowPlayingItem == nil) {
        currentTimeLabel.text = @"00:00";
    } else {
        double currentTime = iPodController.currentPlaybackTime;
        currentTimeLabel.text = [NSString stringWithFormat: @"%02d:%02d",
                (int) currentTime/60,
                (int) currentTime%60];
        currentTimeSlider.value = (float) currentTime;
    }
}
```

Of course, if the music is playing when that method is called, it will be accurate only for an instant. We need to constantly update it. Much like we did for the clocks in Chapter 9, *Preferences*, on page 163, we can use an NSTimer to repeatedly call updateCurrentiPodItemTime (for a reminder on how this works, flip back to Section 9.4, *Side Trip: Updating the Clock Label Every Second*, on page 172). Declare NSTimer *currentTimeUpdateTimer; in the .h, and set it up in the .m's viewDidLoad.

MediaLibrary/MusicLibraryClient/Classes/MainViewController.m

```
// create timer to update clock
currentTimeUpdateTimer = [NSTimer scheduledTimerWithTimeInterval:0.1
        target:self selector:@selector(updateCurrentiPodItemTime)
        userInfo:NULL repeats:YES];
```

Figure 15.2: VIEWING THE CURRENTLY PLAYING IPOD ITEM

The last thing to do in viewDidLoad is to do a one-time call to our three GUI-updating methods so that when the view loads, it will get the playback state, current time, and current item metadata from the iPod application:

MediaLibrary/MusicLibraryClient/Classes/MainViewController.m

```
[self updatePlayState];
[self updateCurrentiPodItemMetadata];
[self updateCurrentiPodItemTime];
```

And now we have everything we need to display what's going on in the system music player. Launch iPod (or Music), and start playing some music, a podcast, or an audiobook. Then, while it's still playing, build and run your application on the device. When it comes up, you'll see the current item in the main view, as shown in Figure 15.2. Note that although we have yet to take control of the playback from our own application, you can do so with the headset clicker provided with the iPhone: click once to pause, again to resume.

Getting Notifications from the Player

Right now, the only metadata component that's being repeatedly updated is the current time label, serviced by the NSTimer that we set up. If your song ends and the media player moves on to the next one, most of the display will be wrong. The same will happen if you skip to the next song by double-clicking your headset clicker or going back by triple-clicking it.

Fortunately, there's a systemwide service for notifying applications of asynchronous events, the *notification center*. Applications can send notifications to the notification center, which are then sent to all registered *observers*. The MPMusicPlayerController provides two notifications, which are identified by NSString constants:

- MPMusicPlayerControllerPlaybackStateDidChangeNotification indicates that the current playback state (playing, paused, stopped, skipping forward, and so on) has changed.

- MPMusicPlayerControllerNowPlayingItemDidChangeNotification notifies observers that the current item has changed, either as a result of the user skipping forward or back or as one item ending and the next in the queue beginning.

For the first of these, we can simply call our updatePlayState method to update the Play/Pause button state. For the latter, we'll want to update all the metadata components.

You add an observer for a given notification by getting the default instance of the NSNotificationCenter and calling addObserver:selector:name:object:. The parameters to this method are the object that will receive callbacks, a method selector, the notification name, and an optional "sender" argument that restricts notifications to be only those sent by a specific object. We'll set up the observers in viewDidLoad.

MediaLibrary/MusicLibraryClient/Classes/MainViewController.m
```
[[NSNotificationCenter defaultCenter] addObserver:self
            selector: @selector (playbackStateChanged:)
            name:@"MPMusicPlayerControllerPlaybackStateDidChangeNotification"
            object:nil];
[[NSNotificationCenter defaultCenter] addObserver:self
            selector: @selector (nowPlayingItemChanged:)
            name:@"MPMusicPlayerControllerNowPlayingItemDidChangeNotification"
            object:nil];
[[MPMusicPlayerController iPodMusicPlayer] beginGeneratingPlaybackNotifications];
```

Now we have to write the callback methods that handle the notifications. These methods need to take an NSNotification argument, which provides the name of the notification, the sending object, and (for some notifications) a "user info" dictionary of additional information. In this case, we don't need any of that; we can just call our update methods.

MediaLibrary/MusicLibraryClient/Classes/MainViewController.m
```
-(void) playbackStateChanged: (NSNotification*) notification {
      [self updatePlayState];
}

-(void) nowPlayingItemChanged: (NSNotification*) notification {
      [self updatePlayState];
      [self updateCurrentiPodItemMetadata];
      [self updateCurrentiPodItemTime];
}
```

With this addition, you can now skip forward and back by double- or triple-clicking your headset clicker, and you'll see the metadata change in the main view. It's more impressive if you do this with a playlist with songs from different albums, such as a Genius playlist. Try it.

15.2 Controlling iPod Playback

You don't just have to be a passive client to the iPod application. The MPMusicPlayerController provides a number of methods to affect playback: play, pause, stop, skipToNextItem, skipToBeginning, and so on. Some of the playback mode properties are also writable, such as repeatMode, shuffleMode, and currentPlaybackTime.

This makes implementing our Play/Pause button very simple. Declare handlePlayPauseTapped as an IBAction, connect the Play/Pause button to it, and implement it like so:

MediaLibrary/MusicLibraryClient/Classes/MainViewController.m
```
- (IBAction) handlePlayPauseTapped {
      MPMusicPlayerController *iPodController =
            [MPMusicPlayerController iPodMusicPlayer];
      if (iPodController.playbackState == MPMusicPlaybackStatePlaying) {
            [iPodController pause];
      } else {
            [iPodController play];
      }
}
```

Notice that this handler doesn't update the selected state of the button. It doesn't need to do so. When the iPod application starts playing, it will send a notification of the state change, which will be received by the observer registered in the previous section, which updates the button status as necessary.

Setting the iPod Player's Current Time

We can also set the current time within the playing track and thereby get a nice "scrubbing" behavior; instead of imprecise and slow "rewind" and "fast-forward" buttons (which you could certainly implement with MPMusicPlayerController's beginSeekingForward, beginSeekingBackward, and endSeeking methods), we'll let the user drag the slider and set the current time as they do.

The only thing we have to be careful of is the fact that the slider's value is already being updated with every call to updateCurrentiPodItemTime, which is being called several times a second by the NSTimer. So, we need to stop doing that while the user is performing a scrub gesture.

Start by defining the instance variable BOOL userIsScrubbing; in the header file, and set its value to **NO** in viewDidLoad. We can set this flag value when the user initiates a touch gesture on the slider and when he or she ends it. So, declare two IBActions to manage userIsScrubbing, and implement them:

MediaLibrary/MusicLibraryClient/Classes/MainViewController.m

```
-(IBAction) handleScrubberTouchDown {
       userIsScrubbing = YES;
}

-(IBAction) handleScrubberTouchUp {
       userIsScrubbing = NO;
}
```

Next, in IB, connect the Touch Down event to handleScrubberTouchDown, and connect both Touch Up Inside and Touch Up Outside to handleScrubberTouchUp. Finally, rewrite the last line of updateCurrentiPodItemTime to not update the slider if the user is scrubbing:

```
if (! userIsScrubbing)
       currentTimeSlider.value = (float) currentTime;
```

Now we're not going to be fighting over the slider, because the timer tries to set it to one value and the user to another. The final step is to tell the MPMusicPlayerController to reset the current time in response to a drag gesture. Fortunately, the slider also generates a Value Changed event as the user slides. Declare - (IBAction) handleScrub; in the header file, and in IB connect the Value Changed event to it. The implementation is easy: just reset the player's currentPlaybackTime property to the value of the slider.

```
MediaLibrary/MusicLibraryClient/Classes/MainViewController.m
-(IBAction) handleScrub {
    MPMusicPlayerController *iPodController =
            [MPMusicPlayerController iPodMusicPlayer];
    iPodController.currentPlaybackTime = currentTimeSlider.value;
}
```

With this task completed, you should now be able to scrub backward and forward within songs, even while they're playing.[3] The slider sends out continuous Value Changed events throughout the drag gesture, so you'll hear immediate results and see the current time update as you scrub back and forth.

15.3 Using the iPod Library

Thus far, we have depended on the user already playing music in the iPod or Music application for us to have a currently playing song to work with. We can also take control of the player by sending it new songs to play, but to do that, we need to use the Media Player framework to discover just what's available in the user's media library.

The Media Player framework offers two means of identifying items in the user's media library, which are represented as MPMediaItems (just like the now-playing item returned by the MPMusicPlayerController is):

- The MPMediaQuery class allows you to search for items in the library by various criteria, such as kind (song, podcast episode, audiobook), item title, artist, album title, and so on.

- The MPMediaPickerController class manages a GUI that allows you to browse through the media library. Depending on the mode you choose (albums, artists, podcasts, and so on), the presentation may be almost identical to what you see in the iPod or Music application.

To examine and exercise these APIs, we'll use the flip side of the MusicLibraryClient to find songs to play and then send those to the native music player application when the user dismisses the flip-side view with the Done button.

---

3. At least in the 3.0 betas, scrubbing the iPod player is somewhat "jumpier" than scrubbing through a local file with the AVAudioPlayer, as we'll do in Section 16.6, *Playing Audio with the AVFramework*, on page 348.

> **Limits of the Media Player Framework**
>
> Although there's a lot of great functionality in the Media Player framework, there are significant limits to what you can find in the library and what you can do with it. Here are a few of the big ones:
>
> - *No access to video*: Video items such as movies, TV shows, and video podcasts cannot be discovered with the query API or the picker and can never be a currently playing item from your application's point of view, since the iPod application stops playing such items when it quits.
> - *No access to audio data*: The MPMediaItem does not let you access the audio files themselves, their compressed data, or their decoded audio streams.
> - *No write access*: You can't edit, delete, or write new audio metadata from your application.
> - *No access to arbitrary metadata*: Metadata standards like ID3 for MP3 files and user data atoms in AAC allow for any kind of user-defined metadata, but the Media Player framework provides just a standard set of common property names.

Designing the Flip Side

The MPMusicPlayerController, whether it represents the system's media player ([MPMusicPlayerController iPodMusicPlayer];) or one allocated for your own application ([MPMusicPlayerController applicationMusicPlayer];), accepts new audio for playback in the form of a *queue*. You can send the player an MPMediaItemCollection that you get from a query or build yourself or an MPMediaQuery to perform and play the results of. So, for our flip side, let's develop a GUI that lets the user build a queue by performing one or more queries. The user will select results from those queries, which will be used to build up an array of selected items. These items will be displayed on-screen as a UITable. Finally, when the user clicks Done, we'll send the items to the native iPod player; when we return to the main view, we should see the first of the user's selections start playing.

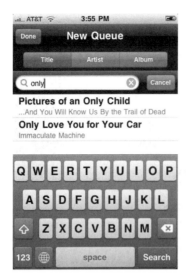

Figure 15.3: QUERYING THE IPOD LIBRARY WITH A UISEARCHDISPLAYCON-TROLLER

The utility application project template has already provided us with a rudimentary flip side, consisting of the classes FlipsideView and FlipsideViewController and a FlipsideView.xib file. The flipping functionality is already wired up, so all we need to do to get started is to build our GUI. Open FlipsideView.xib in Interface Builder, and you'll see the default elements: a navigation bar with a Done button and a navigation item called Title. Here's what we'll add:

- A segmented control right below the navigation bar, with three segments: Title, Artist, and Album.

- Below that, a search bar and search display controller. This is a new and remarkable addition to IB's library in iPhone OS 3.0. Previously, you could add a search bar but were responsible for wiring all its behavior yourself. This new feature adds not just a UISearchBar to your view but also creates a UISearchDisplayController, to which the search bar is prewired. When the search bar gets user input, it slides in a search results table over the current view, at the same time as it brings in the keyboard. In Figure 15.3, we can see what this will look like in our finished example, with the search results table and keyboard covering the rest of the view.[4]

---

4. Note that for this screenshot, we've implemented a search-on-keystroke behavior,

- Fill the rest of the flip-side view with a UITable for the queue of songs that the user will be creating.

As usual, we have some IBOutlets to declare. Add the following instance variables to FlipsideViewController.h, and then declare them as properties with **@property** and **@synthesize**:

MediaLibrary/MusicLibraryClient/Classes/FlipsideViewController.h
```
UISegmentedControl *searchTypeControl;
UISearchDisplayController *searchController;
UITableView *queueTable;
```

We'll also have to declare that you handle three protocols: delegates for the table and search bar, as well as the data source for the table:

MediaLibrary/MusicLibraryClient/Classes/FlipsideViewController.h
```
@interface FlipsideViewController : UIViewController <UITableViewDataSource,
           UITableViewDelegate, UISearchBarDelegate> {
```

In IB, connect the outlets from File's Owner to the appropriate items in the nib. You'll also want to connect the table's delegate and dataSource to File's Owner. Notice, by the way, that the UISearchDisplayController has already made four connections of its own to File's Owner: delegate, searchContentsController, searchResultsDataSource, and searchResultsDelegate. Obviously, we'll have some work to do to service those roles, but at least the connections are prewired.

Finally, change the navigation bar's title to New Queue. When you're done laying out the GUI, it should look like Figure 15.4, on the next page.

Implementing Media Library Search

Our search GUI should let the user type some text into the search bar; query the media library for songs, artists, or albums with that text in the relevant metadata property; and populate the search controller's slide-in table with the results. When the user picks a song, we'll add that to a queue of songs to eventually send to the media player application. We're going to need to declare two arrays as instance variables: an array to hold the most recent set of search results and a mutable array that the user can add selections to.

MediaLibrary/MusicLibraryClient/Classes/FlipsideViewController.h
```
NSArray *searchResults;
NSMutableArray *newPlaybackQueue;
```

---

while the final app won't actually search until the user taps Search.

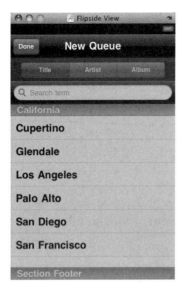

Figure 15.4: LAYING OUT A GUI FOR BUILDING MEDIA QUEUES

We'll also clear out the queue every time the flip-side view comes up, since any previous queue will have already been sent to the system media player. You can do this in viewDidLoad.

`MediaLibrary/MusicLibraryClient/Classes/FlipsideViewController.m`
```
[newPlaybackQueue release];
newPlaybackQueue = [[NSMutableArray alloc] init];
```

These arrays will be used by the data sources of both tables. In fact, we have a rather interesting situation in that our view controller is set to be the delegate and data source for two tables: the slide-in search results table provided by the UISearchDisplayController and the queue table we added to the view ourselves. But it won't be a problem, as you'll see.

To provide the search, we need to implement some of the delegate methods declared by the UISearchBarDelegate protocol. You can choose to search on every keystroke by implementing searchBar:textDidChange: or only when the search button is pressed, with searchBarSearchButtonClicked:. In writing this chapter, we found the search performance sluggish with just 500 songs in the library of a first-gen iPhone, so we recommend waiting for the search button before doing the query. You'll also want to handle the delegate method called when the user cancels the search.

MediaLibrary/MusicLibraryClient/Classes/FlipsideViewController.m

```
- (void)searchBarSearchButtonClicked:(UISearchBar *)searchBar {
    [self updateSearchResults];
    [searchController.searchResultsTableView reloadData];
}

- (void) searchBarCancelButtonClicked: (UISearchBar *) searchBar {
    searchResults = nil;
}
```

Obviously, this is a simple implementation of searchBarSearchButtonClicked:, because it defers the query work to a yet-to-be-written updateSearchResults method. The reason you want to put the query in its own method is that you'll want to requery if the user taps the segmented control to change the query type.

So, let's look at how we're going to perform the query. The Media Player framework provides an MPMediaQuery class to handle querying the media library. It comes with several "canned" queries exposed as class methods, such as songsQuery and podcastsQuery, which return all items of a given type.

Using one of these queries or starting with an empty query, you can add *filter predicates* to narrow down the results to those matching a certain criteria. Most of the typically user-visible properties, such as song titles and artist names, can be used in this way as search criteria. (Check the MPMediaItem documentation to see which properties are marked as **filterable**.)

To search the media library, we'll start with an MPMediaQuery for all the songs and then filter this down to just those that contain the text in the search bar, in either the song title, artist name, or album name. We can do this by **switch**ing on the value of the segmented control and building an MPMediaPropertyPredicate to add to the query. With the predicate added, we can retrieve the matching items as an NSArray from the query's items property.

MediaLibrary/MusicLibraryClient/Classes/FlipsideViewController.m

```
- (void) updateSearchResults {
    NSString *searchText = searchController.searchBar.text;
    MPMediaQuery *query = [MPMediaQuery songsQuery];
    switch (searchTypeControl.selectedSegmentIndex) {
        case 0: {
            MPMediaPropertyPredicate *titlePredicate =
            [MPMediaPropertyPredicate predicateWithValue:searchText
                forProperty:MPMediaItemPropertyTitle
```

```
                                comparisonType: MPMediaPredicateComparisonContains];
                        [query addFilterPredicate:titlePredicate];
                        break;
                }
                case 1: {
                        MPMediaPropertyPredicate *artistPredicate =
                        [MPMediaPropertyPredicate predicateWithValue:searchText
                                forProperty:MPMediaItemPropertyArtist
                                comparisonType: MPMediaPredicateComparisonContains];
                        [query addFilterPredicate:artistPredicate];
                        break;
                }
                case 2: {
                        MPMediaPropertyPredicate *albumPredicate =
                        [MPMediaPropertyPredicate predicateWithValue:searchText
                                forProperty:MPMediaItemPropertyAlbumTitle
                                comparisonType: MPMediaPredicateComparisonContains];
                        [query addFilterPredicate:albumPredicate];
                        break;
                }
                default: {
                        // unknown segment type - just return everything
                        query = [MPMediaQuery songsQuery];
                }
        }

        [searchResults release];
        searchResults = query.items;
        [searchResults retain];
}
```

As you can see, you create an MPMediaPropertyPredicate with class methods that take a value to filter on, which property it applies to, and an optional comparisonType that can be either MPMediaPredicateComparisonEqualTo or MPMediaPredicateComparisonContains. Textual comparisons are case insensitive, and there is no public way to make them case sensitive.

Note that the results come back unsorted. You could get a sorted array with NSArray's various sorting methods or set the query's groupingType property and then retrieve the collections property, which returns results as an array of MPMediaItemCollections, grouped by the specified grouping type. To keep things simple in this example, we'll just leave the individual results unsorted.

So, our updateSearchResults method populates searchResults with an array of MPMediaItems matching the string tapped into the search bar. We also want to reperform this search if the user changes the search type by tapping the segmented control after doing a search (in other words, changing from searching for songs with a given term to artists or albums with that term). This is easy: declare - (IBAction) searchType-Changed; in FlipsideViewController.h, connect the segmented control's Value Changed event to this action, and write an implementation that just calls our updateSearchResults and refreshes the search controller's table.

MediaLibrary/MusicLibraryClient/Classes/FlipsideViewController.m

```
- (IBAction) searchTypeChanged {
      [self updateSearchResults];
      if (searchController.active == YES) {
            [searchController.searchResultsTableView reloadData];
      }
}
```

Implementing the Search Result and Media Queue Tables

We've now performed the search, but the results are just sitting in the searchResults array. We need to implement the UITableViewDelegate and UITableDataSource methods to present them to the user.

As mentioned, the FlipsideViewController is wired up as the delegate to two different tables: the slide-in table owned by the UISearchDisplayController for search results and our own table that we want to use for the queue that the user builds by selecting search results. So, our delegate methods will have to pay attention to which table is calling them.

Let's start with the delegate method tableView:didSelectRowAtIndexPath:. When a user taps an item in the search results table, we want to get the selected MPMediaItem and put it in the newPlaybackQueue array. Then we update the queue table and dismiss the search GUI.

MediaLibrary/MusicLibraryClient/Classes/FlipsideViewController.m

```
- (void)tableView:(UITableView *)tableView
            didSelectRowAtIndexPath:(NSIndexPath *)indexPath {
      if (tableView != queueTable) {
            // add item from search results to queue
            [newPlaybackQueue addObject:
                  [searchResults objectAtIndex:indexPath.row]];
            [queueTable reloadData];
            [tableView deselectRowAtIndexPath:indexPath animated:NO];
            [searchController setActive:NO animated:YES];
            searchResults = nil;
      } else {
            [tableView deselectRowAtIndexPath:indexPath animated:NO];
      }
}
```

Notice that selecting a row in the queue table does nothing.

That covers the most important action: getting a selected search result into the new playback queue array. Now we just have to provide the other essential table methods, providing the number of sections and size of the tables.

MediaLibrary/MusicLibraryClient/Classes/FlipsideViewController.m
```
- (NSInteger)numberOfSectionsInTableView:(UITableView *)tableView {
    return 1;
}

- (NSInteger)tableView:(UITableView *)tableView
            numberOfRowsInSection:(NSInteger)section {
        if (tableView == queueTable) {
            return [newPlaybackQueue count];
        } else {
            return [searchResults count];
        }
}
```

Using a similar strategy, we could also provide a section header for the queue table to show the number of items the user has added.

MediaLibrary/MusicLibraryClient/Classes/FlipsideViewController.m
```
- (NSString *)tableView:(UITableView *)tableView
            titleForHeaderInSection:(NSInteger)section {
        if (tableView == queueTable) {
            return [NSString stringWithFormat:
                    @"Queue (%d songs)", [newPlaybackQueue count]];
        } else {
            return nil;
        }
}
```

Finally, we need to provide cells. The boilerplate implementation provided by Xcode for other table implementations (such as those shown in Section 5.4, *Cell Styles*, on page 81) will do fine. Just change the style to UITableViewCellStyleSubtitle, and then set textLabel and detailTextLabel to property values retrieved from an MPMediaItem retrieved from the appropriate array (newPlaybackQueue for the queueTable, searchResults for the search display controller's table).

MediaLibrary/MusicLibraryClient/Classes/FlipsideViewController.m
```
- (UITableViewCell *)tableView:(UITableView *)tableView
            cellForRowAtIndexPath:(NSIndexPath *)indexPath {
    static NSString *CellIdentifier = @"Cell";
    UITableViewCell *cell =
            [tableView dequeueReusableCellWithIdentifier:CellIdentifier];
```

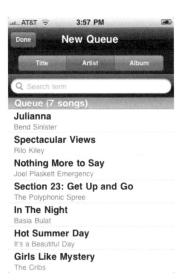

Figure 15.5: BUILDING A PLAYBACK QUEUE OF MPMEDIAITEMS

```
    if (cell == nil) {
        cell = [[[UITableViewCell alloc]
                    initWithStyle:UITableViewCellStyleSubtitle
                    reuseIdentifier:CellIdentifier] autorelease];
}
    // Configure the cell
    MPMediaItem *anItem = nil;
    if (tableView == queueTable) {
            anItem = [newPlaybackQueue objectAtIndex:indexPath.row];
    } else {
            anItem = [searchResults objectAtIndex:indexPath.row];
    }
    cell.textLabel.text =
            [anItem valueForProperty:MPMediaItemPropertyTitle];
    cell.detailTextLabel.text =
            [anItem valueForProperty:MPMediaItemPropertyArtist];
    return cell;
}
```

With these table methods implemented, you can now select search results and build a queue, as shown in Figure 15.5.

Playing a Queue

The final step is to take the queue of songs selected by the user and send them to the native media player application. The utility application template has already wired the navigation bar's Done button to a `done` method that flips back to the main view, so we can add code in that method to set up the playback queue and then flip.

As we discussed before we even started building the flip-side GUI, the `MPMusicPlayerController` has several methods that accept a queue of items to play. Since we've already assembled the items we want to play, we'll call `setQueueWithItemCollection:`, providing an `MPMediaItemCollection` built from our array.

`MediaLibrary/MusicLibraryClient/Classes/FlipsideViewController.m`

```
- (IBAction)done {
    if ([newPlaybackQueue count]) {
        MPMusicPlayerController *iPodController =
            [MPMusicPlayerController iPodMusicPlayer];
        [iPodController stop];
        MPMediaItemCollection *queueCollection =
            [MPMediaItemCollection collectionWithItems:
                newPlaybackQueue];
        [iPodController setQueueWithItemCollection: queueCollection];
        [iPodController play];
    }
    // provided by xcode template to flip back
    [self.delegate flipsideViewControllerDidFinish:self];
}
```

15.4 Browsing the iPod Library

In the previous section, we provided our own search GUI for searching for items to add to the playback queue. Along with `MPMediaQuery`, the Media Player framework gives us another means of finding items from the media library: `MPMediaPickerController`.

This picker is a `UIViewController` whose view is, depending on how you configure it, almost identical to the native iPod or Music applications. You initialize it with a bit field of `MPMediaTypes` to display. For example, to create a browse GUI to show only podcasts and audiobooks, you'd create the picker like this:

```
MPMediaPickerController *pickerController =
    [[MPMediaPickerController alloc]
        initWithMediaTypes: MPMediaTypePodcast | MPMediaTypeAudioBook];
```

This creates a tabbed GUI with two tabs, one each for podcasts and audiobooks, each highly similar to the native application. The biggest difference comes from whether you choose to have the picker allow multiple selections; with this option set to **YES**, the items will have + characters to allow the user to add them to the picker's own queue of selections.

Let's add a browse option to our flip-side GUI. Add a tab bar button to the right side of the navigation item, and give it the title Browse. Declare - (IBAction) handleBrowseTapped; in FlipsideViewController.h, and implement it as follows:

MediaLibrary/MusicLibraryClient/Classes/FlipsideViewController.m

```
- (IBAction) handleBrowseTapped {
    MPMediaPickerController *pickerController =
            [[MPMediaPickerController alloc]
                    initWithMediaTypes: MPMediaTypeMusic];
    pickerController.prompt = @"Add songs to queue";
    pickerController.allowsPickingMultipleItems = YES;
    pickerController.delegate = self;
    [self presentModalViewController:pickerController animated:YES];
    [pickerController release];
}
```

We start by creating the MPMediaPickerController and initializing it to show the music in the user's library. Along with setting properties for a user-readable prompt and whether to allow multiple selections, we provide a delegate implementing the MPMediaPickerControllerDelegate protocol. Since we want that to be the FlipsideViewController itself, be sure to go over to the header file and add MPMediaPickerControllerDelegate to the list of implemented protocols. With the delegate set, we can show the picker controller in a modal view, which slides in from the bottom, as shown in Figure 15.6, on the following page.

When the user taps Done or Cancel (which may not be present, depending on the controller's media types and multiple-selection properties), the MPMediaPickerController calls the delegate method mediaPicker:didPickMediaItems: or mediaPickerDidCancel:, as appropriate. For our purposes, all that's necessary is to add any selected items to the newPlaybackQueue and to dismiss the modal view controller.

MediaLibrary/MusicLibraryClient/Classes/FlipsideViewController.m

```
- (void)mediaPicker: (MPMediaPickerController *)mediaPicker
            didPickMediaItems:(MPMediaItemCollection *)mediaItemCollection {
    [self dismissModalViewControllerAnimated:YES];
    [newPlaybackQueue addObjectsFromArray:mediaItemCollection.items];
    [queueTable reloadData];
}
```

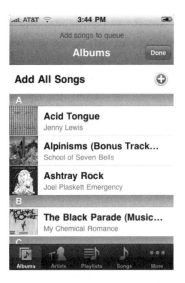

Figure 15.6: BROWSING THE MEDIA LIBRARY WITH MPMEDIAPICKERCONTROLLER

```
- (void)mediaPickerDidCancel:(MPMediaPickerController *)mediaPicker {
    [self dismissModalViewControllerAnimated:YES];
}
```

So with that, you've given your user the same level of playback control they enjoy in the iPod application: the user can browse the music library, search it, pick one or more items to play, and play them, complete with metadata display and scrubbing. What you do with this is up to you. Rather than just offering a differently styled player, consider the possibilities of playing the user's favorite music (perhaps based on properties such as MPMediaItemPropertyPlayCount or MPMediaItemPropertyRating) as background music in your application. Build a social application that analyzes their library and compares it to other users' libraries.[5] Combine music library analysis with the location API to help them find nearby record stores and concert venues with music they'd enjoy. Or you can be cute and unlock features in your application based on whether the user has artists and songs you like.

---

5. With their permission, of course, because sharing the user's data has serious privacy implications.

Chapter 16
Playing and Recording Audio

Although support for video is pretty simplistic in iPhone 3.0, the depth of the audio APIs is remarkably extensive. The Core Audio framework provides multiple layers of abstraction, from a high-level Objective-C API that makes it exceedingly simple to set up a basic audio recorder or player that works with files in your Documents directory to low-level C APIs for streaming audio that allow you to work with raw audio samples directly.

That said, the "80-20 rule" applies: a large percentage of developers will use a small, critical subset of this framework. For the bulk of this chapter, we'll focus on the high-level Objective-C classes provided for the most common tasks: recording and playing local files.

16.1 Creating an AVAudioRecorder

Prior to iPhone SDK 3.0, developers who wanted to record audio needed to work with the Audio Queue Services API (described later in Section 17.3, *Audio Queue Services*, on page 367), a procedural C API that uses a streaming metaphor. You'd get callbacks with newly captured audio samples and have to write them to a file with another C API for writing audio files. Lots of people complained that this was too much work for a simple task.

In response, iPhone SDK 3.0 provides a much easier-to-use option in the form of the AVAudioRecorder class, part of the AVFoundation framework. This class allows you to create a recorder object with a destination URL and some settings to define the format you want to record into and then just start recording audio data into the file with self-explanatory methods like record, pause, and stop. Of course, nothing is ever *that* easy with audio, as we'll see when we consider the options for compressed audio and how few combinations of settings are actually valid, but for many people, this is going to be much more productive.

Setting Up the Sample Application

To exercise AVFoundation's recording and playback features, we'll build a sample application that can record from whatever audio input is available (the phone's mic or the headset mic) into files and that can then play back the recorded files. We'll give the user a wide variety of options in terms of recording formats, which will help us explore the ins and outs of audio formats.

The sample app we've prepared is one of the most ornate in the book. We'll describe the major elements of it, but since we don't want to take the focus off audio for the sake of a four-page trip into Interface Builder, this is one case in which you'd really be best served by looking in the downloadable sample code, where we've provided AudioRecorderPlayer-Starter as a stub project with the GUI built out and wired up. If you're determined to work through building it from scratch, then assume that any **#import**s, properties, helper methods, IBActions, and IBOutlets that we haven't explicitly explained are things we expect you can handle on your own.

For our sample, we'll set up a tab-based application, AudioRecorderPlayer, with two tabs: one for recording and one for playing back the recorded files. This means we'll eventually be writing two view controllers, RecordViewController and PlayViewController, and setting these as the classes for the two view controller objects created for you in MainView.xib. The tab bar template defaults to putting your first view in MainView.xib; in the download, you'll find we've chosen to create a RecordView.xib for the first view and to have the first tab's view controller load its view from this nib. We've also used the refactoring tool to change the FirstViewController to RecordViewController.

In Figure 16.1, on the facing page, we can see what we've set up for the main view in RecordView.xib. The key elements are a filenameField to enter a file to record into, a UISegmentedControl formatSegments that offers a choice between PCM and Encoded formats, a formatButton to bring up format-details modal views, a currentTimeLabel, a disabled recordPauseButton, and a stopButton. Below these are two wide and short views for the leftLevelMeter and rightLevelMeter, which will be implemented with a

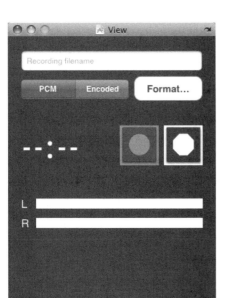

Figure 16.1: LAYOUT FOR AUDIORECORDERPLAYER'S FIRST TAB

custom class later. Of course, all of these will need to be declared as IBOutlets and connected in Interface Builder.

The AVAudioRecorder class is in the optional AVFoundation framework, so remember to add that to your project's Frameworks folder and do an #import <AVFoundation/AVFoundation.h>; in your header file. The class has only one viable initializer method, initWithURL:settings:error:, which takes a file to write to (as an NSURL), an NSDictionary of format settings, and a pointer to an NSError to report any error that prevented the recorder object from being created. We'll need both a location to write to and the settings values in order to create the recorder object. We could either create the recorder when the user taps the record button or record it any time the file location or settings change. Let's opt for the latter approach: if we get an error setting up the recorder (say, from invalid format settings), we'll be able to "fail early" and let the user know before they even attempt to record. So, any time the text field ends editing or the format changes, we'll call a createAVAudioRecorder to set up a new audioRecorder object, our instance of AVAudioRecorder.

Let's start with the text field. Set its delegate to the RecordViewController, and provide the usual textFieldShouldReturn implementation that dismisses the virtual keyboard with [textField resignFirstResponder];. The other thing we should do is to call createAVAudioRecorder when the user is done editing. More than that, we could make some effort to validate the field. There are a lot of restrictions on audio file formats, something we'll discuss in Section 16.2, *Uncompressed Audio Formats*, on the next page and Section 16.3, *Encoded Audio Formats*, on page 335. For example, uncompressed PCM audio in a .wav file must be little-endian, while the same data in an .aif file has to be big-endian. And the *only* legal content in an .mp3 file is MPEG-1 Layer 3 data. We'll save ourselves a lot of headaches by defaulting to the Core Audio Format (.caf) format, which is content-agnostic. Anything Core Audio can play can be stored in a .caf file. For this and other reasons, it's the recommended format for iPhone audio, and we'll honor that by defaulting to .caf if no file extension is provided.

MediaPlayback/AudioRecorderPlayer/Classes/RecordViewController.m
```
- (void)textFieldDidEndEditing:(UITextField *)textField {
    [textField resignFirstResponder];
    // verify that there's a legitimate filename extension
    if ([[textField.text pathExtension] length] == 0)
        textField.text =
            [NSString stringWithFormat: @"%@.caf", textField.text];
    [self createAVAudioRecorder];
}
```

To get the NSURL needed to create the AVAudioRecorder, we'll need the path to the Documents directory. Here's a utility method that lazily instantiates a _documentsPath variable the first time it's needed:

MediaPlayback/AudioRecorderPlayer/Classes/RecordViewController.m
```
- (NSString*) documentsPath {
    if (! _documentsPath) {
        NSArray *searchPaths =
            NSSearchPathForDirectoriesInDomains
                (NSDocumentDirectory, NSUserDomainMask, YES);
        _documentsPath = [searchPaths objectAtIndex: 0];
        [_documentsPath retain];
    }
    return _documentsPath;
}
```

By using that, we can begin to write our createAVAudioRecorder method by creating an NSURL that represents the given filename in the Documents directory.

```
MediaPlayback/AudioRecorderPlayer/Classes/RecordViewController.m
- (NSError*) createAVAudioRecorder {
        [audioRecorder release];
        audioRecorder = nil;

        NSString *destinationString = [[self documentsPath]
                        stringByAppendingPathComponent:filenameField.text];
        NSURL *destinationURL = [NSURL fileURLWithPath: destinationString];
```

Now let's look at the second argument to AVAudioRecorder's initWithURL:settings:error:, which is settings. The documentation describes this as an NSDictionary of key-value pairs, with four general sets of keys: general audio format settings, linear PCM settings, encoder settings, and sample rate conversion settings.

16.2 Uncompressed Audio Formats

We'll start with linear pulse code modulation (PCM). This is what we usually mean when we talk about "uncompressed audio." To really understand it, it helps to step back and think about how digital audio works. Sound is oscillating waves of pressure moving through some medium, within a range of frequencies that humans can perceive. When you speak, your larynx creates waves of sound that move through the air and vibrate someone else's eardrum, which is how they hear it.

Analog sound systems, like the original radio, telephone, and records, transmit sound by representing these sound waves as electrical signals. A microphone's vibration is transmitted as an electrical signal over a wire, which vibrates a speaker to reproduce the sound. With digital audio, we represent a sound wave by *sampling* the amplitude of signal thousands of times a second and assessing the signal strength at that instant as a numeric value. *Pulse code modulation* is the term we use for representing a sound wave as a set of sample values, with *linear PCM* meaning that the sample values are directly proportional to the signal amplitude (as opposed to having some other relationship, like logarithmic). For example, what we generally think of as "CD quality" audio is 44,100 samples a second,[1] stored in 16-bit values.

---

1. 44,100 is not a magic number. The *Nyquist-Shannon sampling theorem* tells us that we can reproduce frequencies by sampling at double the rate of the highest frequency we need to reproduce. Since human hearing generally tops out around 20,000Hz, we need to sample at double that rate.

Linear PCM has the benefits of being universal—every digital audio system is based on it—and of being the highest possible quality. If you're creating and editing audio, you'll want to work with PCM throughout your workflow and convert to a compressed format only at the end. After all, PCM sounds great, but the data gets really big.

For our recording application, let's start with supporting recording into PCM. In the GUI, we have a segmented control to switch between PCM and Encoded, along with a Format button. The idea here is that with so many settings for each kind of audio, we'll need to bring up a whole new view to let the user configure their recording in that format. We'll create separate views for each kind of format, with separate view controllers: PCMSettingsViewController and EncodingSettingsViewController. Create these empty classes, give the RecordViewController an IBOutlet for each, and then go into RecordView.xib and drag two view controller objects into the nib document, using the Identity inspector to change their classes to the new classes you created. Then you should be able to connect the outlets to these new VC objects. Also, be sure to create nib files for their views—PCMSettingsView.xib and EncodingSettingsView.xib—and use the Properties inspector to point each view controller to its view nib.

So, the first order of business is to bring up the appropriate configuration view when the user taps Format. This is a pretty simple job for an event handler wired up to that button:

MediaPlayback/AudioRecorderPlayer/Classes/RecordViewController.m
```
-(IBAction) handleFormatButtonTapped {
        if (formatSegments.selectedSegmentIndex == 0) {
                [self presentModalViewController:pcmSettingsViewController
                        animated:YES];
        } else {
                [self presentModalViewController:encodingSettingsViewController
                        animated:YES];
        }
}
```

When the user has selected PCM and taps the Format button, they'll navigate to the pcmSettingsViewController, which slides in the PCM configuration view that is defined in PCMSettingsView.xib and that looks more or less like Figure 16.2, on the next page.

The five fields in this view come from the settings defined in the AVAudioRecorder documentation. The first two come from the General settings: sample rate and the number of channels (which we'll handle by providing an on/off "stereo" switch). The keys for these settings are

Figure 16.2: LAYOUT FOR THE PCM SETTINGS VIEW

AVSampleRateKey and AVNumberOfChannelsKey. Supporting stereo is a nice touch, though it's worth noting that the default input devices—the phone mic and the headset mic—are mono.

The next three settings are unique to PCM. For sample bit depth, the AVLinearPCMBitDepthKey has four legal values: 8, 16, 24, or 32. 16 is a good choice for a default, because that's what's used internally in Core Audio's "canonical" audio format. The next switch is used for the AVLinearPCMIsBigEndianKey, which affects how values larger than 8 bits are interpreted. As mentioned earlier, WAV files can be little-endian only, AIFFs can only be big-endian, and CAFs can be either. A more elaborate application might want to preset this switch based on the current file type. Finally, there's a switch to set AVLinearPCMIsFloatKey, signaling whether the samples are integers or floating-point values. Integer PCM is more common and is the canonical format on iPhone OS.

Provide properties for the five UI controls in the PCM settings view controller: sampleRateField, stereoSwitch, sampleDepthField, bigEndianSwitch, and floatingSamplesSwitch. You'll also want to have an IBOutlet property to hold on to the RecordViewController, a connection you can make in IB.

The reason for this is that when the user taps the Done button,[2] you need to tell the RecordViewController to try to create an AVAudioRecorder based on the changed values in this modal view controller.

MediaPlayback/AudioRecorderPlayer/Classes/PCMSettingsViewController.m

```
-(IBAction) handleDoneTapped {
        [self dismissModalViewControllerAnimated:YES];
        [recController createAVAudioRecorder];
}
```

Back in RecordViewController, we can continue the implementation of createAVAudioRecorder, using the values from this PCMSettingsViewController:

MediaPlayback/AudioRecorderPlayer/Classes/RecordViewController.m

```
Line 1  NSMutableDictionary *recordSettings =
    -           [[NSMutableDictionary alloc] initWithCapacity:10];
    -   if (formatSegments.selectedSegmentIndex == 0) {
    -           // force pcm settings view to load so its fields can be read
    5           pcmSettingsViewController.view;
    -           [recordSettings setObject:
    -           [NSNumber numberWithInt: kAudioFormatLinearPCM] forKey: AVFormatIDKey];
    -           float sampleRate =
    -                   [pcmSettingsViewController.sampleRateField.text floatValue];
   10           [recordSettings setObject:
    -           [NSNumber numberWithFloat:sampleRate] forKey: AVSampleRateKey];
    -           [recordSettings setObject:
    -                   [NSNumber numberWithInt:
    -                           (pcmSettingsViewController.stereoSwitch.on ? 2 : 1)]
   15                   forKey:AVNumberOfChannelsKey];
    -           int bitDepth =
    -                   [pcmSettingsViewController.sampleDepthField.text intValue];
    -           [recordSettings setObject:
    -           [NSNumber numberWithInt:bitDepth] forKey:AVLinearPCMBitDepthKey];
   20           [recordSettings setObject:
    -                   [NSNumber numberWithBool:
    -                                   pcmSettingsViewController.bigEndianSwitch.on]
    -                   forKey:AVLinearPCMIsBigEndianKey];
    -           [recordSettings setObject:
   25                   [NSNumber numberWithBool:
    -                                   pcmSettingsViewController.floatingSamplesSwitch.on]
    -                   forKey:AVLinearPCMIsFloatKey];
    -   }
```

---

2. If you use a navigation bar and a tab bar button, as we've done here, you'll get a selector action in IB's Connections inspector. Ctrl+drag this to the File's Owner's handleDoneTapped method, and the connection will change from "selector" to "handleDoneTapped." This is analogous to the default flip-side view provided by Xcode's utility application template.

Before looking at the key-value pairs, notice line 5. This handles a sneaky problem with nib loading: we want to get values from the PCM settings view, perhaps the default values set in IB if they haven't been changed. But if the user has never brought up the settings view (for example, we're calling createAVAudioRecorder because the user has set only the filename), then all the properties will be unassigned, because the view won't have even been loaded. So, line 5 references the view, forcing it to load and connecting the properties.

Next we provide a key-value pair to provide the format setting. Obviously in this case, the format is PCM, and we use the constant kAudioFormatLinearPCM, defined in the file CoreAudioTypes.h and wrapped by an NSNumber object, as the value for the AVFormatIDKey.

Although the rest of this **if** block is a lot of code, it's mostly involved with converting the types used by the UI widgets into the NSNumbers expected for values in the NSDictionary used for the settings argument. For example, based on whether the stereoSwitch is on, we set the AVNumberOfChannelsKey to 1 or 2. You can look up more information on the legal values of each key in the AVAudioSettings.h header file that defines the settings keys. For text fields with numeric contents, we convert the text to an **int** or **float** and wrap them with an NSNumber, and the UISwitch values are converted to NSNumbers via numberWithBool:.

16.3 Encoded Audio Formats

We've handled recorder settings for PCM; now we need to consider encoded formats. These are the various forms of audio compression that allow us to get audio files down to a size that's more appropriate for sending over networks or storing on small devices...with the trade-off that most encoded formats are "lossy," meaning that the decoded audio is not technically identical to the original source. Many formats are defined in CoreAudioTypes.h, but not all are supported on the iPhone, and not all are available for recording. Apple has said the following formats are available for use with the AVAudioRecorder:

- Linear PCM (kAudioFormatLinearPCM): As shown in the previous section.

- AAC (kAudioFormatMPEG4AAC): The MPEG-4 Audio codec popularized by iTunes and the iPod. As of this writing, only the second-generation iPod touch and the iPhone 3GS can record into this format.

- **ALAC** (kAudioFormatAppleLossless): An encoder introduced in iTunes 4.5 that shrinks files by about 50 percent while exactly reproducing the source audio.

- **IMA4** (kAudioFormatAppleIMA4): Also known as ADPCM in the Windows world; this is a simple and not very CPU-intensive 4:1 compression for 16-bit audio.

- **u-law and a-law** (kAudioFormatULaw and kAudioFormatALaw): Very old codecs originally developed for telephony.

- **Internet Low-Bitrate Codec** (kAudioFormatiLBC): A codec designed for Voice over IP, streaming audio, and other network audio tasks. Among iLBC's appealing traits is that it handles dropped frames better than many other audio codecs, making it well suited for unreliable transmission protocols like UDP.

For our example, we have a second view controller, EncodingSettingsViewController, that we'll use to manage a view that lets the user choose one of these formats and adjust encoding settings. Note that it's beyond the scope of this chapter to enforce the limitations of each format; for example, many of these have a fixed set of valid encoding bitrates. When you use these encoding formats in your own applications, you'll want to study the details of your chosen format(s) and what you're allowed to do with them.

The Interface Builder layout for the EncodingSettingsView.xib file is shown in Figure 16.3, on the facing page. The first component is a segmented view offering a choice of several of the supported formats. These map to an **enum** in our EncodingSettingsViewController.h that lists the formats in the segmented control:

MediaPlayback/AudioRecorderPlayer/Classes/EncodingSettingsViewController.h
```
enum ENCODED_FORMAT_SEGMENT_VALUES {
        ENCODED_FORMAT_AAC = 0,
        ENCODED_FORMAT_ALAC,
        ENCODED_FORMAT_IMA4,
        ENCODED_FORMAT_ILBC,
        ENCODED_FORMAT_ULAW
};
```

After this, the sample rate and stereo fields are the same as in the PCM settings; in fact, you can drag and drop the fields from that view to this one. The last three settings are for keys that are used only by encoded formats. The AVEncoderBitRateKey defines the encoded bitrate in hertz. You've probably used this in iTunes when you decide whether

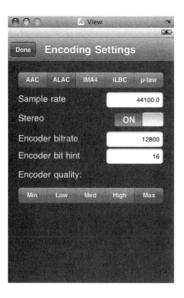

Figure 16.3: LAYOUT FOR THE ENCODING SETTINGS VIEW

to rip your CDs at 128Kbps, 160, 192, or higher. Next, the AVEncoderBitDepthHintKey is a value from 8 to 32, analogous to PCM's AVLinearPCMBitDepthKey. Finally, there's an AVEncoderAudioQualityKey that takes one of five values enumerated in AVAudioSettings.h.

Declare properties for each of these fields and one for the RecordViewController, as before, with the Done button again wired up to dismiss the modal view controller, and call recController createAVAudioRecorder];. Back in RecordViewController.m, we continue with the **else** block that creates an NSDictionary for the recorder settings in this case, the case of using an encoded format:

MediaPlayback/AudioRecorderPlayer/Classes/RecordViewController.m

```
else {
        // force pcm settings view to load so its fields can be read
        encodingSettingsViewController.view;
        NSNumber *formatObject;
        switch ([encodingSettingsViewController.formatSegments
                    selectedSegmentIndex]) {
            case (ENCODED_FORMAT_AAC):
                formatObject =
                    [NSNumber numberWithInt: kAudioFormatMPEG4AAC];
                break;
```

```objc
            case (ENCODED_FORMAT_ALAC):
                    formatObject =
                            [NSNumber numberWithInt: kAudioFormatAppleLossless];
                    break;
            case (ENCODED_FORMAT_IMA4):
                    formatObject =
                            [NSNumber numberWithInt: kAudioFormatAppleIMA4];
                    break;
            case (ENCODED_FORMAT_ILBC):
                    formatObject =
                            [NSNumber numberWithInt: kAudioFormatiLBC];
                    break;
            case (ENCODED_FORMAT_ULAW):
                    formatObject =
                            [NSNumber numberWithInt: kAudioFormatULaw];
                    break;
            default:
                    formatObject =
                            [NSNumber numberWithInt: kAudioFormatLinearPCM];
    }
    [recordSettings setObject:formatObject forKey: AVFormatIDKey];
    float sampleRate =
            [encodingSettingsViewController.sampleRateField.text floatValue];
    [recordSettings setObject: [NSNumber numberWithFloat:sampleRate]
            forKey: AVSampleRateKey];
    [recordSettings setObject:
            [NSNumber numberWithInt:
              (encodingSettingsViewController.stereoSwitch.on ? 2 : 1)]
            forKey:AVNumberOfChannelsKey];
    int encoderBitrate =
            [encodingSettingsViewController.encoderBitrateField.text intValue];
    [recordSettings setObject:[NSNumber numberWithInt:encoderBitrate]
            forKey:AVEncoderBitRateKey];
    int bitDepthHint =
            [encodingSettingsViewController.encoderBitHint.text intValue];
    [recordSettings setObject:
                [NSNumber numberWithInt:bitDepthHint]
            forKey:AVEncoderBitDepthHintKey];
    int encoderQuality;
    switch ([encodingSettingsViewController.qualitySegments
                    selectedSegmentIndex]) {
            case (0) : encoderQuality = AVAudioQualityMin; break;
            case (1) : encoderQuality = AVAudioQualityLow; break;
            case (2) : encoderQuality = AVAudioQualityMedium; break;
            case (3) : encoderQuality = AVAudioQualityHigh; break;
            case (4) : encoderQuality = AVAudioQualityMax; break;
    }
    [recordSettings setObject:
                [NSNumber numberWithInt: encoderQuality]
            forKey: AVEncoderAudioQualityKey];
}
```

This is a lot of code, but again, there's not that much happening. It gets the selected value from the format UISegmentedControl and fetches the constant for that format, wrapping it in an NSNumber. The handling of the sample rate and number of channels is copied directly from the PCMSettingsViewController shown earlier, and the other settings are handled analogously, with numeric strings converted to NSNumbers.

16.4 Using the AVAudioRecorder

Now that we have a place to record the audio data and the format to record in, we're ready to create the AVAudioRecorder. The final parameter to its initWithURL:settings:error: initializer is a pointer to an NSError, which it will populate with an error object if the recorder object can't be initialized with the provided URL and settings dictionary. We'll use this error object to create an alert to the user if initialization fails.

```
MediaPlayback/AudioRecorderPlayer/Classes/RecordViewController.m
NSError *recorderSetupError = nil;
audioRecorder = [[AVAudioRecorder alloc] initWithURL:destinationURL
            settings:recordSettings error:&recorderSetupError];
[recordSettings release];

if (recorderSetupError) {
        UIAlertView *cantRecordAlert =
        [[UIAlertView alloc] initWithTitle: @"Can't record"
                message: [recorderSetupError localizedDescription]
                delegate: nil
                cancelButtonTitle:@"OK"
                otherButtonTitles:nil];
        [cantRecordAlert show];
        [cantRecordAlert release];
        return recorderSetupError;
}
[audioRecorder prepareToRecord];
recordPauseButton.enabled = YES;
audioRecorder.delegate = self;
```

After all the setup work, it takes just a one-line alloc and initWithURL:settings:error: to create the recorder object. Assuming it succeeds, we can call prepareToRecord to allocate needed resources for recording; this will eliminate a pause if those tasks need to be performed upon calling record. We also set the view controller as the recorder's delegate, which means we'll get activity callbacks from the recorder, as described by the AVAudioRecorderDelegate protocol. On the other hand, if recorderSetupError is non-**nil**, then we show a UIAlertView and bail.

> ### Joe Asks...
> #### What the Heck Does 1718449215 Mean?
>
> More than perhaps any other framework on the iPhone, Core Audio relies on the use of *four-character codes* (4cc's) for data identifiers and response codes. If you get an error back when creating the AVAudioRecorder, it's likely that you'll find that the error's domain is **NSOSStatusErrorDomain** and the code is a 4-byte int like 1718449215. These values are actually created as four-char C strings, as you can see by exploring Core Audio's header files.
>
> There are several techniques to convert the value back to a readable string. The string substitution sequence %s treats its argument as a null-terminated char*, so you can log error codes like this:
>
> MediaPlayback/AudioRecorderPlayer/Classes/RecordViewController.m
> ```
> int errorCode = CFSwapInt32HostToBig ([recorderSetupError code]);
> NSLog(@"Error: %@ [%4.4s])",
>   [recorderSetupError localizedDescription], (char*)&errorCode);
> ```
>
> This prints the NSError's localized error message and a string representation of the 4cc, which it gets by dealing with endianness (the 4cc's are all big-endian), and setting the C string's format to be *exactly* four characters. The output looks like this:
>
> ```
> Error: Operation could not be completed. (OSStatus
>         error 1718449215.) [fmt?])
> ```
>
> Another technique to get 4cc's is to use the Calculator application. Switch to programmer view (⌘ 3), click the Decimal and ASCII buttons, and paste in your return code number.
>
> So, what is fmt? Now that you have some text, you can search Core Audio's header files and find it in AudioFile.h, where it's the value of the kAudioFileUnsupportedDataFormatError constant.

Assuming the AVAudioRecorder has been created successfully, using it is very straightforward. Its three essential methods are record, pause, and stop. The difference between the latter two is that pause indicates that you intend to resume recording into the file with a subsequent call to record, while stop finishes the recording session, writing the data out to the file and closing it. Another call to record after a stop would actually overwrite the file. With those rules in mind, it's pretty easy to write record/pause/stop utility methods:

MediaPlayback/AudioRecorderPlayer/Classes/RecordViewController.m

```
-(void) startRecording {
        [audioRecorder record];
        recordPauseButton.selected = YES;
        formatButton.enabled = NO;
}

-(void) pauseRecording {
        [audioRecorder pause];
        recordPauseButton.selected = NO;
}

-(void) stopRecording {
        [audioRecorder stop];
}
```

Then you call these utility methods from event handlers on the buttons:

MediaPlayback/AudioRecorderPlayer/Classes/RecordViewController.m

```
-(IBAction) handleRecordPauseTapped {
        NSLog (@"handleRecordPauseTapped");
        if (audioRecorder.recording) {
                [self pauseRecording];
        } else {
                [self startRecording];
        }
}

-(IBAction) handleStopTapped {
        NSLog (@"handleStopTapped");
        if (audioRecorder.recording) {
                [self stopRecording];
        }
}
```

One other thing to deal with is the delegate callbacks from the AVAudioRecorder. Two of these deal with the end of the recording session, either normally or abnormally. We can use this message to reset the GUI for the next recording.

MediaPlayback/AudioRecorderPlayer/Classes/RecordViewController.m
```
- (void)audioRecorderDidFinishRecording:(AVAudioRecorder *)recorder
            successfully:(BOOL)flag {
    NSLog (@"audioRecorderDidFinishRecording:successfully:");
    recordPauseButton.selected = NO;
    recordPauseButton.enabled = NO;
    formatButton.enabled = YES;
    [audioRecorder release];
    audioRecorder = nil;
    filenameField.text = @"";
}

- (void)audioRecorderEncodeErrorDidOccur:(AVAudioRecorder *)recorder
            error:(NSError *)error {
    recordPauseButton.selected = NO;
    recordPauseButton.enabled = NO;
    [audioRecorder release];
    audioRecorder = nil;
    filenameField.text = @"";
}
```

The other two delegate methods don't require specific handling for our application but are well worth knowing about. audioRecorderBeginInterruption: signals that the application is being interrupted, typically by an alarm or by an incoming phone call. In either case, the user may switch to another application as a response to the interruption, such as answering the call:

If the user takes the call, your application delegate will get a call to applicationWillTerminate: and will end shortly thereafter. However, if the user declines the call, the recorder's delegate will get the audioRecorderEndInterruption: callback, and your application will continue running.

The only thing our application would really need to do in the case of an interruption is to pause the recording, and as it turns out, the AVAudioRecorder automatically pauses when interrupted anyway.[3]

16.5 Monitoring Recording Levels

In the example, we've provided two features to provide feedback to the user while he or she is recording. The first is a current time label. This is identical to the time display for the music library player in Section 15.1, *Working with Time Properties*, on page 308. In short, you create an NSTimer to periodically call an update method, in which you get the currentTime property (an NSTimeInterval) from the AVAudioRecorder object and use a formatted string to create an MM:SS-style string that you can then set as the label's text. Adding this feature to the recorder is simply a matter of cutting and pasting from the earlier project and changing the iPodController reference to audioRecorder. For good measure, you might want to check whether the audioRecorder is **nil**, meaning we failed to create it, and blank out the currentTimeLabel in that case.

Another interesting thing we can do to provide feedback is to get power levels from the recorder object. A set of methods in AVAudioRecorder allows us to do just that. It actually provides two level-metering values: the *average power level* is what we usually think of as the instantaneous loudness or softness of the channel, and the *peak power level* is the greatest power level recorded over the most recent period (usually about a second). You've probably seen both of these on audio displays: average power is the "jumpy" value that goes up and down with the loudness of the speaker's voice, while the peak power is a value that's almost always greater than the average power, representing the loudest instant in the last second or so.

In fact, we'll create just such a level meter for our recorder. You'll need to create a new UIView subclass, called LevelMeterView. In IB, use the Identity inspector to change the class of the two short and wide views (next to the "L" and "R" labels) to this new class.

As shown in Chapter 19, *Drawing in Custom Views*, on page 385, we'll use custom drawing to render these level meters. To keep it simple, we'll just draw a white bar whose width is proportional to the average power and a two-pixel-wide red line for the peak power. Your headers for

---

3. Actually, as of iPhone SDK 3.0, the recorder *stops* instead of pausing, despite what the documentation says.

this class will need instance variables for the average and peak power levels, as well as CGRects to represent the areas we'll draw in. The only method we need is one to set the levels for a given instant.

MediaPlayback/AudioRecorderPlayer/Classes/LevelMeterView.h

```objc
@interface LevelMeterView : UIView {
    float power;
    float peak;
    CGColorRef levelColor;
    CGColorRef peakColor;
    CGRect levelRect;
    CGRect peakRect;
}
- (void) setPower: (float) pow peak: (float) pk;
@end
```

We'll start in the view controller, with the code to call the view's setPower:peak:, and then go back to handle the drawing. After creating the AVAudioRecorder, use its settings property to determine whether it's set to record in stereo or mono. If there's only one channel, we'll disable the view for the right channel and just use left as our mono view. We also tell the recorder at this point that we want to use level monitoring by setting the meteringEnabled property.

MediaPlayback/AudioRecorderPlayer/Classes/RecordViewController.m

```objc
audioRecorder.meteringEnabled = YES;
if ([[audioRecorder.settings
        objectForKey:AVNumberOfChannelsKey] intValue] > 1) {
    leftLevelLabel.text = @"L";
    rightLevelLabel.hidden = NO;
    rightLevelMeter.hidden = NO;
} else {
    leftLevelLabel.text = @"M";
    rightLevelLabel.hidden = YES;
    rightLevelMeter.hidden = YES;
}
```

Now, in our updateAudioDisplay method, repeatedly called by the timer to update the current time label, we'll add some code to get the levels from the recorder and set them on our custom views. You get the levels with the instance methods averagePowerForChannel: and peakPowerForChannel:, but before calling them, you need to call updateMeters to refresh the average and peak values across all channels.

MediaPlayback/AudioRecorderPlayer/Classes/RecordViewController.m

```objc
[audioRecorder updateMeters];
[leftLevelMeter setPower: [audioRecorder averagePowerForChannel:0]
                   peak: [audioRecorder peakPowerForChannel: 0]];
```

```
if (! rightLevelMeter.hidden) {
        [rightLevelMeter setPower: [audioRecorder averagePowerForChannel:1]
              peak: [audioRecorder peakPowerForChannel: 1]];
}
```

But what are the values? We'll need to understand that to draw an appropriate level view. You might expect them to range from 0.0 to 1.0, like the values of the MPMusicPlayerController's volume property, but if you were to NSLog() the values, you'd find they range from -160.0 for silence to 0.0 for maximum power. Moreover, the scale clearly isn't linear; if there's any sound at all, you'll quickly find the values ranging from, say, -30.0 to 0.0.

What's happening is that the power levels returned by the AVAudioRecorder are in *decibels*. The decibel is a logarithmic unit of measurement relative relative to a reference level. Since decibels use base-10, a difference of 10 decibels is an order of magnitude difference in terms of power. It doesn't *sound* ten times louder, because human hearing is logarithmic in nature, which is why decibels are a good system for representing power levels.

For the sake of displaying a graphic level, though, we want to convert from a logarithmic scale to a linear one, so the visual change with the level meter will seem appropriate to the change we're hearing. We can convert from the logarithmic -160.0 to 0.0 scale to a linear 0.0 to 1.0 scale with the formula widthPercentage = $10^{0.5*level}$. We can then multiply that **float** by the width of the view to figure out how wide the level bar should be, with 0.0 creating an empty bar, 1.0 creating a full one, and values in between being proportionally wide.

So, start by initializing the drawing colors and the constant parts of the drawing rectangles. A good place to do this is in LevelMeterView's initWithCoder:, called when the view is loaded from the nib:

MediaPlayback/AudioRecorderPlayer/Classes/LevelMeterView.m

```
- (id) initWithCoder: (NSCoder*) decoder {
        if (self = [super initWithCoder: decoder]) {
                // Iniitialization code
                levelColor = [UIColor whiteColor].CGColor;
                levelRect.origin.x=0;
                levelRect.origin.y=0;
                peakColor = [UIColor redColor].CGColor;
                peakRect.size.width=2;
                peakRect.origin.y = 0;
        }
        return self;
}
```

Next, implement the setPower:peak: method that the updater calls. All you need to do here is to remember the levels and request a redraw.

`MediaPlayback/AudioRecorderPlayer/Classes/LevelMeterView.m`

```objc
-(void) setPower: (float) pow peak: (float) pk {
    power = pow;
    peak = pk;
    // request redraw
    [self setNeedsDisplay];
}
```

The call to setNeedsDisplay will result in a call to drawRect: at some point, and that's where we do our custom rendering. We just need to erase the current view, calculate the dimensions of levelRect and fill it, and then do the same for peakRect.

`MediaPlayback/AudioRecorderPlayer/Classes/LevelMeterView.m`

```objc
- (void)drawRect:(CGRect)rect {
    // Drawing code
    CGContextRef context = UIGraphicsGetCurrentContext();
    
    // erase view
    CGColorRef undrawColor = self.backgroundColor.CGColor;
    CGContextSetFillColorWithColor (context, undrawColor);
    CGContextFillRect (context, rect);
    
    // figure out how far to draw
    levelRect.size.height = rect.size.height;
    levelRect.size.width = pow (10, (0.05 * power)) * rect.size.width;
    
    // fill with color
    CGContextSetFillColorWithColor(context, levelColor);
    CGContextFillRect(context, levelRect);
    
    // draw peak as 2-pixel wide bar
    CGContextSetFillColorWithColor(context, peakColor);
    peakRect.size.height = rect.size.height;
    peakRect.origin.x = pow (10, (0.05 * peak)) * rect.size.width;
    if (peakRect.origin.x >= (rect.size.width - 2))
        peakRect.origin.x = rect.size.width - 2;
    CGContextFillRect(context, peakRect);
}
```

The crucial parts of this code are the calculations based on the power levels. On line 12, we calculate a percentage width for the average power level, multiply that by the view's width, and make that the width of the levelRect, which we then fill with a white color. For the peak indicator, we use a similar calculation on line 21, though in this case we use the calculated value as the x value for the left side of peakRect, whose width is always 2.

Figure 16.4: RECORDING AUDIO WITH AUDIORECORDERPLAYER

So, with the level meters set up, you can try recording with the application. As you record, watch the level meters move in response to the loudness of your voice in the mic. Note that most input devices are mono, so you may see levels only on the left channel, as is the case in Figure 16.4.

On the simulator,[4] you'll be able to find your recorded files in ~/Library/ Application Support/iPhone Simulator/User/Applications/; sort by date to figure out which of the folders (named as arbitrary application ID hex codes) is your application, and then look in its Documents folder. Or

---

4. Note that on the simulator, the AVAudioRecorder seems to not work with FireWire input devices, like the old external iSight. Use the Sound system preference to set your audio input to a built-in mic or a USB device.

you can press on to the next section, in which we'll add a player tab to play our recorded files.

16.6 Playing Audio with the AVFramework

The AVFoundation framework also provides a convenient Objective-C class for playing audio, AVAudioPlayer. In fact, this class was introduced in iPhone OS 2.2, so it's more backward compatible than the 3.0-only AVAudioRecorder.

Take a look at its documentation, and you'll find it's very similar to the AVAudioRecorder. It has simple play, pause, and stop methods for playback control, as well as duration, numberOfChannels, and currentTime properties, the last of which is writable, meaning you can use it to jump around a file (just like the MPMusicPlayerController's currentPlaybackTime, shown in Section 15.2, *Setting the iPod Player's Current Time*, on page 313). It even has a level-metering API identical to the one supplied by the AVAudioRecorder.

Building a player for our recorded files will largely be a matter of borrowing from the recorder tab and the music library player's code. Since the AudioRecorderPlayer stores all its recordings in the Documents directory, let's have our player let the user pick a file from that directory and play it, with a current time display, scrubbing, and level meters...all of which you've already seen.

Create a PlayViewController class (with New File and the UIViewController template, of course) and a PlayView.xib file for its GUI. Open MainWindow.xib in IB, and find the tab bar controller's second tab controller; use its Attributes inspector to set its NIB Name to PlayView and the Identity Inspector to set its class to PlayViewController. Now the tab knows to use your new view controller and connect it to your new view.

Next, you can open the PlayView.xib file in IB to lay out the playback GUI. Once you set File's Owner's class to PlayViewController, add components from the library to the view to get the interface shown in Figure 16.5, on the next page.

The UIPickerView is somewhat overwhelming, but it does have the advantage of being very easy to code. It's also worth noting that Interface Builder lets you drag and drop between nibs: in this screenshot, we've dragged the level meters and their labels from the RecordView.xib file

Figure 16.5: LAYOUT FOR AUDIORECORDERPLAYER'S SECOND TAB

directly into this view, maintaining the level meters' identity as LevelMeterViews (note that they also have transparent backgrounds, which is why we've selected the left meter for the sake of this screenshot, because they'd otherwise be invisible).

You'll need to set up IBOutlets for filePicker, playPauseButton, stopButton, currentTimeLabel, currentTimeSlider, durationLabel, leftLevelMeter, rightLevelLabel, rightLevelMeter, and volumeSlider. Remember to also connect the picker's dataSource and delegate connections to File's Owner.

Let's start by handling the file-selection picker. You'll need to edit the header file to declare that you implement the UIPickerViewDelegate and UIPickerViewDataSource methods, which will be backed up with an NSArray* filenames that you can populate in viewDidLoad, where we'll also populate a _documentsPath instance variable.

MediaPlayback/AudioRecorderPlayer/Classes/PlayViewController.m

```
[_documentsPath release];
NSArray *searchPaths =
        NSSearchPathForDirectoriesInDomains
        (NSDocumentDirectory, NSUserDomainMask, YES);
_documentsPath = [searchPaths objectAtIndex: 0];
[_documentsPath retain];
[filenames release];
filenames = [[NSFileManager defaultManager] directoryContentsAtPath:_documentsPath];
[filenames retain];
```

With the array of filenames, we can easily implement the picker's data source and delegate:

MediaPlayback/AudioRecorderPlayer/Classes/PlayViewController.m

```
#pragma mark UIPickerViewDataSource methods
-(NSInteger) pickerView: (UIPickerView*) pickerView
            numberOfRowsInComponent: (NSInteger) component {
    return [filenames count];
}

- (NSInteger) numberOfComponentsInPickerView:(UIPickerView *)pickerView {
    return 1;
}

#pragma mark UIPickerViewDelegate methods
- (NSString *)pickerView:(UIPickerView *)pickerView titleForRow:(NSInteger)row
                    forComponent:(NSInteger)component {
    return [filenames objectAtIndex:row];
}
```

Now, every time the user selects a different file, we'll try to create a new AVAudioFilePlayer object with the file, which we'll do in a yet-to-be-written createAVAudioPlayer helper method:

MediaPlayback/AudioRecorderPlayer/Classes/PlayViewController.m

```
- (void)pickerView:(UIPickerView *)pickerView didSelectRow:(NSInteger)row
            inComponent:(NSInteger)component {
        // stop if playing
        [audioPlayer stop];
        playPauseButton.selected = NO;
        [self createAVAudioPlayer];
}
```

To create an AVAudioPlayer, we have two options. There is an initWithData:error: method that takes an NSData object referring to some kind of audio data in memory. That might be useful if you've downloaded some audio into an in-memory buffer. What's more useful to us now is the initializer method initWithContentsOfURL:error:. It's important to note that while this method takes an NSURL, it can load audio data *only* from file://-style URLs; the AVAudioPlayer does not play RTSP streaming audio, Shoutcast-style http streams, or any other form of network audio. If you want to build a streaming audio player, you'll need to use the frameworks introduced in Section 17.3, *A Core Audio Overview*, on page 366.

Here's the complete implementation of createAVAudioPlayer:

MediaPlayback/AudioRecorderPlayer/Classes/PlayViewController.m

```
Line 1  - (NSError*) createAVAudioPlayer {
    -       [audioPlayer release];
    -       audioPlayer = nil;
    -       currentTimeSlider.value = 0;
    5
    -       NSString *filename = [filenames objectAtIndex:
    -               [filePicker selectedRowInComponent:0]];
    -       NSString *playbackPath =
    -               [_documentsPath stringByAppendingPathComponent: filename];
    10      NSURL *playbackURL = [NSURL fileURLWithPath: playbackPath];
    -       NSError *playerSetupError = nil;
    -       audioPlayer = [[AVAudioPlayer alloc]
    -               initWithContentsOfURL:playbackURL error:&playerSetupError];
    -
    15      if (playerSetupError) {
    -           NSString *errorTitle =
    -               [NSString stringWithFormat:@"Cannot Play %@:", filename];
    -           UIAlertView *cantPlayAlert =
    -           [[UIAlertView alloc] initWithTitle: errorTitle
    20                      message: [playerSetupError localizedDescription]
    -                      delegate:nil
    -                      cancelButtonTitle:@"OK"
    -                      otherButtonTitles:nil];
    -           [cantPlayAlert show];
    25          [cantPlayAlert release];
    -           audioPlayer = nil;
    -           durationLabel.text = @"--:--";
    -           return playerSetupError;
    -       }
    30
    -       audioPlayer.delegate = self;
    -       audioPlayer.meteringEnabled = YES;
    -       audioPlayer.volume = volumeSlider.value;
    -       currentTimeSlider.maximumValue = audioPlayer.duration;
    35      durationLabel.text = [NSString stringWithFormat: @"%02d:%02d",
    -                   (int) audioPlayer.duration/60,
    -                   (int) audioPlayer.duration%60];
    -       return playerSetupError;
    -   }
```

The method appends the current filename to the _documentsPath, converts that path to an NSURL, and creates the AVAudioPlayer on lines 12–13, displaying an alert if there's an error.[5]

As you can see, we have a few tasks at the end related to the current time slider and level metering. Building out the play view from this point is entirely a matter of doing things that you've already done before:

- The current time display is updated with an NSTimer that repeatedly calls an updateAudioDisplay method to update the currentTimeLabel with the player's currentTime property, reformatted to an MM:SS-style string. You saw this in the recorder tab, the music library chapter, and the clock in Chapter 9, *Preferences*, on page 163.

- The updateAudioDisplay can also update the currentTimeSlider. You made a slider like this draggable in Section 15.1, *Working with Time Properties*, on page 308, and the same approach works here: repeatedly update its value from the player's currentTime property, and set that property when you get a Value Changed event from the slider (you'll also want to use a isScrubbing value to tell updateAudioDisplay to not automatically update the slider's value if you've received a Touch Down event, just like in the iPod player example).

- As mentioned earlier, the level meters work *exactly* as they do for the recording tab, described in Section 16.5, *Monitoring Recording Levels*, on page 343. The level-metering API in the AVAudioPlayer is identical to that in AVAudioRecorder.

- The AVAudioPlayerDelegate defines methods that are largely the same as those in AVAudioRecorderDelegate: audioPlayerDidFinishPlaying:successfully: and audioPlayerDecodeErrorDidOccur:error: can be used to update the GUI when playback ends either normally or because of an error, while audioPlayerBeginInterruption: and audioPlayerEndInterruption: notify you of interruptions from incoming calls or other system events that lead the user to terminate your application.

By applying the techniques learned earlier, copying and tweaking code as necessary, you can finish out the PlayViewController with little trouble, or just check out the downloadable example code.

---

5. You'll often get an error when you switch to this tab on the simulator, because the first file the picker lands on will be the nonaudio .DS_Store file. You could improve this app by filtering the filenames array to include only files with extensions appropriate to audio files, like .caf and .aif.

16.7 Interacting with Audio Sessions

One other task is required for any application that uses audio: setting up an *audio session*. This is an object that represents how your application will interact with the rest of the audio system.

It's important that you declare an *audio category*, which describes how your application uses audio and how it will (or will not) interact with other audio on the system. For example, silent applications shouldn't mind if the user has music playing in the background from the iPod application. But a game with its own soundtrack, a music-making application, a net radio client, and other sorts of applications would want to make sure that their sound is the only thing the user hears. And in other cases, it may be desirable to mix your application's audio with that from the iPod application.

The AVAudioSession offers an Objective-C class for working with the audio session, and we'll start with the setCategory:error: method. This is how you declare your audio intentions to the system, and the method should be called early in your application's life cycle. For our application, it's easy enough to call it in AudioRecorderPlayerAppDelegate's applicationDidFinishLaunching method.

MediaPlayback/AudioRecorderPlayer/Classes/AudioRecorderPlayerAppDelegate.m
```
AVAudioSession *audioSession = [AVAudioSession sharedInstance];
NSError *audioSessionError = nil;
[audioSession setCategory: AVAudioSessionCategoryPlayAndRecord
                    error: &audioSessionError];
if (audioSessionError)
       NSLog (@"Error setting audio category: %@",
              [audioSessionError localizedDescription]);
```

The only way to get a AVAudioSession object is via the class method sharedInstance. Using this object, we declare our category to be AVAudioSessionCategoryPlayAndRecord, which means the application both plays and records audio. This has the effect of silencing the iPod or other background audio, as well as reserving the audio input hardware for our application's use.

Five categories are defined in AVAudioSession, each representing a certain class of audio application behavior and with defined effects on the iPhone's audio system in terms of access to audio input or output hardware, mixing with the iPod or other background audio, and whether the ring/silent switch is obeyed.

Here's an overview:

Category	Gets Input Hardware	Gets Output Hardware	Mixes with iPod	Obeys Ring/Silent
AVAudioSessionCategoryAmbient	No	Yes	Yes	Yes
AVAudioSessionCategorySoloAmbient	No	Yes	No	Yes
AVAudioSessionCategoryPlayback	No	Yes	No	No
AVAudioSessionCategoryRecord	Yes	No	No	No
AVAudioSessionCategoryPlayAndRecord	Yes	Yes	No	No

With the category declared, you set the audio session "active" to acquire the audio input and/or output hardware, potentially silencing any system audio in the background:

MediaPlayback/AudioRecorderPlayer/Classes/AudioRecorderPlayerAppDelegate.m

```
[audioSession setActive:YES error:&audioSessionError];
```

The AVAudioSession can also be used to inspect the current audio system. You can get the hardware's sampling rate and I/O buffer duration and set preferred values for these.[6] You can also inspect the number of channels on the input and output hardware and whether audio input is available.

This last point is an important one for applications that hope to record audio, since not all iPhone OS devices are capable of recording audio. All iPhones always have an input device: the headset mic, the phone mic, or possibly some other input device. On the other hand, the first-generation iPod touch does not support audio input of any kind. And on the second-generation iPod touch, audio input is available only if the user has plugged in an appropriate device, like the standard iPhone headset with the clicker. Since the availability of input depends not on model but by what the user has attached to it, a robust application will use the AVAudioSession to check for audio input hardware at runtime.

---

6. It's a little strange to see the I/O buffer information exposed by the high-level AVFoundation, since it's only in using the low-level Core Audio APIs that you'd have low enough audio software latency to care about hardware latency.

For example, you can add an alertIfNoAudioInput method to RecordView-Controller and call it when the record button is tapped:

MediaPlayback/AudioRecorderPlayer/Classes/RecordViewController.m

```
-(BOOL) alertIfNoAudioInput {
        AVAudioSession *session = [AVAudioSession sharedInstance];
        BOOL audioHWAvailable = session.inputIsAvailable;
        if (! audioHWAvailable) {
                UIAlertView *cantRecordAlert =
                [[UIAlertView alloc] initWithTitle: @"Can't record"
                        message: @"No audio input hardware is available"
                        delegate: nil
                        cancelButtonTitle:@"OK"
                        otherButtonTitles:nil];
                [cantRecordAlert show];
                [cantRecordAlert release];
        }
        return audioHWAvailable;
}
```

So, with a high-level API for recording and playing audio in a variety of formats, iPhone OS 3.0's AVFoundation framework gives you a lot of capability in a pretty straightforward Objective-C API, one that isolates the developer from a lot of the intricacies and challenges of working with audio.

On the other hand, those details are what some developers are interested in working with. Although AVFoundation makes audio simple for the developer who wants to play flat files, it hides the harder issues of streaming, mixing, and processing. There are lower-level frameworks that let you work at that level if you choose, and in the next chapter, we'll take a look at how they work.

Chapter 17
Core Audio

In Chapter 16, *Playing and Recording Audio*, on page 327, we learned how to use the Objective-C AVFoundation framework to record and play audio files in a variety of compressed and uncompressed formats. That framework is designed to provide an easy-to-use API for developers with fairly basic audio needs—the developer who doesn't want to mess with details and says, "I just want to play this MP3 file!" In the process, it trades capability for simplicity. To keep things easy, AVFoundation doesn't allow you to do any of the following:

- Play audio from the network.

- Mix multiple sounds.

- Access the audio data, either in encoded or decoded form.

- Access audio metadata, other than duration and number of channels. In fact, the AVAudioPlayer does not expose the format of the audio it's playing.

- Do anything with recorded audio other than saving it to a file.

You've probably seen iPhone applications that do all of these things, so clearly they're all possible. The key is that they're using lower-level APIs, which (unlike AVFoundation) have been available in the iPhone SDK since its original introduction to the public. The various audio APIs on the iPhone are all part of Core Audio, a comprehensive, stream-based infrastructure for digital audio. Core Audio consists of a number of interrelated APIs, which range from the low-level processing of small buffers of audio samples all the way up to the high-level audio player and recorder classes in AVFramework. By choosing to move to one of several lower levels of abstraction, you can access more functionality for your audio application.

However, to do so, we're going to have to leave Objective-C behind....

17.1 Using the Procedural-C APIs

As introduced way back in Section 3.1, *The iPhone Software Architecture*, on page 24, most of the media support in iPhone OS is in the Core Media layer, which exists at a lower level of abstraction than Cocoa Touch. Once you go down to Core Media or the other lower levels of the stack, you find that the APIs are C functions, not Objective-C classes. That means we'll be using C not only for audio but also for several other APIs in the Core Services and Core Media layers, including Quartz/Core Graphics and Address Book. We've already called into these layers a few times, such as when we called Core Foundation's NSLog() function.

Apple commonly refers to C as *Procedural C*, to distinguish it from Objective-C, and has employed a number of design patterns and programming conventions throughout its APIs to provide an OO-like experience when using procedural C. You'll want to make sure you understand these before you start digging into the Core APIs, because the concepts are fundamental to using the APIs correctly.

Opaque Types and Objects

While the Core APIs offer objects, there's still no such thing as a class in C. Instead, Core Foundation defines *opaque types*, which are similar to classes in that they hide an implementation from you but lack some of the other traits of OO classes, like polymorphism. For example, the CFString opaque type represents a string of Unicode characters, just like an NSString does in Cocoa. The internal data structures of the CFString are completely unavailable to you; in fact, your code will only ever deal with references to this type, defined (and documented) as the pointer CFStringRef.

To simulate the idea of calling a method on an object but within procedural C's functional idiom, the Core Foundation APIs use a naming convention. Functions that operate on an object begin with the object's opaque type and take the object as their first argument. For example, you would get the length of a CFStringRef named myCFString like this:

```
length = CFStringGetLength (myCFString);
```

Similarly, functions that take parameters put those parameters after the object, as in this example of getting the Unicode character at index 3 (that is, the fourth character) of the string:

```
fourthUniChar = CFStringGetCharacterAtIndex (myCFString, 3);
```

The CFStringRef is immutable, just like NSString. As you might've guessed, there is another opaque type, CFMutableStringRef, that allows you to change the string's contents.

As you may have noticed already, there's a high degree of correlation between some of the opaque types in Core Foundation and their Objective-C equivalents in Cocoa's Foundation framework. It's so high, in fact, that in some cases, a Cocoa object and a Core Foundation object are the same thing. The documentation for some of the key Cocoa classes indicates that they're *toll-free bridged* to Core Foundation equivalents, and vice versa. What this means is that you can cast between Cocoa and Core Foundation references, at zero cost. If you wanted to use myCFString with a Cocoa method, you'd just cast it like this:

```
NSString *myNSString = (NSString*) myCFString;
```

Similarly, you can use your NSStrings with Core Foundation functions by just casting in the other direction:

```
CFStringRef myCFString = (CFStringRef) myNSString;
```

You have to be careful, though, because similar names are no promise that two types can be toll-free bridged. For example, NSBundle is not toll-free bridged to the similar-sounding CFBundleRef; therefore, you can't cast between them. Your best source of information is Xcode's documentation viewer, because the overview for a class or opaque type will typically indicate whether it's toll-free bridged to an equivalent.[1]

Memory Management in Core Foundation

When working with Core Foundation objects, you employ a system of memory management that's much more like Cocoa than the malloc() and free() scheme of traditional C programming.

In Cocoa, you create objects with the alloc method. In Core Foundation, the functions that create objects take a reference to an *allocator* object that allocates and deallocates memory for objects. Although you can

---

1. Apple's *Carbon-Cocoa Integration Guide* [App07a] also lists all the toll-free bridgeable relationships, in the "Interchangeable Data Types" section. It's obviously meant for the Mac, but all the classes and opaque types listed there exist in iPhone OS as well.

customize this behavior, you will probably never want to—just pass **NULL** or kCFAllocatorDefault to object-creating functions to get the default allocator.

The allocator is associated with the object throughout its lifetime, relocating it in memory if necessary and freeing its memory when the object is no longer needed. It does this via a reference-counting scheme that is nearly identical to Cocoa's. We hope that, throughout this book, you've kept in mind the *fundamental rule* about memory management introduced in Section 3.7, *Managing Application Memory*, on page 43: you own any object you obtain by way of a method with alloc, new, or copy in its name, and you must eventually release it; conversely, you don't own objects obtained through other means and must not release them, unless you become an owner by means of retain.

The rules for Core Foundation are highly analogous:

- You own any object you create by means of a function with Create or Copy in its name, and you must eventually CFRelease() any such object.

- You don't own any object you obtain by other means (notably, functions with Get in their names), and you must *not* CFRelease() them.

- If you need to hold on to an object reference and ensure it is not freed, become an owner by calling CFRetain(). As an owner, you must eventually call CFRelease() on the object.

17.2 Playing System Sounds

With these concepts in mind, let's take on the simplest of the C-based audio APIs, *System Sound Services*.

This API is provided for playing simple, in-memory sounds of 30 seconds or less and is meant for user interface sounds, such as key clicks or event alert beeps, or discrete action sounds in simple games, such as the tap of a Go stone being placed on the board. System Sound Services is not appropriate for more sophisticated or immersive uses of audio: the API can't play longer sounds and offers no control over volume, stereo positioning, or a means of stopping a sound that has started playing. Also, system sounds cannot be compressed. You need to use linear PCM or IMA/ADPCM in an .aif, .wav, or .caf container.

We'll exercise this API by creating a view-based application, SystemSoundsDemo. You'll need to create some short sounds for this example.[2] For the downloadable sample application, we used the Mac's speech synthesizer to speak the numbers 1–9 with different voices and saved each of these as CAF files named 1.caf through 9.caf, adding all of them to the project's Resources group.[3]

To use the System Sound Services API, add AudioToolbox.framework to the project's Resources group, and add the header #import <AudioToolbox/AudioToolbox.h> to SystemSoundsDemoViewController.h.

For the user interface, we'll need just a single table and a label. The table will let the user tap to play sounds, and the label will be used to show status. Start by editing SystemSoundsDemoViewController.h to append <UITableViewDataSource, UITableViewDelegate> to the **@interface** declaration, and add IBOutlet UILabel* statusLabel; inside the **@interface** block. Also declare the label as a **@property**, and **@synthesize** the property in the .m implementation file. Now, open SystemSoundsDemoViewController.xib in Interface Builder, and add the table and label to the view. After laying them out, wire the table's dataSource and delegate outlets to the File's Owner, and connect the label to the statusLabel outlet.

Assuming you use nine sounds like we did for the sample, it's simple enough to implement the table data source methods by providing some constants at the top of the **@implementation** block:

MediaPlayback/SystemSoundsDemo/Classes/SystemSoundsDemoViewController.m
```
NSString *soundNames[] = {
        @"Vibrate", @"One", @"Two", @"Three", @"Four",
        @"Five", @"Six", @"Seven", @"Eight", @"Nine"
};
NSInteger SOUND_COUNT = 10;
```

Notice how we're going to treat row 0 of the table as Vibrate. Now you implement the table data source methods by having tableView:numberOfRowsInSection: return SOUND_COUNT and by adding a single line to tableView:cellForRowAtIndexPath:'s default behavior.

---

2. Ironically, despite the name System Sound Services, the API offers no access to the various alert beeps installed with the system or the user's custom ringtones.
3. Since a reader asked, we created the files with Automator's "Ask for Text" and "Text to Audio File" actions and then converted from AIFF to CAF with Core Audio's afconvert command-line utility.

```
// Set up the cell...
cell.textLabel.text = soundNames[indexPath.row];
return cell;
```

Creating System Sounds

When the application gets the tableView:didSelectRowAtIndexPath: callback, it needs to react to the row selection by finding one of the CAFs and playing the sound. We can create a system sound with the C function AudioServicesCreateSystemSoundID(). This function takes a CFURLRef parameter providing a **file://**-style URL to the sound and the address of a SystemSoundID[4] that it can populate. The function returns an **OSStatus** to indicate success or failure; if the return value is kAudioServicesNoError, then you can take the SystemSoundID and pass it to AudioServicesPlaySystemSound() to play the sound. You could also pass the system sound to the similar function AudioServicesPlayAlertSound(), which plays the sound with vibration on an iPhone and plays an alert jingle (ignoring the specific sound you passed in) on iPod touch.

So, here's how we'll create a system sound, given the indexPath passed to tableView:didSelectRowAtIndexPath:.

MediaPlayback/SystemSoundsDemo/Classes/SystemSoundsDemoViewController.m

```
Line 1  // create a system sound id for the selected row
    -   SystemSoundID soundID;
    -
    -   OSStatus err = kAudioServicesNoError;
    5   if (indexPath.row == 0) {
    -           // special case: vibrate
    -           soundID = kSystemSoundID_Vibrate;
    -   } else {
    -           // find corresponding CAF file
   10           NSString *cafName = [NSString stringWithFormat: @"%d", indexPath.row];
    -           NSString *cafPath =
    -                   [[NSBundle mainBundle] pathForResource:cafName ofType:@"caf"];
    -           NSURL *cafURL = [NSURL fileURLWithPath:cafPath];
    -           err = AudioServicesCreateSystemSoundID((CFURLRef) cafURL, &soundID);
   15   }
```

---

4. Note that this type is defined as a 32-bit pointer and is not an opaque type (which would be a pointer to some **struct** and would have Ref at the end of its name). Therefore, you don't have to CFRelease() it, though it has its own cleanup routine that we'll attend to shortly.

There are a couple of interesting things to point out here. First, if the row is 0, then line 7 uses a System Sound Services constant, kSystem-SoundID_Vibrate, to get vibration (or the iPod touch jingle) as a system sound. Otherwise, we get the filename from the row (for example, 3.caf) on line 10, find the CAF in the bundle on line 12, and convert the path to an NSURL on line 13.[5] Finally, line 14 uses toll-free bridging to cast the NSURL to a CFURLRef as needed by AudioServicesCreateSystemSoundID() to create the system sound. A key thing to note here—and you're going to see this a *lot* in these C-based APIs—is how we provide the address of our SystemSoundID pointer, which will be populated with a reference to the system sound if the call is successful.

Playing, Monitoring, and Disposing of System Sounds

We're basically ready to play the sound, but there's a detail we haven't accounted for: managing the sound's life cycle. If you're done with a sound, you need to dispose of it with AudioServicesDisposeSystemSoundID(), but when to do so is a matter of timing. If we wanted to keep the sound in memory because we intended to use it repeatedly, then we could just make it an instance variable and dispose of it in dealloc. But for this app, we will want to dispose of the sound once we're done playing it.

The trick is knowing when it's OK to dispose of the sound, since the application will keep going while the sound plays. If you dispose of the sound immediately after calling AudioServicesPlaySystemSound(), you won't hear the sound, since it will be destroyed just after you start playing it. What you need is to know when the sound is done playing, so you can clean it up.

To do this, we can register a *completion process*, a C function that will be called when the sound is done playing. You register a process with AudioServicesAddSystemSoundCompletion(). Here's the rest of our tableView:didSelectRowAtIndexPath: implementation, which registers a completion process, plays the sound, and alerts the user if any of our System Sound Services calls returned an error code.

---

5. Note that system sounds can be created only from file:// URLs, not from http:// or any other network URL.

MediaPlayback/SystemSoundsDemo/Classes/SystemSoundsDemoViewController.m

```
Line 1  if (err == kAudioServicesNoError) {
            // set up callback for sound completion
            err = AudioServicesAddSystemSoundCompletion
                (soundID,                       // sound to monitor
     5          NULL,                           // run loop (NULL==main)
                NULL,                           // run loop mode (NULL==default)
                SystemSoundsDemoCompletionProc, // callback function
                self                            // data to provide on callback
            );
    10      statusLabel.text = @"Playing";
            AudioServicesPlaySystemSound (soundID);
        }
        if (err != kAudioServicesNoError) {
            CFErrorRef error = CFErrorCreate(NULL, kCFErrorDomainOSStatus, err, NULL);
    15      NSString *errorDesc = (NSString*) CFErrorCopyDescription (error);
            UIAlertView *cantPlayAlert =
                [[UIAlertView alloc] initWithTitle:@"Cannot Play:"
                             message: errorDesc
                             delegate:nil
    20                       cancelButtonTitle:@"OK"
                             otherButtonTitles:nil];
            [cantPlayAlert show];
            [cantPlayAlert release];
            [errorDesc release];
    25      CFRelease (error);
        }
```

The completion process is registered on lines 3–9. This function takes five parameters, as described in its documentation. AudioServicesAddSystemSoundCompletion() takes the system sound to play, the run loop and run loop mode to use (just use **NULL**s to get the default behavior), the name of the C function to call back, and finally a "user data" pointer, for which we'll just pass in our view controller object with **self**.

With the completion process registered, we can set the label to Playing (line 10) and finally start playing the sound on line 11.

If anything has gone wrong—specifically, if the OSStatus returned by any of the System Sound Services calls has returned a value other than **kAudioServicesNoError**, then an error-handling block (lines 13 to 26) presents the user with an alert message. Line 14 creates a CFErrorRef from the returned OSStatus value, and line 15 copies its description string and (via toll-free bridging) casts it to an NSString for use in a UIAlertView.[6]

---

6. CFErrorRef is toll-free bridged to NSError, so you could also cast to NSError and use its

Notice how, since we create new objects with CFErrorCreate() and CFErrorCopyDescription() on lines 14 and 15, we are responsible for releasing them on lines 24 and 25.

Now the last thing we need is to actually supply the callback function. This code (or a C forward declaration) should appear earlier in the source file than the AudioServicesAddSystemSoundCompletion() that references it.

The function header is defined by AudioServicesSystemSoundCompletionProc(), which you can find by way of a link from AudioServicesAddSystemSoundCompletion()'s documentation. The two arguments are the system sound that has finished playing and the client data pointer you registered. Here's a simple implementation:

MediaPlayback/SystemSoundsDemo/Classes/SystemSoundsDemoViewController.m
```
void SystemSoundsDemoCompletionProc (
            SystemSoundID   soundID,
            void            *clientData)
{
    AudioServicesDisposeSystemSoundID (soundID);
    ((SystemSoundsDemoViewController*)clientData).statusLabel.text = @"Stopped";
};
```

As you can see, the first thing we do is to use the C function AudioServicesDisposeSystemSoundID() to dispose of the system sound and any resources it is using. Next, we set the text of the label to Stopped. Even though the code is inside SystemSoundsDemoViewController.m, it is a C function and therefore has no awareness of the view controller object or its instance variables. So, we cast the client data object to an instance of SystemSoundsDemoViewController and access the statusLabel property to set the text.

With this, you're finally ready to use your sounds. You've provided code to create SystemSoundIDs, play them, and dispose of them in a registered completion process. Do a Build and Go to run the application in the simulator.

---

methods for handling the error.

As you tap around each of the table cells, as well as Vibrate, you should hear each of the various system sounds play. Though the use of procedural C makes it seem difficult—and it is difficult, if you don't do much C—this is a pretty simple API, compared to the rest of Core Audio. Certainly for playing one-off sounds, it's preferable to using the more involved APIs of the next section. Of course, coding your UI sounds with the Objective-C AV Foundation would be easier still, though System Sound Services does have the advantage of being somewhat less resource-intensive.

17.3 A Core Audio Overview

By using the System Sounds Services API, we've had a brief taste of what's in Core Audio. As you push deeper into its APIs, you get access to more functionality, at the price of greater complexity and challenge.

With its tricky C-based API and prerequisite knowledge of digital signal processing, really getting into Core Audio would pretty much require its own book, so in lieu of that, we'll use this section to provide an overview of the Core Audio frameworks and what each one provides.

Technically, the Core Audio infrastructure is split into five frameworks: Audio Toolbox, Audio Unit, AVFoundation, and Core Audio (headers used by multiple frameworks),[7] and OpenAL. Within these, there are a number of services provided by Audio Toolbox, which are almost like individual frameworks unto themselves.

Audio Session Services

We introduced the concept of the audio session—an expression of the application's use of audio and its relationship to other audio on the device—in Section 16.7, *Interacting with Audio Sessions*, on page 353. In that section, we worked with AVFoundation's AVAudioSession. That class is an Objective-C abstraction atop the C APIs in this Audio Toolbox service. With the Audio Session API, you can define an audio category as before and inspect traits of the audio hardware, such as the latency of its input and output hardware, the hardware sample rate, and more. In the Audio Toolbox, this is the API that provides notification of interruptions from other applications, like incoming phone calls, something that AVFoundation forwards to the AVAudioPlayer and AVAudioRecorder.

There's additional functionality in the C API not currently exposed by AVAudioSession. One particularly useful thing you can do with Audio Sessions is to inspect the current audio path, that is, whether the output is going to speakers or headphones, by retrieving the session's kAudioSessionProperty_AudioRoute. You can also set up a listener on this property so you'll get a callback when the route changes. With this, you could automatically stop an audio player when the user removes his or her headphones.

Audio Queue Services

Audio Queue Services is a high-level (relative to the rest of Core Audio, anyway) API for recording and/or playing audio from any source that can provide audio samples. It works by exchanging buffers of audio data. For example, a playback audio queue repeatedly calls back to your application with empty buffers that you fill with audio data (either

---

7. From here out, we'll use "Core Audio" to refer to the entire infrastructure and "Core Audio framework" for this tiny framework.

compressed formats or PCM samples). For recording, the queue delivers buffers with data captured from the microphone or other input device.

Since Audio Queue Services doesn't tie you into storing recorded audio into files or playing audio only from files or memory, you might use this API for network audio applications, such as web radio players or chat applications. Since you provide the audio data directly to the queue, you could also support encrypted or DRM'ed media with this API: you'd decrypt your source into audio data in memory and provide that directly to the queue.

Audio Queue Services also provides a level metering API, as well as an "offline rendering" function, which would allow you to export audio to a buffer instead of to the audio output hardware.

Audio File Services

Another part of the Audio Toolbox, Audio File Services simplifies working with the various container formats for audio. With Audio File Services, an application that's working with an audio stream can write that stream or read it from various file formats that support the stream format, without having to worry about the many differences between .caf, .mp3, .wav, .aif, and so on.

For those interested in such details, Audio File Services' property API also lets you work with many forms of audio file metadata in a format-agnostic way. For example, this API lets you get album/artist/title metadata from music files, in addition to working with metadata that would be useful in sound editing, like regions, markers, and SMPTE time codes.

Audio File Stream Services

The Audio File Stream Services API helps you deal with audio streams that don't necessarily have a beginning or end and therefore can't be accessed in a random-access fashion like a file's contents can. This is another API you would use for network-based audio applications. As you read data from a stream, you'd have Audio File Stream Services parse the stream and determine its format, eventually delivering packets of audio data that you could then pass to an audio queue or process in your own code.

Audio Conversion Services and Extended Audio Files

Audio Toolbox's Audio Conversion Services is a set of APIs for converting encoded data to PCM, or vice versa (but not between two different encoded formats, unless you use PCM as an intermediate representation). It can also be used to convert between different PCM representations. It works on bufferfuls of data of in-memory audio data. You don't have to use this service if you're using an Audio Queue, because the queue performs these kinds of conversions for you automatically.

A related service, Extended Audio File Services combines the Audio File and Audio Conversion services, allowing you to convert formats and read from or write to a file in one action.

Audio Units and Audio Unit Graphs

An entire framework unto itself, Audio Units is the lowest publicly accessible level of the Core Audio infrastructure. Audio Units are software objects that process audio samples and can be combined in a chain to achieve complex audio processing streams.

The most important of these is the I/O unit, retrieved via the Audio Component API with the constant kAudioUnitSubType_RemoteIO. The I/O unit is an abstraction over the audio input and output hardware. You can connect a "renderer callback" to the I/O unit to get callbacks to provide audio samples, which will be sent directly to the speaker or headphones. Unlike the AVAudioPlayer and Audio Queues, which use buffers and therefore introduce latency, connecting directly to the I/O unit is very low-latency. You can get your audio out in less than 30 milliseconds—equivalent to the screen refresh rate—by sending PCM directly to the I/O unit. Going in the other direction, you can also get low-latency audio capture by working with the I/O unit. Low-latency is important if you're writing an application in which you want to your user to interact directly with sound, such as virtual instruments that need to play or modulate sound immediately in response to touch, shake, or voice. iPhone SDK 3.0 also adds a Voice Processing I/O audio unit, which performs *echo cancellation*: if your app performs two-way voice chat, this unit prevents user A's voice from being picked up by user B's microphone, and coming back to user A as a sort of digital "echo." The Voice Processing I/O unit is used by the Game Kit voice chat API, described back in Section 13.9, *Voice Chat*, on page 290.

Aside from the I/O units, the iPhone also comes with several mixer units to combine multiple audio sources and comes with an iPod Equalizer Unit to perform the audio processing offered by the native music player. As of iPhone SDK 3.0, you can write your own audio units to add custom audio processing to your application.

You can connect audio units together, such that one unit's output is another's input. The output unit at the end "pulls" audio by calling a render method on any units connected to its input, which in turn pull data from any of their upstream connections. However, these connections can be somewhat burdensome to set up and maintain, because they work with a property-style API. The Audio Unit Graph API greatly simplifies working with multiple audio units. You create an AUGraph and then fill it with AUNodes, which are just wrappers around audio units. The graph makes it easy to connect and coordinate the nodes in a graph. You connect nodes with AUGraphConnectNodeInput() and start the audio processing with AUGraphStart().

OpenAL

The last Core Audio framework we'll discuss isn't an Apple framework at all. OpenAL is an industry-standard framework for 3D positional audio. Its design is highly analogous to the OpenGL graphics API, and the two are meant to be used together for game programming. On the iPhone, OpenAL is implemented atop the rest of Core Audio, using a 3D mixer audio unit to represent the spatialized sound in the left-right soundfield of the headphones. Since the OpenAL data is sent to a mixer unit and then onto the I/O unit, it's another means of achieving very low-latency audio on the iPhone. This is mostly of use to game developers, who can count on in-game sound keeping up with the on-screen action.

As a third-party API, the OpenAL instruction set is nothing like the rest of Core Audio and has more in common with the idioms of OpenGL. OpenAL is hosted by Creative Technology, and as of 2009, its website (http://openal.org) redirects to a Creative site, where you can find documentation, examples, and a wiki.

We hope this chapter has given you an idea of the capabilities of the audio frameworks on the iPhone and which ones will be useful to you for a given application. For more information, you can consult Apple's *Core Audio Overview* [App07b] and go from there to the reference or programming guide for each framework.

There has been an ongoing effort on Apple's part to make things easier: playing audio files on iPhone SDK 2.0 required the use of Audio Queues, which although high-level relative to Core Audio (certainly easier than expecting developers read files themselves, convert encoded packets to PCM, and feed those to the I/O unit), was still too challenging for many developers. iPhone 2.2 added AVFramework's AVAudioPlayer, and 3.0 adds AVAudioRecorder and AVAudioSession. If you're in the majority of developers that simply needs to play from or record into an encoded format, those classes will serve you well. But if you have more sophisticated needs, you can determine what level of Core Audio suits your needs—do you want to hand buffers of encoded data to an Audio Queue, or do you want to process PCM data in audio units?—and use those functions.

Chapter 18
Events, Multi-Touch, and Gestures

Multi-Touch is one of the features that just blows people away. The ability to pinch and rotate photos makes people smile the first time they see it. The great thing is that with UIKit it's straightforward to add these types of interactions to our applications. In this chapter, we'll cover the event system and how our applications fit into it and respond to the events to make a fantastic application.

The UIKit defines a different interaction model than a desktop model. The most obvious is the touch instead of the mouse click. But even more interesting is the addition of multiple touches at the same time. In a typical desktop environment, there is a single mouse and only one event type to deal with at a time. On the iPhone, there can be multiple touches going on at the same time with some touches going away while new ones come in. The touches have context and can indicate user intent (like rotating photos). The event system on the iPhone gives us some great opportunities to make amazing user interactions possible. Let's dig into how the events work.

18.1 Event Model

Events are the life blood of an application; in order to make great apps, you have to understand how your app gets the events and what is possible once you get them. This section is all about events and the iPhone events model.

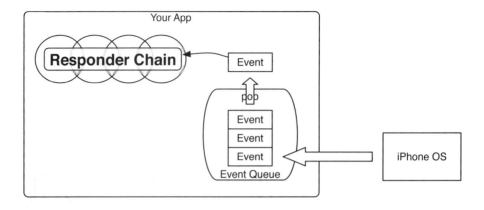

Figure 18.1: EVENT DELIVERY SYSTEM

Events come from the operating system via the responder chain. We briefly discussed responders and their role in event processing in Chapter 2, *Hello iPhone*, on page 9 and in a few places since. Let's talk about them again in the context of event delivery. In Figure 18.1, we can see the events as they are processed by the OS and then sent to our application's event queue. The event loop dequeues events off the queue and sends them to the responder chain.

The responder chain is simply a list of objects connected together via their nextResponder property. As a responder receives an event, it either processes the event or passes it on to its nextResponder. Many of the objects you use to build your application are responders. The application object, all views (including windows), and all UI widgets are responders. Being in a chain allows the objects deepest in the chain to respond to events they understand but then delegate up the chain for events they don't understand.

If you have done much UI or event-based programming, you have probably run into a responder chain of one type or another. The pattern community calls this pattern the "chain of responsibility." It's simply a way to organize *processing objects* (the responders here) so they can pass off any *command objects* (events in this context) that they don't know how to use to other processing objects in a consistent way.

The application object is the keeper of the first responder and the object in your application that receives all events as they are dequeued from the event queue. So, as each event comes to the application, it is sent to the first responder and then travels its way back up the responder chain, ending eventually in either being processed or being discarded.

When the user's finger touches the screen over a view, the view is sent the touchesBegan:withEvent: method. If the view is a button or other control, it processes the event by invoking its actions. If the view is one of your views and you don't implement the method, then the default implementation on UIView is invoked, and the responder chain is sent the message until either something responds or the event is discarded. As the user moves the touch around on the screen, a series of touchesMoved:withEvent: methods are sent and are processed in the same way. When the user lifts their finger(s) off the screen, the touchesEnded:withEvent: is sent and again is processed by the responder chain. If the system interrupts the application (because of a system event like an incoming phone call), the application is sent the touchesCancelled:withEvent: to give your application the chance to clean up any information it was keeping related to the current event processing happening.

In each of the event methods, the first argument is a set of the touches that have changed since the last event method was called. The second argument is the event that contains all the touches (via the allTouches property) or all the touches for a particular view (via allTouchesForView:). Each touch knows which view it started in and its current and previous location in that view.

Implementing these event-processing methods and using the touches information is key to making your application Multi-Touch. In the remainder of this chapter, we will be looking at the details of how to make applications do many of the cool Multi-Touch things that other iPhone apps do, like swipe and rotate. Throughout the remainder of this chapter, we are going to implement several examples of handling events. Let's get started with tracking touches.

18.2 Tracking Touches

To provide a good Multi-Touch experience to your users, you have to understand how to track touches. As the user's finger moves across elements, the direction and duration are used to discern intent in interesting ways. For example, when you touch an email cell in the mail application and slide your finger across the cell, the Delete button appears so you can delete the email. This event type is known as a *swipe*. In

Figure 18.2: DOTS FROM TOUCHES

the Photos application, *flicking* takes the user to the next or previous photo, and *pinching* and *stretching* cause the photo to zoom in or out. Tracking touches is what makes Multi-Touch gestures possible.

Enough talk about events—let's dig into some code and build an example. Create a new view-based project, and call it Draw. In this example, we'll build an application that tracks touch events in a view. For each touchesMoved:withEvent: method, we are going to capture the location and then draw a dot at that location. The example is shown in Figure 18.2.

To make this application work, the view tracks both the beginning of the touch as well as the touch moving. Create a new view subclass, and call it DrawView. Open the DrawViewController.xib file in Interface Builder, and change the class of the view to DrawView.

As with all event streams, we begin in the touchesBegan:withEvent: method. Here is the code:

MultiTouch/Draw/Classes/DrawView.m

```
- (void)touchesBegan:(NSSet *)touches withEvent:(UIEvent *)event {
    UITouch *touch = [touches anyObject];
    CGPoint location = [touch locationInView:self];
    Dot *dot = [[[Dot alloc] init] autorelease];
    dot.x = location.x;
    dot.y = location.y;
    [self.dots addObject:dot];
    [self setNeedsDisplay];
}
```

In this code, we're creating the first Dot for the sequence of events. On line 3, we get the location of the touch in the view, create a Dot for that location, and add that to the list and then mark the view for display. The Dot class is very simple and is only a holder of the x and y values.

The list of Dots are used to draw circles (that is, the dots) in the view, one for each instance at the location captured in the object. Keep in mind that this code is meant to give us a visual of what tracking our events can do for us, not be an example of how to do the most efficient drawing on the iPhone.

Next is the touchesMoved:withEvent: method. Here is the code:

MultiTouch/Draw/Classes/DrawView.m

```
- (void)touchesMoved:(NSSet *)touches withEvent:(UIEvent *)event {
    UITouch *touch = [touches anyObject];
    CGPoint location = [touch locationInView:self];
    Dot *dot = [[[Dot alloc] init] autorelease];
    dot.x = location.x;
    dot.y = location.y;
    [self.dots addObject:dot];
    [self setNeedsDisplay];
}
```

In this method, we are doing more or less the same thing as in the touchesBegan:withEvent: except that this method is called many more times than the previous method.

The Dot is a really simple subclass of NSObject. It plays the role of our Model class. To create it, make a new subclass of NSObject, and add two properties of type CGFloat named x and y. Add the instance variables and the @synthesize statements to the implementation file, and you're done with the Dot class.

18.3 Tapping

Tapping is another fundamental way users interact with their iPhones. Taps go through the same cycle of touchesBegan, touchesMoved, and touchesEnded as other events, so in order to make our applications respond to one or more taps, we implement one or more of these methods. In the simple case where we want to provide a "double tap" action for one of our responders, we can implement touchesBegan:withEvent:. If the tapCount property is equal to two, then we invoke the "double tap" functionality. If not, then we ignore the event.

An interesting case comes up if we want a responder to do one action on a single tap and do a different action on a double tap. The system will send the responder the first (single tap) event and then send the second tap as a new event. The responder gets what looks like a single tap and a double tap.

It turns out that making this work is not difficult, but it can be confusing if you haven't done something like it before. The trick is to delay the performance of the single tap action until you know whether it's really a single tap or the first of a double tap sequence. The way we do that on the iPhone is with the performSelector:withObject:afterDelay: method. This method queues up the execution of a method (a selector is the data structure the ObjC runtime uses to represent methods) to happen some time later.

Once we queue up the response to the initial single tap, we can then wait for the double tap, and if it does come, we can cancel the initial single tap invocation via another method called cancelPreviousPerformRequestsWithTarget:. All this can seem a bit abstract, so let's look at the code. In this example, the view turns red on a single tap and turns blue on a double tap. Here is part of the code:

MultiTouch/MultiTap/Classes/MultiTapView.m
```
- (void)touchesEnded:(NSSet *)touches withEvent:(UIEvent *)event {
  UITouch *touch = [touches anyObject];
  if(touch.tapCount == 1) {
    [self performSelector:@selector(turnRed) withObject:nil afterDelay:0.10f];
  }
  if(touch.tapCount == 2) {
    [self turnBlue];
  }
}
```

When the tap ends (the user picks up their finger) and the tapCount is 1, the single tap method is queued up to be executed in 0.1 seconds. If the tapCount is 2, then the double tap method is invoked right away. This is relatively straightforward code. However, we are not canceling the single tap method, so if this was all we had, then both would happen, and the view would turn blue then red. Let's look at the code to fix that:

MultiTouch/MultiTap/Classes/MultiTapView.m

```
- (void)touchesBegan:(NSSet *)touches withEvent:(UIEvent *)event {
  UITouch *touch = [touches anyObject];
  if(touch.tapCount == 2) {
    [[self class] cancelPreviousPerformRequestsWithTarget:self
                                              selector:@selector(turnRed)
                                                object:nil];
  }
}
```

If the second tap comes in, then the single tap method is canceled. The delay we put in of 0.1 seconds gives the user that amount of time to tap a second time and register the double tap event; otherwise, it will be another single tap. Using this technique, we can give multiple actions to our responders. For example, you might want to provide a control that selects on a single tap but edits on a double tap.

18.4 Multi-Touch Gestures

To this point we have been looking at scenarios where we consider only one finger down on the screen at a time. Although that covers most of the event handling needed for most applications, there are cases where more than one finger can be used to indicate more complex interactions. The canonical case in the built-in apps is the pinch gesture. It is hard to imagine a better user interaction model that could be done with one finger. Who wants to tap buttons even if they fade in and out or whatever when the two finger gesture is so natural? We should strive for the same type of natural interaction model in our apps. Make something amazing that the users already know how to use.

Let's return to the simple draw application we did back in Section 18.2, *Tracking Touches*, on page 375 and modify it to draw dots for each touch on the screen. It's surprisingly simple to add Multi-Touch drawing to this application.

The first thing we need to do is turn on Multi-Touch for the view. We could do this in code by setting the multipleTouchEnabled property to

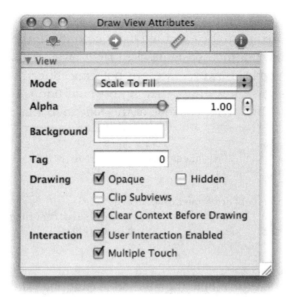

Figure 18.3: MULTIPLE TOUCH TURNED ON

YES, but it's more straightforward to do the configuration in IB. Open DrawViewController.xib, and select Draw View (DrawViewController.xib is part of the Draw project we built earlier). Switch on the Multiple Touch checkbox in the Attributes inspector (Command-1). The inspector with the checkbox turned on is shown in Figure 18.3.

Now that we have changed the view so that it tracks multiple touches, we need to take advantage of that in our code. Here is the code for the touchesBegan:withEvent: method. Let's look at it in detail.

MultiTouch/DrawMulti/Classes/DrawView.m
```
- (void)touchesBegan:(NSSet *)touches withEvent:(UIEvent *)event {
    UITouch *anyTouch = [touches anyObject];
    if(anyTouch.tapCount > 1) {
        self.dots = nil;
        [self setNeedsDisplay];
        return;
    }
    BOOL needsDraw = NO;
    for(UITouch *touch in event.allTouches) {
        if(UITouchPhaseBegan == touch.phase) {
            CGPoint location = [touch locationInView:self];
            Dot *dot = [[[Dot alloc] init] autorelease];
```

```
            dot.x = location.x;
            dot.y = location.y;
15          [self.dots addObject:dot];
            needsDraw = YES;
          }
        }
        if(needsDraw) {
20        [self setNeedsDisplay];
        }
      }
```

This code is not all that different from the code we discussed previously except that when adding dots, we consider every touch (starting on line 9) that is part of the event instead of any touch in the touches parameter. Also notice on line 10 that only the touches just beginning are considered. In this way, the application does not add a new dot for existing touches as a new touch joins the event cycle. Next up, let's look at the touchesMoved:withEvent: method. Here is the code:

MultiTouch/DrawMulti/Classes/DrawView.m

```
Line 1  - (void)touchesMoved:(NSSet *)touches withEvent:(UIEvent *)event {
          BOOL needsDraw = NO;
          for(UITouch *touch in event.allTouches) {
            if(UITouchPhaseStationary != touch.phase) {
5             CGPoint location = [touch locationInView:self];
              Dot *dot = [[[Dot alloc] init] autorelease];
              dot.x = location.x;
              dot.y = location.y;
              [self.dots addObject:dot];
10            needsDraw = YES;
            }
          }
          if(needsDraw) {
            [self setNeedsDisplay];
15        }
        }
```

Recall that touches are active while the finger the touch represents is on the screen. Whenever any of the touches moves, the touchesMoved:withEvent: method is called. As the events are processed, the application will avoid looking at any stationary touches on 4. That will prevent the application from adding dot upon dot on any touches that are not moving.

One other thing to notice in this code. The view is not keeping any persistent information about the events and touches it receives, so there is nothing to be removed or cleaned up when the events are done. Therefore, there is no implementation of touchesCanceled:withEvent:.

Figure 18.4: TOUCH ZOOM

With just a bit of math, we could determine whether the touches are getting closer together or further apart and thus provide a pinch or zoom gesture for the application. Being able to process multiple touches and know whether they are moving or not is key to adding gestures to your application. Let's look at an example that provides the pinch gesture.

In Figure 18.4, we can see a simple application that displays the robot in a layer and allows you to zoom in or out with the pinch gesture. The two gray circles are from the simulator; you can simulate Multi-Touch by holding down the Alt key.

The pinch gesture tracks two fingers on the display. If they are moving together, you zoom out; when moving apart, you zoom in. The view needs to keep track of the distance between the two touches. Let's look at the code to track that. Here is the code for touchesBegan:withEvent::

MultiTouch/PinchZoom/Classes/PinchZoomView.m

```
- (void)touchesBegan:(NSSet *)touches withEvent:(UIEvent *)event {
  if(event.allTouches.count == 2) {
    pinchZoom = YES;
    NSArray *touches = [event.allTouches allObjects];
    CGPoint pointOne = [[touches objectAtIndex:0] locationInView:self];
    CGPoint pointTwo = [[touches objectAtIndex:1] locationInView:self];
    previousDistance = sqrt(pow(pointOne.x - pointTwo.x, 2.0f) +
                            pow(pointOne.y - pointTwo.y, 2.0f));
  } else {
    pinchZoom = NO;
  }
}
```

In this code, we first check to make sure we have a two-finger gesture happening, and then if so, we turn the flag on that indicates a pinch gesture and then calculate the distance between the two touches. We keep the flag and the distance in instance variables so we can use them in the touchesMoved:withEvent: method. Let's look at that code now:

MultiTouch/PinchZoom/Classes/PinchZoomView.m

```
Line 1  - (void)touchesMoved:(NSSet *)touches withEvent:(UIEvent *)event {
   -      if(YES == pinchZoom && event.allTouches.count == 2) {
   -        NSArray *touches = [event.allTouches allObjects];
   -        CGPoint pointOne = [[touches objectAtIndex:0] locationInView:self];
   5        CGPoint pointTwo = [[touches objectAtIndex:1] locationInView:self];
   -        CGFloat distance = sqrt(pow(pointOne.x - pointTwo.x, 2.0f) +
   -                                pow(pointOne.y - pointTwo.y, 2.0f));
   -        zoomFactor += (distance - previousDistance) / previousDistance;
   -        zoomFactor = fabs(zoomFactor);
  10        previousDistance = distance;
   -        self.robotLayer.transform =
   -        CATransform3DMakeScale(zoomFactor, zoomFactor, 1.0f);
   -      }
   -    }
```

As either or both fingers are moved around on the screen, the view is sent the touchesMoved:withEvent: method. If a pinch gesture is in progress, we calculate the current distance between the two touches and use that to calculate the zoom factor. Once the zoom factor is calculated, we get the absolute value on line 9 so that we are never using a negative zoom. If you don't do this, the image will flip when the factor turns negative. When one or more touches end, we reset the flag and the distance.

Here is the code:

```
MultiTouch/PinchZoom/Classes/PinchZoomView.m
- (void)touchesEnded:(NSSet *)touches withEvent:(UIEvent *)event {
  if(event.allTouches.count != 2) {
    pinchZoom = NO;
    previousDistance = 0.0f;
  }
}
```

We reset the values only if we don't have two touches. That is important because someone might place a third finger on the screen and then pick it up. We don't want to reset the pinch until the two fingers are off the screen.

The PinchZoom example uses a CALayer to achieve the pinch zoom effect. Although it is not necessary to use a layer, it was the easiest way to demonstrate the principles without getting deep into the internals of Core Graphics. So, please forgive the forward usage of layers and believe that it is much easier to understand pinch zoom here with a layer. For more detail about layers, see Chapter 21, *Core Animation*, on page 409.

In this chapter, we have learned about the event model and how it works in the Multi-Touch environment on the iPhone. We also went over approaches to making our applications respond to multiple touches in interesting ways.

Chapter 19

Drawing in Custom Views

So far, we have created our application's views using off-the-shelf components and Interface Builder. Sometimes you will want to build something special—something you can't create with the standard components. In this chapter, you will build custom user interfaces by extending the UIView class and drawing your own UI via the Core Graphics library.

The Core Graphics library provides the primitives that you will use to assemble into custom drawings. Specifically, in this chapter, we'll focus on using the vector-drawing features of Core Graphics. You will learn about the image-drawing features of Core Graphics in Chapter 20, *Drawing Images and Photos*, on page 397.

One word of caution—it takes a long time to get drawing just right. If you don't need custom drawing, then stick with the built-in components.

19.1 Drawing Model

Drawing on the iPhone is a somewhat indirect affair. You never just start drawing; instead, you have to tell Cocoa Touch what needs to be drawn. Then, when it is a good time for the drawing to be done (during the event processing cycle), Cocoa Touch will do all the necessary configuration and then tell your view to draw by invoking the drawRect: method.

In addition, iPhone applications don't usually redraw the whole screen. Instead, individual views (or even regions of the views) are marked as needing to be updated. We write the code that updates the view in the

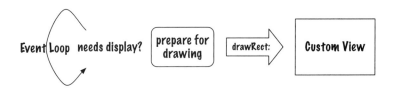

Figure 19.1: DRAWING CYCLE

drawRect: method. Then during the event cycle, Cocoa Touch tells the view to update by invoking drawRect:. Figure 19.1 graphically describes the drawing cycle.

During each pass through the event cycle, the drawing machinery checks to see whether any views have been marked as needing to be drawn. Each view that needs to be redrawn will be sent the drawRect: method after Cocoa Touch takes care of all the configuration necessary to draw. We will look at how to mark a view as needing an update shortly.

Once in the drawRect: method, you can use any of the Core Graphics routines to draw any content that needs to be drawn. Your freshly drawn pixels are then pushed to the screen after drawRect: finishes.

19.2 Vector Drawing

Let's make a simple app that fills a rectangle with blue. As you know from the drawing cycle, you update the contents of the rectangle when the view's drawRect: method is called. So, that is where we'll put the code to fill the rectangle.

In this example, we will start a new project named Filling; use the view-based template for this project. Once you have the project, create a new subclass of UIView called FillingView. In the New File dialog box, choose "Objective-C class," and then in the "Subclass of" pop-up choose UIView.

Now comes the fun part. Open FillingView.m, and move to the drawRect: method. It has the inviting comment "Drawing code" placed there by the template begging us to fill it in. There are, of course, an almost limitless number of options here, but we are going to start simple. We are going to use a convenience function to fill the rectangle.

Here is the code:

`Drawing/Filling/Classes/FillingView.m`

```
- (void)drawRect:(CGRect)rect {
    [[UIColor blueColor] setFill];
    UIRectFill(rect);
}
```

The first line of code sets blue as the color to fill the rectangle with (more on setting graphics context state such as fill color soon). The second line fills the CGRect (that is, the rect parameter) passed into the drawRect: method.

The UIRectFill() function takes a rectangle and creates a Core Graphics object called a CGPath and fills it. This function hides the details from us of how paths are created (thus making it easy). These kinds of functions are called *convenience functions* because they make some of our drawing easier. However, if you want to do more than fill a rectangle, you need to move a step deeper to understand the underlying path objects. We will cover paths in a moment; for now let's look at what we need to get our newly created view into our user interface.

Open the FillingViewController.xib file. Select the view, and change its class by clicking ⌘-4. Then type FillingView into the Class field. Click the Build and Go button, and you should see the entire screen filled with blue.

Congratulations! You have successfully built your first custom view that draws to the screen. It was a fairly straightforward endeavor. Now that we have seen the basic mechanisms that allow us to draw to the screen, let's look in more detail at paths and how they work.

19.3 Paths

We use paths to describe rectangles, circles, and other shapes we want to draw. Paths can be stroked, filled, or both stroked and filled. Stroking a path causes the path to be drawn according to the state in the graphics context. Filling causes the area inside the path to be painted. In Core Graphics, there are four primitives used to represent paths, points, lines, arcs, and curves.

A point is a single location in two-dimensional space. Don't think of it as a pixel. You need to think of it the way you did back in geometry class. A point takes up no space at all, so drawing one will not show anything on the screen. You can add as many points as you'd like to your path. To get something to show up, you'd use one of the other primitives, which are composed with points.

A line is defined by two points: the starting point and ending point. A line can be stroked, which means that whatever properties are set on the graphics context (such as stroke width or color) will be applied to the line between the two points. Lines occupy no area, so they can't be filled. Instead, you would use a set of lines or curves together to make a shape and then close the path and fill that.

An arc is defined by single point (the center), a radius, a start angle, and a stop angle. To get a circle, you'd make an arc with an angle starting at zero radians and stopping at 2π radians. Since an arc is a path and marks an area, it can be filled or stroked or both. When you fill a path that is not closed (any path that is not explicitly closed is open), the filling routines will close it for you.

Finally, curves are represented as a Bezier curve. You define a curve with four points: two points describe the end points, and the other two points describe the tangent to the curve at each end point. A curve is like an line in that it does not describe an area; however, since there is curve in a Bezier curve, a call to the fill routines will close the path (by drawing a line from the start point to the end point) and will thus end up with some area to fill.

Paths can be used to draw some very sophisticated graphics. In fact, you can do with Core Graphics almost everything you can do with advanced vector graphics tools.

The simple single-line rectangle fill that we saw in the previous example is easy to use, but like many other easy-to-use things, it hides a bunch of flexibility from us. We are going to rebuild that example to see what is going on behind that single function invocation so that we can better understand the path primitives.

Here is the code for a new drawRect: method that will fill a 100x100 rectangle with an origin at 10,10. We started a whole new project and created a new view (and changed the view's class in IB), but you can simply change your existing FillingView's drawRect: if you'd prefer. We will be covering graphics contexts in a moment; for now, focus on the path code.

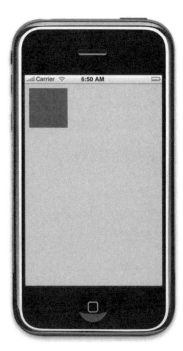

Figure 19.2: BLUE-FILLED VIEW

Drawing/FillingSquare/Classes/FillingView.m

```
- (void)drawRect:(CGRect)rect {
  CGMutablePathRef path = CGPathCreateMutable();
  CGPathMoveToPoint(path, NULL, 10.0f, 10.0f);
  CGPathAddLineToPoint(path, NULL, 100.0f, 10.0f);
  CGPathAddLineToPoint(path, NULL, 100.0f, 100.0f);
  CGPathAddLineToPoint(path, NULL, 10.0f, 100.0f);
  CGPathCloseSubpath(path);
  CGContextRef ctx = UIGraphicsGetCurrentContext();
  CGContextSetFillColorWithColor(ctx, [UIColor blueColor].CGColor);
  CGContextAddPath(ctx, path);
  CGContextFillPath(ctx);
}
```

When you run this application, you should see something that looks like Figure 19.2. Instead of us filling the whole view, we now have a square.

Chapter 19. Drawing in Custom Views

Figure 19.3: BAR GRAPH DRAWING

Now that you have seen how to build a path with a really simple shape, let's move on to a bit more complex example that draws a bar graph with rounded tops. Figure 19.3 is a screenshot of the application in action.

Let's look at the code a piece at a time. Here is the drawRect: method:

Drawing/BarGraph/Classes/BarGraphView.m
```
Line 1  - (void)drawRect:(CGRect)rect {
   -        CGSize size = self.bounds.size;
   -        CGFloat width1 = size.width * 0.75f;
   -        CGFloat width2 = size.width * 0.35f;
   5        CGFloat width3 = size.width * 0.55f;
   -
   -        CGFloat height = size.height * 0.2f;
   -
   -        CGRect one = CGRectMake(0.0f, height + 5.0f,
  10                                width1, height - 10.0f);
   -        CGRect oneText = CGRectMake(10.0f, height + 25.0f,
   -                                    width1, height - 30.0f);
```

```
           CGRect two = CGRectMake(0.0f, 2.0 * (height + 5.0f),
                                   width2, height - 10.0f);
15         CGRect twoText = CGRectMake(10.0f, 2.0 * height + 30.0f,
                                       width2, height - 30.0f);
           CGRect three = CGRectMake(0.0f, 3.0 * (height + 5.0f),
                                     width3, height - 10.0f);
           CGRect threeText = CGRectMake(10.0f, 3.0 * height + 35.0f,
20                                       width3, height - 30.0f);

           CGContextRef ctx = UIGraphicsGetCurrentContext();

           [[UIColor blueColor] setFill];
25         CGPathRef pathOne = [self pathInRect:one];
           CGContextAddPath(ctx, pathOne);
           CGPathRelease(pathOne);
           CGContextFillPath(ctx);
           [[UIColor blackColor] setFill];
30         [@"One" drawInRect:oneText withFont:[UIFont systemFontOfSize:34]];
           [[UIColor redColor] setFill];
           CGPathRef pathTwo = [self pathInRect:two];
           CGContextAddPath(ctx, pathTwo);
           CGPathRelease(pathTwo);
35         CGContextFillPath(ctx);
           [[UIColor blackColor] setFill];
           [@"Two" drawInRect:twoText withFont:[UIFont systemFontOfSize:34]];
           [[UIColor yellowColor] setFill];
           CGPathRef pathThree = [self pathInRect:three];
40         CGContextAddPath(ctx, pathThree);
           CGPathRelease(pathThree);
           CGContextFillPath(ctx);
           [[UIColor blackColor] setFill];
           [@"Three" drawInRect:threeText withFont:[UIFont systemFontOfSize:34]];
45     }
```

The method is somewhat long, but it is really only doing five things:

- On line 3, we calculate the width of the bar.

- On line 7, we calculate the height of the bar.

- On line 9, we create a rectangle for the bar to be drawn in.

- On line 25, we add the path to the context and then fill it.

- On line 30, we draw the label.

Of course, each of these five steps is repeated for each of the bars, so it looks like a ton of code. Don't let the volume of code throw you off, though. Really, only the five things are taking place. Now to see how the path is created, let's take a look at that code.

```
Drawing/BarGraph/Classes/BarGraphView.m
-(CGPathRef) pathInRect:(CGRect)rect {
  CGMutablePathRef path = CGPathCreateMutable();
  CGFloat radius = CGRectGetHeight(rect) / 2.0f;
  CGPathMoveToPoint(path, NULL, CGRectGetMinX(rect), CGRectGetMinY(rect));
  CGPathAddLineToPoint(path, NULL, CGRectGetMaxX(rect) - radius,
                       CGRectGetMinY(rect));
  CGPathAddArc(path, NULL, CGRectGetMaxX(rect) - radius,
               CGRectGetMinY(rect) + radius,
               radius, -M_PI / 2.0f, M_PI / 2.0f, NO);
  CGPathAddLineToPoint(path, NULL, CGRectGetMinX(rect), CGRectGetMaxY(rect));
  CGPathCloseSubpath(path);
  CGPathRef imutablePath = CGPathCreateCopy(path);
  CGPathRelease(path);
  return imutablePath;
}
```

This code also looks a bit complex on first blush, but it's really not. We initially calculate the radius of the rounded end of the bar to be half the height of the bar. We then move to the top-left corner of the bar, place our first point, continue around the shape until we get to the bottom-left corner, and then close the path. Once the path is closed, we make an immutable copy, do some memory cleanup by releasing the mutable path, and then return.

These examples give you a quick overview of what is possible within Core Graphics, but there is a lot more. A big part of the flexibility and power of Core Graphics comes from the graphics context. Let's take a look at that next.

19.4 Graphics Context

Think back to when you drew as a kid. When you wanted red, you picked up the red crayon and drew the lines of your masterpiece. When blue was called for, you switched to the blue crayon. Well, the graphics context in Core Graphics is a lot like your hand. As we switch the colors and other parameters of the graphics context, we are switching between different crayons. In the previous examples, we glossed over the [UIColor blueColor]; line of code. Now we are going to see what is really going on with that color object.

You can assign the value of many different variables in the graphics context to affect the way drawing gets done. We have seen the fill color used in the previous two examples, but there are many other variables that determine what happens when we fill or stroke a path. For example, we can set the width of the "crayon" to be 5, and then when we stroke the path, the area covered by our virtual ink will be 5 units wide. We can set other variables that affect how a path is stroked too, such as the dash pattern (to make dashed lines for example), the stroke color, and many others. Let's go back to our filled square and stroke the path after setting the stroke width. Here is the code to set the stroke width to 10 and the stroke color to yellow and then stroke the path:

Drawing/StrokedFilledSquare/Classes/StrokedFilledSquareView.m

```
- (void)drawRect:(CGRect)rect {
  // create and build the path
  CGMutablePathRef path = CGPathCreateMutable();
  CGPathMoveToPoint(path, NULL, 10.0f, 10.0f);
  CGPathAddLineToPoint(path, NULL, 100.0f, 10.0f);
  CGPathAddLineToPoint(path, NULL, 100.0f, 100.0f);
  CGPathAddLineToPoint(path, NULL, 10.0f, 100.0f);
  CGPathCloseSubpath(path);
  CGContextRef ctx = UIGraphicsGetCurrentContext();
  // configure the fill parameters and fill the path
  CGContextSetFillColorWithColor(ctx, [UIColor blueColor].CGColor);
  CGContextAddPath(ctx, path);
  CGContextFillPath(ctx);
  // configure the stroke parameters and stroke the path
  CGContextSetStrokeColorWithColor(ctx, [UIColor yellowColor].CGColor);
  CGContextSetLineWidth(ctx, 10.0f);
  CGContextAddPath(ctx, path);
  CGContextStrokePath(ctx);
  CGPathRelease(path);
}
```

There are a couple of important things to notice here. First, we set all the values we are interested in before we do the drawing. Remember that the graphics context defines the crayon. If we have a red crayon in our hand when we go to draw a line, we will get a red line. If we want a yellow line, we need the yellow crayon. The second thing to notice here is that we have to add the path to the context again. When we stroke or fill a path, it is cleared from the context. Our newly drawn square looks like Figure 19.4, on the following page.

You can do a lot more with drawing paths and manipulating the graphics context to get just the look you want. We have only begun to scratch the surface of what is possible. For a lot more detail on vector drawing with the Core Graphics library, see *Programming with Quartz* [GL06] or the Apple online documentation *Introduction to Quartz 2D Programming Guide* [App07c].

Figure 19.4: STROKED-FILLED SQUARE

19.5 Redisplaying a View

Now we should take a moment to discuss how to force your view to redraw its content. We pay a performance penalty every time we push new pixels to the screen. The bits that get colored in a drawRect: eventually have to get pushed out to the graphics card (and thus to the screen), and that takes time. Since it costs a lot to get new pixels pushed to the screen, the UIKit is careful to draw only when absolutely necessary. With each pass through the event loop, the drawing machinery of Cocoa Touch checks to see which views (if any) need to be redrawn. By default, nothing needs to be redrawn; we have to explicitly mark views as needing to be redrawn before they will be redrawn. If a view does need to be redrawn, then the drawing machinery focuses on that view, configures a graphics context for drawing, makes it current, and then sends the drawRect: to the view.

You can mark a view that needs to be redrawn by calling the setNeedsDisplay or setNeedsDisplayInRect: method. The first method marks the entire view as needing redisplay; the second method has a rectangle parameter that lets us mark a subregion of the view as needing redisplay. Whichever means you use to mark a view for redisplay, Cocoa Touch will pass the appropriate rectangle (either the whole bounds or the subset of the bounds) into the drawRect: method when it is time to draw.

You never call the drawRect: method directly; instead, mark a view (or part of a view) as needing to be redrawn. Cocoa Touch and the drawing machinery will take care of configuring a graphics context for you. Then, when it is time to draw, Cocoa Touch will call your view's drawRect: method.

If you are doing complex drawing that takes a long time to calculate or set up, then it makes sense in your drawRect: method to use the rectangle parameter to limit what you recalculate and redraw. If your drawing is less sophisticated, then it might make sense to just redraw the whole view again anyway. The only way to really know is to try both ways and do some performance profiling; see Chapter 28, *Performance Tuning*, on page 511 for more details.

In this chapter, we have covered the basics of drawing on the iPhone. From the basics of drawing that you have learned here, you can build some really cool applications that take advantage of the power inherent in Core Graphics.

Chapter 20
Drawing Images and Photos

In this chapter, we will learn how to work with images on the iPhone. The images might be bundled with the application, be part of the photo library, or be new images captured with the iPhone's camera.

Wherever the images come from, we will want to display them, and that is the theme of this chapter. In Chapter 19, *Drawing in Custom Views*, on page 385, we discussed the drawing model and how to use the vector drawing routines. In this chapter, we are going to use our knowledge of the drawing model to draw images. We are going to use images from our projects as well as grab photos from the photo library.

There are many cool ways to use the photos in an application. For example, wouldn't it be cool to have an application that allowed us to take a picture of our kids and turn that photo into an animated bobble head when we shake the phone? Silly but fun! Another great way to integrate photos is found in the Contacts application. You could click where an image would go for your contact and be prompted to attach a photo from your library or even use the camera to take the person's picture. As is often the case, the hard part is deciding what you want the user experience to be. The actual code is fairly straightforward. All you have to do is make a couple of connections in Interface Builder and implement a couple of methods.

Our journey through this chapter starts with using the UIImageView to draw our images. This class makes it really easy for us to get our images onto the screen. Next up we will look at the UIImage class and see how we can draw images with it. Although this class is not as easy to use as UIImageView, it gives us more flexibility in how an image is

> ### Joe Asks...
> #### Can I Use Any Image Format?
>
> Not exactly. Although the iPhone is able to use many different image formats (JPEG, GIF, and so on), it is optimized for PNG files. So, make sure your artwork is in PNG format before shipping your application.

drawn. Finally, we will look at how to integrate with the photo library and camera so we can make our applications all the more personal for our users.

20.1 Basic Image Drawing

To draw images, you have two options:

- The UIImageView provides the simplest method of showing an image on the screen.
- The second option is to use a UIImage object to represent the image and then draw the image into a view.

We'll take start with the simplest approach and put an image into a UIImageView.

If you don't have complex requirements on the way your images are drawn, then the UIImageView is a great way to display the images. All you need to do is set the image property on the image view to get your image drawn to the screen.

Let's build a sample application that uses a UIImageView to display an image that will be contained in your application bundle.

Create a new view-based project called ImageView. Next you need to add an image to the project. Feel free to choose the image photo.png included with the sample code or one of your own images. To add an existing file to your project, choose the group you want to add it to, right-click that group, choose Add > Existing Files, and follow the steps in the wizard. We almost always choose the "Copy items into destination group's folder (if needed)" checkbox on the last page of the wizard so that we get a copy instead of a reference to the file.

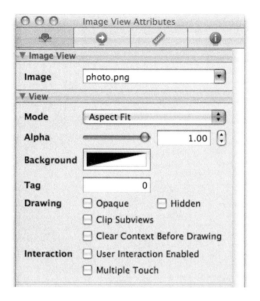

Figure 20.1: CHANGING THE IMAGE

Now that we have an image, we can do the rest of the work for this example in Interface Builder. Open the ImageViewViewController.xib file. From the Library (⌘-⇧-L), choose a UIImageView, and drag it out to the view. Resize the image view (with the Size inspector, select the image view, and hit ⌘-3) to a size that fits comfortably on the screen, such as 300x300 with the origin at 10, 10.

Our image view is ready, we can place an image into the view, and it will be drawn. Open the Attributes inspector with ⌘-1, and in the top pull-down, select the image you added to your project. Our inspector looks like Figure 20.1.

After you get your project to this point, spend some time modifying the Mode pull-down value so that you can see the different ways the UIImageView displays the image. The choice you make depends on how you want the image to draw. Our image drawn with the Aspect Fit option chosen is shown in Figure 20.2, on the following page.

Here is the great part; our image is displayed on-screen, and we have written no code. Simplicity—the beauty of the image view. This makes it very easy for us to get images into our UI. However, the only control we have over the way the image is drawn is the different modes. If these

Figure 20.2: IMAGE DRAWN WITH ASPECT FIT

modes meet your requirements, then you are good to go. If not, it is time to look at the UIImage. Let's look at that next.

20.2 Customizing the Image Display

The UIImage allows us to specify how and where an image is drawn. It gives us the ability to drastically change what the image looks like on the screen.

Let's build an example that takes advantage of this flexibility. Create a new view-based project called Image, add a new UIView subclass to it called ImageView, and then change the class of the view in ImageView-Controller.xib to ImageView. Also, add the same image you used in the last project to this project (if you don't want to create a new project, you can always go back to the last custom view project and modify that). Now that we have the setup done, let's implement the drawRect: method to draw our image.

Images/Image/Classes/ImageView.m

```
Line 1  - (void)drawRect:(CGRect)rect {
          UIImage *image = [UIImage imageNamed:@"photo.png"];
          CGFloat idealSize = 300.0f;
          CGFloat ratio = 1.0f;
    5     CGFloat heightRatio = idealSize / image.size.height;
          CGFloat widthRatio = idealSize / image.size.width;
          if(heightRatio < widthRatio) {
            ratio = heightRatio;
          } else {
   10       ratio = widthRatio;
          }
          CGRect imageRect = CGRectMake(10.0f, 10.0f, image.size.width * ratio,
                                        image.size.height * ratio);
          [image drawInRect:imageRect blendMode:kCGBlendModeDifference alpha:1.0f];
   15   }
```

In this code we are basically implementing the Aspect Fit mode (one of the other modes available from the UIImageView) but with a small twist. We are drawing the image with the difference blend mode just to show off some of the flexibility we get from using the UIImage class. Let's go through the code piece by piece. On line 2, we see the loading of the image with the imageNamed: method. This method will look in the resources of the project for an image with the specified name. After the image is loaded, we can get its size and then calculate the appropriate ratio to preserve the aspect ratio (as shown on line 7).

Now that we have the rectangle that will preserve the aspect ratio, we can draw the image. We draw the image with the drawInRect:blendMode:alpha: method and pass in the rectangle on line 14. The blend mode and alpha parameters give us much more control over the way the image is drawn than what we had when drawing with the UIImageView. In this case, we are only setting the alpha value to 1.0. The blend mode parameter gives us a bunch of flexibility in the way the image will look. The blend mode changes the way the image's pixels are blended with the background pixels. There are many different options; you should try several of them just to get a feel for the ways you can blend images. In Figure 20.3, on the next page, we can see the image drawn in difference mode with an alpha of one. As you can see, we can get some pretty cool-looking effects using the UIImage drawing methods.

Images can also be drawn with the CGContextDrawImage(), but the details of using the underlying core graphics context to draw images are a bit more involved. The default coordinate system would draw the image upside down, so we are required to do a bit of coordinate transformation

Figure 20.3: IMAGE DRAWN WITH DIFFERENCE BLEND MODE

before drawing. There is a sample program (called CGImage) in the code for the book that draws the image properly if you are interested in the details.

20.3 Image Picker

Now that you have seen how to draw an image, it is time to look at how to get an image from the user's photo library. Not only is this a fun and easy feature to implement, but it also brings another level of personalization to your application.

You interact with either the camera or the photo library on the iPhone through an instance of UIImagePickerController. The image picker, as it is known, is a subclass of view controller. As such, it does all you would expect from a view controller. It manages a screen full of information, it has a view that can be animated in or out, and it acts as a controller between the view elements and the underlying model. The image picker is the sole interface we have to the underlying photos in the photo library. Fortunately, using the image picker is as easy as setting a delegate and implementing a couple of methods. Let's look at that now.

We first touched on delegation in Chapter 3, *iPhone Development Fundamentals*, on page 23 where we saw the application's delegate. As you recall, delegation is a way for us to customize the behavior of an object without having to subclass. In this case, the delegate of the image picker gets to make the decisions of what happens with the chosen image. We implement two methods to interact with the image picker; the first one passes the chosen image, and the second informs the delegate when the user cancels.

Let's create a new project that will grab an image out of the photo library and place it into a UIImageView.

Create a new view-based project called PhotoView. We need two outlets in the PhotoViewViewController: one for the image picker and one for the image view. Modify the PhotoViewViewController.h so that it looks like this:

Images/PhotoView/Classes/PhotoViewViewController.h
```
@interface PhotoViewViewController : UIViewController
<UIImagePickerControllerDelegate> {
    UIImageView *imageView;
    UIImagePickerController *imagePicker;
}

@property(nonatomic, retain) IBOutlet UIImageView *imageView;
@property(nonatomic, retain) IBOutlet UIImagePickerController *imagePicker;

@end
```

Notice that we also made this view controller conform to the UIImagePickerControllerDelegate protocol. We will see the methods of that protocol in a moment. For now, just make sure the header declares our intention that the view controller conforms to this protocol. Next let's go to Interface Builder to configure and set up our image picker. Open PhotoViewViewController.xib by double-clicking the file in Xcode. Open the library with ⌘-L, and find Image Picker Controller. Drag an instance of the image picker into the document window. Your document should look like Figure 20.4, on the next page.

Now that we have the image picker in our nib file, we need to connect our view controller to it. Ctrl+click the PhotoViewController (that is, the File's Owner), and drag to the new image picker. When you mouse up,

Figure 20.4: IMAGE PICKER CONTROLLER

select imagePicker to make the connection to the image picker. We also need to connect the UIImagePickerController's delegate to the view controller, so Ctrl+click the image picker and drag to the view controller. When you mouse up, choose delegate. If you see two rows in the pop-up for delegate, just choose the first one.

Next we need to add an instance of UIImageView and make the connection from the PhotoViewController to it. Go to the library (⇧-⌘-L), choose the UIImageView, and drag it into your view. In the Size inspector (⌘-3), set the width and height to 300, and set the x and y coordinates to 10 and 10, respectively. Now drag from the PhotoViewController to the new image view, and connect the imageView outlet to it.

Now that we have the Interface Builder configuration done, we need to return to Xcode to actually implement the UIImagePickerControllerDelegate protocol as well as write the code that makes the image picker show up.

First, here is the code to make the image picker appear:

Images/PhotoView/Classes/PhotoViewViewController.m
```
- (void)touchesEnded:(NSSet *)touches withEvent:(UIEvent *)event {
    if([[touches anyObject] tapCount] > 1) {
        // bring up image grabber
        if([UIImagePickerController isSourceTypeAvailable:
                UIImagePickerControllerSourceTypeCamera]) {
            self.imagePicker.sourceType = UIImagePickerControllerSourceTypeCamera;
        } else {
            self.imagePicker.sourceType =
                UIImagePickerControllerSourceTypePhotoLibrary;
        }
        self.imagePicker.allowsImageEditing = YES;
        [self presentModalViewController:self.imagePicker animated:YES];
    }
}
```

This method brings up the image picker in a modal way if the user has tapped more than once (that is, a double tap). Remember that since the view controller is in the responder chain, it gets a chance to respond to events. When the touchesEnded:withEvent: goes to the view, it forwards it to the view controller; our code responds by bringing up the image picker.

There are a few things to note here in this code. First, on line 5, we see we're checking for the camera type before setting the source type. It is important that you check to see whether the camera is available (that is, some iPhone OS devices do not have a camera) before trying to use it. Next, on line 11, we are setting the controller to allow editing. When this flag is set to YES, the user is allowed to move and zoom in on the image before hitting the Choose button. If the flag is set to NO, then the user can only choose an image without zooming or moving. Finally, on line 12, the view controller animates the image picker into view as a modal view.

In addition to the two source types shown here, you can also specify UIImagePickerControllerSourceTypeSavedPhotosAlbum to allow the user to choose only one of the saved photos from the library.

One other thing to be aware of. In version 2.2 of the iPhone OS SDK, there is a bug. The simulator reports that it has a camera (if you have an iSight camera on your computer) but then fails to attach to the camera properly, so it cannot take a photo. To work around this bug, you can use this code to always set the source type to UIImagePickerControllerSourceTypePhotoLibrary if the application is running in the simulator.

Images/PhotoView/Classes/PhotoViewViewController.m
```
if([[[UIDevice currentDevice] name] rangeOfString:@"Simulator"].location !=
     NSNotFound) {
  self.imagePicker.sourceType = UIImagePickerControllerSourceTypePhotoLibrary;
}
```

OK, now that we have the image picker animating in, we need to deal with the user choosing a photo or canceling. As you recall, we do that through the delegate methods. Here is the code that handles the cancel:

Images/PhotoView/Classes/PhotoViewViewController.m
```
- (void)imagePickerControllerDidCancel:(UIImagePickerController *)picker {
    [self.imagePicker dismissModalViewControllerAnimated:YES];
}
```

All that we are doing here is dismissing the image picker. Next let's look at using the image after the user has chosen it. Here is the code:

Images/PhotoView/Classes/PhotoViewViewController.m
```
Line 1  - (void)imagePickerController:(UIImagePickerController *)picker
     2  didFinishPickingMediaWithInfo:(NSDictionary *)info {
     3    imageView.image = [info objectForKey:UIImagePickerControllerEditedImage];
     4        [self dismissModalViewControllerAnimated:YES];
     5  }
```

Here we are using the info dictionary parameter to get the edited photo. Since we allow editing, the user can crop and scale the image to their liking. Then when they hit Done, the original image, the edited image, and the crop rectangle are put together into the info dictionary. We get the edited image with the UIImagePickerControllerEditedImage key, the original is found under the UIImagePickerControllerOriginalImage key, and the crop rectangle is found with the UIImagePickerControllerCropRect key. We don't need these last two items in our simple example, but if your application can use them, they are available.

20.4 Capturing Video

In addition to being able to get photos either from the user's library or from the camera, in some iPhone OS 3.0 and newer devices the user can also capture video. We can integrate that video into our applications using the same basic flow. Create an image picker, set the source type, present the picker, and then get the captured media via the delegate callback.

To see whether the device your app is running on is capable of video, you ask the image picker for the source types by sending the availableMediaTypesForSourceType: method. The returned array will contain the constant kUTTypeMovie if the device supports video capture.

If your user's device supports video, then you can get at that video in your delegate method by getting the media URL from the info dictionary with the UIImagePickerControllerMediaURL key. Once you have the URL, you can do whatever makes sense for your application with that video. See Chapter 14, *Video Playback*, on page 293 for more information about using video in your application.

In this chapter, we have seen how to display images that we ship as part of the application bundle. We've also learned to use photos, videos, and camera images with our iPhone application. Integration is straightforward, with the only requirement being to implement the UIImagePickerControllerDelegate protocol.

Chapter 21

Core Animation

Every animation on the iPhone—every flipped view, fade-in, scroll, and so on—is done with Core Animation. The fact that we have not had to say much about Core Animation (CA) and still get so much done is one of the best things about it. It underlies everything in the UI on the iPhone but is almost entirely covered by more abstract, and often simpler, APIs. Not that Core Animation is complex—it's just that it has more knobs and switches that we can tweak and therefore has more for us to understand and think about.

In this chapter, we'll cover the breadth of CA and then go into some of the specific things that can be done on the iPhone only by using Core Animation.

21.1 Introduction to Core Animation

Layers are the building blocks that make everything else in Core Animation possible. Picture a layer as a rectangular 2D surface that "lives" in a 3D world. You can place content such as images or PDF to that surface and then move the surface around easily. The best part for performance is that the content does not have to be redrawn for it to move around.

Actually, moving a layer around is just the beginning. You can shrink everything in a layer. You can move a layer around in three-dimensional space, which is how the Cover Flow view works in the iPod application. You can fade the contents of a layer in or out. In other words, just about every property of the layer can be animated. Geometry properties like position and bounds as well as style properties like opacity or contents can all be animated.

There are two basic types of animation. With *implicit animations*, you set a new position for a layer, and it will move there in a nice, smoothly animated fashion. We don't have to set up timers, think about threads, or do any redrawing. It just works, as the saying goes. When you set or change any property that can be animated, you trigger an implicit animation.

If you can't get the effect you are looking for from the implicit animations, then you can easily replace the animations that are used implicitly with your own *explicit* custom animations. Using explicit animations, we can gain exact control over all aspects of the animation, from the timing to the key points along the animation.

Finally, layers can move in 3D space. Core Animation has some really cool (and relatively easy to use) features that allow us to push our content into 3D space. Before we dive into the cool 3D effects, though, let's spend some time getting to know CA by looking at how we have been using it thus far.

Rather than walk you through creating each new project, in this chapter, we are just going to discuss the details of making the animations happen. If you'd like to follow along, you can start a new project for each section in this chapter. If you'd prefer to have the code to look at rather than starting from scratch, grab it from the book's code bundle.

21.2 Animating UIView

We have already seen lots of Core Animation with UIView, but we've been able to ignore the details because of the integration of Cocoa Touch and Core Animation. The visible part of all UIViews is really a layer. The view basically exists to process events and fit into the responder chain. All the drawing, animation, and other visual features of a view are directly or indirectly handled by the layer.

Animation Blocks

Any properties changed on a UIView within an animation block will be animated with the implicit (or explicit, more on that later) animation defined for that property. Creating animation blocks is a snap. The class method +beginAnimations:context: starts a block. Then when you want the animation to start, you call +commitAnimations. All the changes made while the block is active will be animated instead of appearing

Figure 21.1: BASKETBALL IMAGE VIEW

immediately. Let's look at a really simple example to get us started. A UIImageView with a basketball in it perched at the top of the view is shown in Figure 21.1. We are going to move the image from the top of the screen to the bottom and back.

The animation block begins on line 2 with the call to +beginAnimations:context:. Both arguments to this method are more or less arbitrary and for us to give clues to the delegates we set up (more on that shortly).

Here is the code:

CoreAnimation/SimpleMovement/Classes/SimpleMovementViewController.m

```
Line 1  - (void)touchesEnded:(NSSet *)touches withEvent:(UIEvent *)event {
        [UIView beginAnimations:@"movement" context:nil];
        CGPoint center = movingBall.center;
        if(center.y > 85.0f) {
     5    center.y -= 295.0f;
          movingBall.center = center;
        } else {
          center.y += 295.0f;
          movingBall.center = center;
    10  }
        [UIView commitAnimations];
      }
```

If the ball is at the top, place it at the bottom. If it's at the bottom, place it at the top. Notice that we are setting the center property rather than the origin of the frame. The center property is often simpler to work with than the frame, so use it whenever you can.

The numbers in this example were chosen because they looked nice, not because they have any special significance. Take a moment to run the application. Notice that the ball moves from top to bottom quickly (the default implicit animation lasts for 0.25 seconds). One of the ways we can use explicit animation is to make the duration longer (we will look at customizing animations in Section 21.2, *Customizing Animations*, on the facing page).

We can also animate the transition between two subviews with the set-AnimationTransition:forView:cache:. If we set a transition animation, any change to the view hierarchy will be animated in the manner we request. Here is the code for our example:

CoreAnimation/SimpleMovement/Classes/SimpleMovementViewController.m

```
Line 1  - (void)touchesBegan:(NSSet *)touches withEvent:(UIEvent *)event {
        if(event.allTouches.count > 1) {
          [UIView beginAnimations:@"switch" context:nil];
          [UIView setAnimationTransition:UIViewAnimationTransitionCurlUp
     5                          forView:self.view cache:YES];
          if(nil == basketBall.superview) {
            basketBall.center = tennisBall.center;
            [tennisBall removeFromSuperview];
            [self.view addSubview:basketBall];
    10      movingBall = basketBall;
          } else {
            tennisBall.center = basketBall.center;
            [basketBall removeFromSuperview];
```

```
            [self.view addSubview:tennisBall ];
15          movingBall = tennisBall;
          }

        [UIView commitAnimations];
      }
20  }
```

In this example, we set the transition on line 5 to "curl up" when the view is switched out. We can choose from several built-in transition animations. We're using "curl up" here because it looks cool. Try some of the others to see which ones you like best. Also make sure to try them on your device because transitions often look better on the device than they do in the simulator.

Customizing Animations

So far, we've only been experimenting with animations in their default states. There are lots of options, though, for customizing the way the implicit animations work. Changing things like the timing curve on an animation to an ease-in or ease-out can provide subtle clues to users about what the application is doing. These subtle clues can often make the difference between a good UI and a great one. Let's change our bouncing ball application to use timing curves. Here is the code:

CoreAnimation/TimingCurve/Classes/TimingCurveViewController.m
```
Line 1  - (void)touchesEnded:(NSSet *)touches withEvent:(UIEvent *)event {
    -     [UIView beginAnimations:@"movement" context:nil];
    -     [UIView setAnimationCurve:UIViewAnimationCurveEaseIn];
    -     [UIView setAnimationDuration:1.0f];
5         [UIView setAnimationRepeatCount:3];
    -     [UIView setAnimationRepeatAutoreverses:YES];
    -     CGPoint center = basketBall.center;
    -     if(center.y > 85.0f) {
    -       center.y -= 295.0f;
10        basketBall.center = center;
    -     } else {
    -       center.y += 295.0f;
    -       basketBall.center = center;
    -     }
15        [UIView commitAnimations];
    -  }
```

This code is familiar except for the addition of animation customization code starting on line 3. Setting the curve to "ease in" causes the animation to start slowly and then accelerate. An ease-in often adds a feel of realism to the animations because that is how we see things in the real world move. Nothing in the real world starts moving instantaneously.

We also set the duration to 1.0 second on the next line, which as you'd expect causes the duration of the animation to be one second (instead of the default 0.25). We often use the duration for debugging; setting to one or two seconds allows you to see things that you might miss when it's over in a quarter of a second.

Next in this example, we'll set the repeat count and autoreverse. This animation will cause the basketball to bounce up and down three times (along with the reverse animation). Run the application, and watch the difference the ease-in curve makes. The ease-in curve makes the bouncing much more natural.

In running the animation, you probably noticed that after the final repeat cycle of the animation the ball jumps to the bottom of the screen. That happens because of the nature of the animation repeat cycle. Each animation (recall we are repeating three times) proceeds from the starting point (the top) to its end point (the bottom) and reverses (back to the top). Since we changed the center of the ball image to the bottom of the screen and the animation ends at the top of the screen, we get the jump.

The fix for that is to use keyframe animations and take control of the whole animation. Specifying six keyframes, one for each top and bottom bounce point, would make the ball bounce three times and stop when it hits the bottom on the third bounce. We would probably make each "top point" less than the previous one as well. There is an example of key frame animation in Section 21.3, *Layers*, on page 416, but an in-depth discussion is beyond the scope of this chapter. You can get more information from *Core Animation for Mac OS X and the iPhone* [Dud08].

Animation Delegation

You can set one of your objects to be the delegate of the animation so that when the animation starts or stops your delegate will get notification. This is particularly useful for doing things such as playing sounds when the animation starts or stops. Our coverage of audio begins with Chapter 15, *iPod Library Access*, on page 303, so for our example here we will simply write to the log. After you get audio under your belt, you can replace these log messages with your own sounds.

You can set the delegate for your animations with the setAnimationDelegate: method. You also have to set the start and stop selectors so that Core Animation knows which methods to invoke. You set the start selector with the setAnimationWillStartSelector: and the stop selector with setAnimationDidStopSelector:.

Let's set an animation delegate in our bouncing basketball example. Here is the updated code for the touchesEnded:withEvent: method:

CoreAnimation/AnimationDelegate/Classes/AnimationDelegateViewController.m

```
- (void)touchesEnded:(NSSet *)touches withEvent:(UIEvent *)event {
    [UIView beginAnimations:@"movement" context:nil];
    [UIView setAnimationDelegate:self];
    [UIView setAnimationWillStartSelector:@selector(didStart:context:)];
    [UIView setAnimationDidStopSelector:@selector(didStop:finished:context:)];
    [UIView setAnimationCurve:UIViewAnimationCurveEaseIn];
    [UIView setAnimationDuration:1.0f];
    [UIView setAnimationRepeatCount:3];
    [UIView setAnimationRepeatAutoreverses:YES];
    CGPoint center = basketBall.center;
    if(center.y > 85.0f) {
      center.y -= 295.0f;
      basketBall.center = center;
    } else {
      center.y += 295.0f;
      basketBall.center = center;
    }
    [UIView commitAnimations];
}
```

On line 3, we are setting the delegate for the animations that follow. On the following two lines, we are setting the selectors to be called when the animation starts or stops. These methods can have any name that we want but must have the correct signatures. The start selector should have two parameters; the first is an NSString and is the name we gave the animation in the call to beginAnimations:context:. The second is a void pointer and will be the same pointer passed to beginAnimations:context:. The stop selector should take three parameters: an NSString, a BOOL, and a void pointer. The first corresponds to the name, and the third corresponds to the void pointer passed into the beginAnimations:context: method. The second parameter is a flag indicating whether the animation completed. Implementing these delegate methods gives us another level of insight and control over how the animations proceed and what happens when they start and stop.

To gain complete control over the animations, we need to dig a little deeper into Core Animation. Let's get started by covering the relationship between the layer and the view.

UIView and CALayer

Core Animation layers provide delegation methods that allow us to draw and animate in custom ways without having to subclass the layer. For more information on layers and Core Animation in general, see *Core Animation for Mac OS X and the iPhone* [Dud08]. UIView takes advantage of the delegation because the view is always the delegate of its layer. This allows us to add layer delegate methods to our view and gain additional control over the way an animation happens. Let's look at replacing the default animation with one of our own to see how it works. Here is the code:

CoreAnimation/LayerDelegate/Classes/LayerDelegateView.m
```
- (id <CAAction>)actionForLayer:(CALayer *)layer forKey:(NSString *)key {
    CAAnimation *animation = (CAAnimation *)[super actionForLayer:layer
                                                          forKey:key];
    if([animation respondsToSelector:@selector(setDuration:)]) {
        animation.duration = 2.0f;
    }
    return animation;
}
```

This code is setting the duration to 2.0 seconds for all animations. Although we have seen other ways to set the duration of an animation, the UIView methods work only in an animation block. This code will set the duration of any animation for the view whether we have configured a block.[1] The other cool thing we can do here is replace the default with our own animations. A full discussion of what is possible is beyond the scope of this book; check out *Core Animation for Mac OS X and the iPhone* [Dud08] for more information.

You can also draw into the layer from the delegation methods (drawLayer:inContext: and displayLayer:), but that is rarely necessary because the UIView's drawRect: method is almost always a sufficient place to do your drawing in.

21.3 Layers

The tight integration with UIKit often allows you to do what you need with the slightly simpler UIKit approach. However, for doing 3D transformations, customizing the timing of an animation, and doing other

---

1. During the event loop, the iPhone OS creates a block for us.

customizations, you will get more fined-grained control by using the layer-based APIs.

As is often the case, gains in one area come at the cost of losses in another. Layers are not responders, so they do not respond to events. Even if you implement the event methods (touchesBegan:withEvent:, and so on), they won't get called. For the most part, though, you can almost always get the event handling you need by processing the events in the view that contains the layers. You can use the hitTest: method to find the layer that the user touched and then proceed based on that. It is important that, as you build up your layer-based UI, you remember that the layers are not responders.

Since layers are really 2D objects existing in a 3D space, we can animate them around in 3D space. To do that, we use the zPosition and the transform properties. Let's look at an example using the transform property. Here is the code:

CoreAnimation/ThreeDTransform/Classes/ThreeDTransformViewController.m

```
- (void)touchesEnded:(NSSet *)touches withEvent:(UIEvent *)event {
    CAKeyframeAnimation *animation = [CAKeyframeAnimation
                                      animationWithKeyPath:@"transform"];
    NSValue *initial = [NSValue valueWithCATransform3D:
                        CATransform3DMakeRotation(0.0, 1.0f, -1.0f, 0.0f)];
    NSValue *middle = [NSValue valueWithCATransform3D:
                       CATransform3DMakeRotation(M_PI, 1.0f, -1.0f, 0.0f)];
    NSValue *final = [NSValue valueWithCATransform3D:
                      CATransform3DMakeRotation(0.0, 1.0f, -1.0f, 0.0f)];
    animation.values = [NSArray arrayWithObjects:initial,
                        middle, final, nil];
    animation.duration = 2.0f;
    [self.basketBall addAnimation:animation forKey:@"transform"];
}
```

In this method, we are creating a CAKeyframeAnimation and adding three transforms to it: initial is the starting point, middle is the middle of the animation, and final is back to the starting point. This will cause the layer to rotate in 3D space around the vector 45 degrees up from vertical (that is, 1.0, -1.0 , 0.0) 360 degrees. The basketball about one third of the way through its rotation is shown in Figure 21.2, on the next page.

UIViews are drawn in order according to their position in the subviews property of their superview. CALayers are ordered based on two properties: first the order of the layer in its superlayer's sublayers array and second the value of the zPosition property. If two sibling layers share the

Figure 21.2: BASKETBALL IMAGE VIEW

same zPosition value, then they are ordered according to their order in the parent layer's sublayers array. Otherwise, the layers are drawn in the order specified by their zPosition. Let's look at a simple example that has two layers placed offset from the center of the view; one is red and one is blue. When the application starts, the blue layer has its zPosition set to 5.0, and the red has its zPosition set to -5.0. Here is the code for the viewDidLoad::

CoreAnimation/Depth/Classes/DepthViewController.m
```
- (void)viewDidLoad {
    [super viewDidLoad];
    self.blue = [CALayer layer];
    blue.backgroundColor = [[UIColor blueColor] CGColor];
    blue.bounds = CGRectMake(0.0f, 0.0f, 100.0f, 100.0f);
```

```
    blue.position = CGPointMake(160.0f - 20.0f, 240.0f - 20.0f);
    blue.zPosition = 5.0f;
    [self.view.layer addSublayer:blue];
    self.red = [CALayer layer];
    red.backgroundColor = [[UIColor redColor] CGColor];
    red.bounds = CGRectMake(0.0f, 0.0f, 100.0f, 100.0f);
    red.position = CGPointMake(160.0f + 20.0f, 240.0f + 20.0f);
    red.zPosition = -5.0f;
    [self.view.layer addSublayer:red];
}
```

Even though the blue layer is added to the sublayers list first (and thus should be drawn first), it is drawn last (on top) because its zPosition is set to 5.0, while the red layer has its zPosition set to -5.0. To further illustrate how the zPosition affects layer drawing mode, take a look at this code:

CoreAnimation/Depth/Classes/DepthViewController.m

```
- (void)touchesEnded:(NSSet *)touches withEvent:(UIEvent *)event {
    CGFloat oldBlue = self.blue.zPosition;
    self.blue.zPosition = self.red.zPosition;
    self.red.zPosition = oldBlue;
}
```

This simple change in zPosition causes the layers to switch drawing order. In more complex situations where the layers are rotated in 3D space, the effect of changing zPosition will be more dramatic.

There are many many more things that can be done to get just the effect you want or need from the animations in your application. See the Core Animation book for more details (*Core Animation for Mac OS X and the iPhone* [Dud08]).

21.4 OpenGL ES

OpenGL on the iPhone is done via a CAEAGLLayer layer. You subclass UIView and return the CAEAGLLayer class from the +layerClass method. The layerClass method returns CALayer by default and is called when setting up the view. On the iPhone all views are backed by a layer (the layer is referred to sometimes as the *backing store*) and have their drawing and other visual content cached there. The layerClass method gives us the chance to specify our own layer instead of the default.

If the default OpenGL ES options are sufficient for your drawing, you don't need to do anything else to get an OpenGL ES surface to draw on. However, it is often the case that the defaults are not sufficient. To customize them, you only need to set the drawableProperties property on the layer. A full treatment of what is required to properly configure and initialize an OpenGL ES environment is beyond this book; check out *OpenGL SuperBible* [RSWLH07] for a much more thorough treatment.

Let's dive into the example. We are going to use the default rotating square that comes from the OpenGL ES template from Xcode to explain how to set up your OpenGL environment. We are not going to dive into the details of the OpenGL ES, though. Instead, we will focus on the Core Animation aspects of the example.

Let's take a quick look at the code provided by the template, starting with the code to configure the layer:

CoreAnimation/OpenGL/Classes/EAGLView.m
```
Line 1  - (id)initWithCoder:(NSCoder*)coder {
     -        if ((self = [super initWithCoder:coder])) {
     -                CAEAGLLayer *eaglLayer = (CAEAGLLayer *)self.layer;
     -                eaglLayer.opaque = YES;
     5                eaglLayer.drawableProperties =
     -    [NSDictionary dictionaryWithObjectsAndKeys:
     -     [NSNumber numberWithBool:NO], kEAGLDrawablePropertyRetainedBacking,
     -     kEAGLColorFormatRGBA8, kEAGLDrawablePropertyColorFormat, nil];
     -                context = [[EAGLContext alloc]
    10                initWithAPI:kEAGLRenderingAPIOpenGLES1];
     -
     -                if (!context || ![EAGLContext setCurrentContext:context]) {
     -                        [self release];
     -                        return nil;
    15                }
     -
     -                self.animationInterval = 1.0 / 60.0;
     -        }
     -        return self;
    20  }
```

The initWithCoder: method is called as this view is loaded from the nib file.

Note that the opaque flag on the layer is set on line 4. Making your OpenGL ES layers opaque and letting Core Animation know that can considerably improve performance. So, do all your OpenGL ES drawing in an opaque layer if you can.

We also set the drawableProperties on line 8. The configuration here specifies no retained backing and the RGBA 8 color format (via the kEAGLColorFormatRGBA8 constant). There are many, many options for what kind of color format to use. Given the more limited memory bandwidth available on the iPhone, however, it is important to use the smallest types (including color formats) that work for your application.

We set the API level on line 10, which specifies which version of the API we'd like to use. For iPhone OS 2.0, the value must be kEAGLRenderingAPIOpenGLES1, but expect the list to expand to future releases of the iPhone OS. Next we set the context on line 12, which lets the OpenGL ES runtime know which context to use when we start sending OpenGL ES commands.

In layoutSubviews, we set the current context and re-create the frame buffer (the gnarly OpenGL stuff you have to look in the Blue Book to grok: *OpenGL SuperBible* [RSWLH07]). And if all goes well, the application is ready to draw its first frame. layoutSubviews is called automatically for us because the view is placed in the window. Here is the code:

CoreAnimation/OpenGL/Classes/EAGLView.m
```
- (void)layoutSubviews {
        [EAGLContext setCurrentContext:context];
        [self destroyFramebuffer];
        [self createFramebuffer];
        [self drawView];
}
```

After layoutSubviews is called, the application is more or less finished with its startup tasks. When the app finishes startup, recall that the application delegate is sent the applicationDidFinishLaunching: method. In that method, the delegate sends the view the startAnimation method. Let's look at that code next:

CoreAnimation/OpenGL/Classes/EAGLView.m
```
- (void)startAnimation {
        self.animationTimer =
  [NSTimer scheduledTimerWithTimeInterval:self.animationInterval
                                   target:self selector:@selector(drawView)
                                 userInfo:nil repeats:YES];
}
```

In this method, we are creating the timer that will be fired every animationInterval seconds (in our case it's every 1/60 seconds). Each time the timer fires, it invokes the drawView method.

Let's look at that code next:

`CoreAnimation/OpenGL/Classes/EAGLView.m`

```objc
- (void)drawView {
    const GLfloat squareVertices[] = {
        -0.5f, -0.5f,
         0.5f, -0.5f,
        -0.5f,  0.5f,
         0.5f,  0.5f,
    };
    const GLubyte squareColors[] = {
        255, 255,   0, 255,
          0, 255, 255, 255,
          0,   0,   0,   0,
        255,   0, 255, 255,
    };

    [EAGLContext setCurrentContext:context];

    glBindFramebufferOES(GL_FRAMEBUFFER_OES, viewFramebuffer);
    glViewport(0, 0, backingWidth, backingHeight);

    glMatrixMode(GL_PROJECTION);
    glLoadIdentity();
    glOrthof(-1.0f, 1.0f, -1.5f, 1.5f, -1.0f, 1.0f);
    glMatrixMode(GL_MODELVIEW);
    glRotatef(3.0f, 0.0f, 0.0f, 1.0f);

    glClearColor(0.5f, 0.5f, 0.5f, 1.0f);
    glClear(GL_COLOR_BUFFER_BIT);

    glVertexPointer(2, GL_FLOAT, 0, squareVertices);
    glEnableClientState(GL_VERTEX_ARRAY);
    glColorPointer(4, GL_UNSIGNED_BYTE, 0, squareColors);
    glEnableClientState(GL_COLOR_ARRAY);
    glDrawArrays(GL_TRIANGLE_STRIP, 0, 4);

    glBindRenderbufferOES(GL_RENDERBUFFER_OES, viewRenderbuffer);
    [context presentRenderbuffer:GL_RENDERBUFFER_OES];
}
```

Apart from all the OpenGL ES code, this method does two basic things. First it ensures that the current context is the correct context on line 15, and then it pushes the render buffer to the screen on line 36. The rest of the code in this method is basic OpenGL code that you can discover details about in the Blue Book.

Core Animation can be a powerful addition to your application. When you need advanced control over your animations or the ability to move stuff in 3D space, layers are the way to go.

Chapter 22
Accelerometer

It wasn't that long ago that a phone's physical interface consisted of ten-digit buttons, plus * and #. The iPhone radically overhauls this paradigm with its touchscreen, but it's just as important to consider the unseen *accelerometer* as an essential part of the device's UI. By sensing the pull of gravity and motion, the accelerometer can help an application better adapt to the user's needs. If the device is on its side, accelerometer data can let an app know to rotate its UI. If the device is shaken, an application can use that gesture to trigger an undo or an erase. And for many games, tilting and twisting the device has proven to be a highly intuitive control scheme, another feature facilitated by the accelerometer.

In this chapter, we'll take a deep look at how the accelerometer provides data to your application. We'll start with the higher-level APIs that inform you of the overall orientation of the device, and then we'll look at how the iPhone lets your application get the raw data coming from the on-device accelerometer and, more important, how you can do anything useful with it.

> **Simulators Don't Tilt**
>
> The iPhone Simulator provides no meaningful simulation of the accelerometer. When your application asks the simulator for updates from the accelerometer, nothing happens. Because of this, running the sample applications in this chapter require you to be able to put your apps on a device.

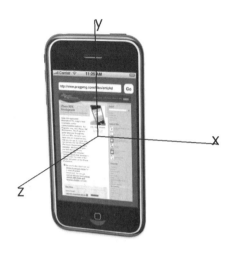

Figure 22.1: VISUALIZING THE ACCELEROMETER AXES

But what is an accelerometer? Basically, it's a tiny device that measures force. Specifically, it reports the force of gravity, user-initiated motion, or both. This motion is reported along three axes: x, y, and z. If you visualize your iPhone in portrait mode, with the home button at the bottom, then x is the axis that goes across your screen (parallel to top and bottom), y is the axis that goes up and down your screen (parallel to the sides), and z is the axis that comes out of the screen (perpendicular to the plane of the screen) right at you. This 3D arrangement as a 2D perspective image is shown in Figure 22.1.

All you ever get from the accelerometer is three floating-point values for the force along these axes, along with a timestamp of when the forces were measured. So... *what the heck do you do with it?*

22.1 Getting Device Orientation

First, let's consider the easy cases of not directly addressing the accelerometer at all. For example, you might be concerned only with adjusting your user interface if the device is rotated. You've known how to do this since way back in Chapter 2, *Hello iPhone*, on page 9. You opt in to rotation notifications by implementing shouldAutorotateToInterfaceOrientation: to return **YES** when it agrees to rotate to the proposed orientation (one of UIInterfaceOrientationPortrait, UIInterfaceOrientationPortraitUpsideDown, UIInterfaceOrientationLandscapeLeft, or UIInterfaceOrientationLandscapeRight). If you support a given rotation, the view controller will get various callbacks such as willRotateToInterfaceOrientation:duration: to inform it of the progress of the rotation animation; check the UIViewController class reference, under the "Handling Rotation" topic, for a complete list of callback methods and their meanings. You can also inspect your current orientation by inspecting the inherited property interfaceOrientation.

Then again, the UIViewController orientation APIs are all about the user interface and, specifically, a UI that uses UIKit. Although we've focused on UIKit throughout the book, it's not the only way to build an iPhone GUI. For example, a game might take over the entire screen with an OpenGL view and make little or no use of UIKit, and such an application would need some other means of getting orientation data.

To get device orientation data without regard for whether the current UI rotates to that orientation, or even uses UIKit at all, you can use the UIDevice class directly. Get the systemwide shared instance with [UIDevice sharedInstance], and then you can inspect its orientation property. One caveat here: you must first call the method beginGeneratingDeviceOrientationNotifications in order to power up the accelerometer. If you don't, the orientation property will always return a meaningless 0. And, having powered up the accelerometer, you need to let go of it at some point in your application with a call to endGeneratingDeviceOrientationNotifications.

22.2 Getting Shakes from the UIResponder Chain

In iPhone OS 3.0, the system provides a limited level of shake detection for your application. The UIResponder class, which is a superclass of UIView, has added three new methods for motion detection: motionBegan:withEvent:, motionEnded:withEvent:, and motionCancelled:withEvent. The first argument is a UIEventSubtype enumerated constant whose only meaningful value in 3.0 is UIEventSubtypeMotionShake.

To get these events, your view needs to be the first responder (see Chapter 18, *Events, Multi-Touch, and Gestures*, on page 373 for more on UIKit event delivery and responders) and then implement at least one of the motion... methods. We'll show an example of this later in the chapter as we build out an extensive accelerometer-based application, as well as noting how you can use raw accelerometer data to do your own shake detection, a technique that is backward compatible with iPhone OS 2.0.

iPhone SDK 3.0 adds one more acceleration API you should be aware of: if you use the NSUndoManager to support undo in your application, you can set the applicationSupportsShakeToEdit property on the shared instance of UIApplication to bring up the undo manager's Undo or Redo button when the user shakes the device.

22.3 Getting Raw Accelerometer Data

Orientation and shake detection are nice, but there are a number of interesting things you can do when you have access to the raw accelerometer data. You can determine your orientation in three dimensions (and not assume that the device is upright), and you can use that tilt as a user interface. Or, you can filter out gravity and just detect user motions: whether the user has turned, shaken, swung, or dropped the device.

For a first example, let's let the accelerometer's detection of gravitational force work for us. The BalanceBall example will let you roll a ball around the screen by looking straight down at the device and tilting it. As you tilt toward one side, the pull of gravity will appear on the x- and/or y-axes, and we'll apply that acceleration to the x and y components of the ball's velocity.

For technical reasons, we decided to use Xcode's Utility Application template (introduced in Section 9.1, *Displaying a Flippable Preference View in Your Application*, on page 163) for this sample. First, it provides us with the view-flipping Info button, which we'll use to add a flip-side preferences screen to enable or disable some filtering logic to improve our use of the accelerometer data. Second, it provides us with a custom UIView subclass that we can customize to draw our playfield.

The Utility Application provides us with a MainViewController to handle the logic of the front-side view, and it's here that we'll keep track of the ball's location and velocity, as well as update the view. So, this is the class where we want to opt into updates from the accelerometer. Let's ask for updates once the view is loaded, that is, in viewDidLoad.

`Accelerometer/BalanceBallFor3.0/Classes/MainViewController.m`
```
// ask for accelerometer events
[[UIAccelerometer sharedAccelerometer] setUpdateInterval: 0.03]; // 30 fps
[[UIAccelerometer sharedAccelerometer] setDelegate: self];
```

The first of these two lines indicates how frequently we want to get updates from the accelerometer, expressed as the time interval between callbacks, in seconds. Nothing is guaranteed, but within certain sensible ranges, you can expect to get more or less the frequency you ask for. Choosing the appropriate frequency is largely a function of what you're going to do with the data. If you wanted to figure out orientation only, you would get acceptable resolution polling 10 or 20 times a second, which would be an interval of 0.10 to 0.05 seconds. For games and other cases where you need the accelerometer to provide you with a higher-resolution user-input device, you generally want to run closer to frame rate, so an interval of 0.033 (30 updates per second) or even 0.02 (50 updates per second) is appropriate.

The second line sets a delegate to get callbacks with the accelerometer data. You can have only one UIAccelerometerDelegate in your application, so if multiple parts of your application wanted updates from the accelerometer, it would be up to your application to distribute them as needed. As soon as you set the delegate, you'll start getting callbacks with accelerometer data, and they'll continue until you **nil**-out the delegate, which is your signal that you're done with the accelerometer. In ball rolling app, we want to always be getting accelerometer data, so we won't **nil** out the delegate until the dealloc method.

Accelerometer/BalanceBallFor3.0/Classes/MainViewController.m

```
- (void)dealloc {
        [[UIAccelerometer sharedAccelerometer] setDelegate: nil];
        [super dealloc];
}
```

The UIAccelerometerDelegate protocol requires us to declare in MainViewController.h that we implement the protocol—so add <UIAccelerometerDelegate> to the @interface as you would for any other protocol implementation—and to implement its one method, accelerometer:didAccelerate:. This method provides us with the UIAccelerometer that provides the data and a UIAcceleration object that provides the acceleration data.

But what are we going to do with this data? We don't have our rolling ball to apply the gravitational acceleration to yet, so let's attend to that.

We'll keep track of the ball's position and velocity in the view controller, leaving the custom view class to just paint the ball. For the sake of this example, we'll use really simple graphic primitives, introduced in Chapter 19, *Drawing in Custom Views*, on page 385. To draw the ball, we'll need to know its current drawing rectangle, represented as a CGRect, and the previous drawing rectangle, which we'll erase before drawing the ball in its new location. So, we need a couple of CGRects and methods to set the ball's location and its size. Here's the MainView.h header file:

Accelerometer/BalanceBallFor3.0/Classes/MainView.h

```
@interface MainView : UIView {
      CGRect ballRect;
      CGRect oldBallRect;
}
-(void) setBallX: (CGFloat) newX Y: (CGFloat) newY;
-(void) setBallWidth: (CGFloat) newW height: (CGFloat) newH;
```

Implementing the setBallWidth:height: method is trivial:

Accelerometer/BalanceBallFor3.0/Classes/MainView.m

```
- (void) setBallWidth: (CGFloat) newW height: (CGFloat) newH {
      ballRect.size.width = newW;
      ballRect.size.height = newH;
      oldBallRect.size.width = newW;
      oldBallRect.size.height = newH;
}
```

Setting the ball's location requires three simple tasks: updating the ballRect's origin point, requesting an update of the on-screen view, and saving off the rectangle we just drew as an erase rectangle for the next update. Erase-and-draw is more efficient than drawing the whole view, and we make it more efficient by setting a clipping rectangle that tells Quartz that the only "dirty" part of the view, that is, the only areas that could possibly require repainting, are in the rectangle defined by the union of the old and new drawing rectangles. We calculate this union rectangle with CGRectUnion().

Accelerometer/BalanceBallFor3.0/Classes/MainView.m

```
-(void) setBallX: (CGFloat) newX Y: (CGFloat) newY {
      ballRect.origin.x = newX;
      ballRect.origin.y = newY;

      // update view
      CGRect clipRect = CGRectUnion (oldBallRect, ballRect);
      [self setNeedsDisplayInRect:clipRect];

      // update oldBallRect
      oldBallRect.origin.x = ballRect.origin.x;
      oldBallRect.origin.y = ballRect.origin.y;
}
```

The setNeedsDisplayInRect: call will result in drawRect: getting called. Our custom implementation of this method draws the view by getting the current graphics context of the view, fills the old drawing rectangle with the background color (which we set as green in IB; your taste may vary), and draws a plain white circle at the ball's current location.

`Accelerometer/BalanceBallFor3.0/Classes/MainView.m`

```
- (void)drawRect:(CGRect)rect {
    // Drawing code
    CGContextRef context = UIGraphicsGetCurrentContext();
    
    // undraw ball at old location
    CGColorRef undrawColor = self.backgroundColor.CGColor;
    CGContextSetFillColorWithColor (context, undrawColor);
    CGContextFillRect (context, oldBallRect);
    
    // draw ball at new location
    CGContextSetGrayFillColor(context, 1.0, 1.0);
    CGContextFillEllipseInRect(context, ballRect);
}
```

With the drawing done, we can return our attention to the view controller and, ultimately, to the accelerometer. In the MainViewController headers, we'll set up variables for the ball's location and position. We'll also want to keep track of the timestamp of the last accelerometer callback in order to compute how much time has elapsed and therefore how far the ball has moved. Finally, wire up an IBOutlet to the MainView so you'll be able to call its setBallX:Y: method.

`Accelerometer/BalanceBallFor3.0/Classes/MainViewController.h`

```
CGFloat ballX;
CGFloat ballY;
CGFloat ballVelocityX;
CGFloat ballVelocityY;
NSTimeInterval lastAccelTimestamp;
```

Along the way, we'll also define some constants for the ball height and width (both 20.0) and a MAX_ACCEL_PER_SEC that represents how many units per second we want an acceleration of 1.0 to represent (that is, the pull of gravity if the device is standing all the way on one edge). With all of this set up, we can provide a method to set the ball near the center of the view, which we'll do when the application starts, and another to check whether the ball has gone out of the view, which will require us to reset the ball.

Accelerometer/BalanceBallFor3.0/Classes/MainViewController.m

```
- (void) resetBall {
        ballVelocityX = 0.0;
        ballVelocityY = 0.0;
        filteredAccelX = 0.0;
        filteredAccelY = 0.0;
        ballX = self.view.frame.size.width / 2;
        ballY = self.view.frame.size.height / 2;
}

- (void) checkBallInPlay {
        if      ((ballX + BALL_WIDTH < 0) ||
                 (ballY + BALL_HEIGHT < 0) ||
                 (ballX > self.view.frame.size.width) ||
                 (ballY > self.view.frame.size.height)) {
                NSLog (@"reset ball out of bounds");
                [self resetBall];
        }
}
```

Now, in viewDidLoad, you can set up the initial state of the game by calling the methods you've created to set the ball's size, reset its location, and update the view. Notice that we have to cast the view property to MainView in order to call setBallWidth:height: and setBallX:Y: without a warning.

Accelerometer/BalanceBallFor3.0/Classes/MainViewController.m

```
[(MainView*) self.view setBallWidth: BALL_WIDTH height:BALL_HEIGHT];
[self resetBall];
[(MainView*) self.view setBallX:ballX Y:ballY];
```

Our rolling ball simulation now has a ball (complete with on-screen appearance and a model of its location and velocity) but no actual rolling. That's because we haven't yet done anything with the accelerometer data that our delegate (the view controller) is going to be receiving. Let's finally deal with that. Each time we get a callback with accelerometer data, we'll apply the x and y components to our ball's velocity, which is measured in units per second. So, if the device is straight up in portrait mode, it should get a 1.0 acceleration on its y-axis, and in landscape mode, it should get 1.0 on the x-axis. Typically, tilting it will result in some amount of pull on both axes, as well as on z, which we don't use in this application. We defined a constant earlier to represent how much acceleration one second of 1g will provide in our game world, so we can implement accelerometer:didAccelerate: to apply that force to the ball's velocity and then use the velocity and the elapsed time to figure out how far to move the ball.

Accelerometer/BalanceBallFor3.0/Classes/MainViewController.m

```
- (void)accelerometer:(UIAccelerometer *)accelerometer
          didAccelerate:(UIAcceleration *)acceleration {
    NSTimeInterval elapsedTime = acceleration.timestamp - lastAccelTimestamp;
    ballVelocityX = ballVelocityX +
        (acceleration.x * MAX_ACCEL_PER_SEC * elapsedTime);
    ballVelocityY = ballVelocityY -
        (acceleration.y * MAX_ACCEL_PER_SEC * elapsedTime);
    if (lastAccelTimestamp > 0) {
    // recalc ball position
    ballX += (ballVelocityX * elapsedTime);
    ballY += (ballVelocityY * elapsedTime);
    // update view
    [mainView updateBallX:ballX Y:ballY];
    // check for ball out of bounds
    [self checkBallInPlay];
    }
    lastAccelTimestamp = acceleration.timestamp;
```

The callback method gives us a UIAcceleration object with four properties: an NSTimeInterval value for the timestamp and three UIAccelerationValue (defined as double) properties: x, y, and z. In this implementation, we take the acceleration on one axis and determine how much of it to apply to the velocity for this callback interval. For example, if we received a 1.0 value and hadn't updated for an entire second, then we'd add or subtract 1.0 seconds times 1.0*g* times MAX_ACCEL_PER_SEC from the velocity. In practice, of course, we'll get much smaller values much more frequently. Another thing to notice is that while we add the accelerator value along the x-axis, we *subtract* it for the y value. This is because the accelerometer reports upward acceleration as positive values and downwards as negative, while the coordinate system of the iPhone has y values increasing as they go down the screen. Therefore, positive y acceleration (a pull toward the top of the screen) needs to produce negative velocity so that the ball's origin point goes "down" relative to the coordinate system.[1]

And with that, the BalanceBall application is ready to play. Assuming you have the credentials to sign an application and put it on your device, go ahead and do so, and roll the ball back and forth. You'll notice it takes a great deal of tilt to turn the ball around, because in part we haven't applied any friction to slow the ball down once it gets going. The application in all its graphic simplicity is shown in Figure 22.2, on the next page.

---

1. It might have been cleaner to simply apply a transformation matrix to the view's

Figure 22.2: MOVING AN ON-SCREEN BALL WITH THE ACCELEROMETER

22.4 Filtering Accelerometer Data

So far, we've made fairly trivial use of the raw data provided to us by the accelerometer, simply mapping the forces measured by the accelerometer onto a virtual ball that we model. But the real art of working with the accelerometer involves making smart decisions about the data we get from the callbacks, such as smoothing out erroneous data and distinguishing user input from the pull of gravity.

Basic Accelerometer Filters

As written, our ball-rolling application takes whatever force it measures and applies it to the ball's velocity, whether that force comes from gravity or from the user's own motions. Many apps will want one or the other of these forces, and a pair of simple *filter* algorithms can suffice

---

drawing code, thereby flipping its coordinate system to match the accelerometer's interpretation of the y-axis.

in many cases. These simple filters, which you see in Apple's documentation as well as many of its examples, are the *low-pass filter* and *high-pass filter*. Each one uses the fact that gravity is constantly present as the key to how it works.

The low-pass filter takes a single accelerometer reading and applies it to previously calculated accelerometer readings using only a small fraction of the new value. This approach devalues sudden movement in favor of long-term force, with the practical result that it's useful for ignoring twitchy fingers in favor of gravity. The high-pass filter is the opposite: it subtracts the calculated low-pass filter value from the current accelerometer value, meaning it discards long-term forces in favor of instantaneous ones. Therefore, the high-pass filter is better suited to detecting the user's movement.

Let's try this with our ball-rolling app. We'll put a segmented control on the flip-side view to allow the user to pick what kind of filtering they want to use. And to further test quick motion detection, we'll use the high-pass filter to detect device shakes.

Since we chose Xcode's utility application template, you have an info button on your main view, which takes you to a flip-side view for setting preferences. Add a UISegmentedControl to FlipsideView.xib (as shown in Figure 22.3, on the following page) and wire it up to a property filterControl. It needs to be a property, so that you can read its value when the flipside view is dismissed. In MainViewController's flipsideViewControllerDidFinish, collect the value from the filterControl, and use it as the new value of filterPref, which is an instance variable for remembering the chosen filter type.

Accelerometer/BalanceBallFor3.0/Classes/MainViewController.m

```
- (void)flipsideViewControllerDidFinish:(FlipsideViewController *)controller {
    [self dismissModalViewControllerAnimated:YES];
    // set low-level filter use to switch state
    filterPref = controller.filterControl.selectedSegmentIndex;
}
```

MainViewController.m is where we're actually going to do the filtering, so just inside the **@implementation**, we'll add a few needed constants, starting with an **enum** for the possible filter values passed in from the segmented control. Next, we'll define a "filtering factor" to represent just how much weight to give each new value we get from the accelerometer. With a factor of 0.1, the low-pass filter will give a 10 percent weight to the new value and 90 percent to its previously filtered value. Finally,

Figure 22.3: PREFERENCE INTERFACE FOR ACCELEROMETER FILTER

we'll want a constant to represent just how much force will count as a "shake"; 2*g*s is strong enough to not be mistaken for anything but a good shake.

`Accelerometer/BalanceBallFor3.0/Classes/MainViewController.m`

```
// filter types, as values from the multiswitch
enum SORT_TYPES {
        NO_FILTER = 0,
        LOW_PASS_FILTER,
        HIGH_PASS_FILTER
};

// how much weight we give to a single accelerometer value
float kFilteringFactor = 0.1;

// how much acceleration (high-pass filtered) counts as a shake
CGFloat SHAKE_THRESHOLD = 2.0;
```

Now we can do the filtering in the accelerometer:didAccelerate: method where we've already been handling the accelerometer data. We'll need instance variables declared in the header file for filteredAccelX and filteredAccelY, because the algorithms for high- and low-pass filters need to use the previously filtered acceleration values to compute the new acceleration. Here are the simple formulas for the low-pass and high-pass filters on the x- and y-axes:

Accelerometer/BalanceBallFor3.0/Classes/MainViewController.m

```
CGFloat lowPassFilteredX = (acceleration.x * kFilteringFactor) +
    (filteredAccelX * (1.0 - kFilteringFactor));
CGFloat lowPassFilteredY = (acceleration.y * kFilteringFactor) +
    (filteredAccelY * (1.0 - kFilteringFactor));
CGFloat highPassFilteredX = acceleration.x - lowPassFilteredX;
CGFloat highPassFilteredY = acceleration.y - lowPassFilteredY;
```

If you keep in mind that we defined kFilterFactor as 0.1, you can see that the low-pass filters basically amount to "add 10 percent of the current accelerometer value to 90 percent of the previously filtered value." And if you think about it, you'll see that this will tend to negate one-time-only jolts in favor of force that shows up in repeated callbacks (that is, gravity). Meanwhile, the high-pass filter takes the computed low-pass value and subtracts it from the current accelerometer value.

All that's left is to use the user's choice of filtered acceleration value to modify the ball's velocity, replacing the earlier computations of ballVelocityX and ballVelocityY as follows:

Accelerometer/BalanceBallFor3.0/Classes/MainViewController.m

```
switch (filterPref) {
        case NO_FILTER: {
                filteredAccelX = acceleration.x;
                filteredAccelY = acceleration.y;
                break;
        }
        case LOW_PASS_FILTER: {
                filteredAccelX = lowPassFilteredX;
                filteredAccelY = lowPassFilteredY;
                break;
        }
        case HIGH_PASS_FILTER: {
                filteredAccelX = highPassFilteredX;
                filteredAccelY = highPassFilteredY;
                break;
        }
}

// apply acceleration to velocity
NSTimeInterval elapsedTime = acceleration.timestamp - lastAccelTimestamp;
ballVelocityX = ballVelocityX + (filteredAccelX * MAX_ACCEL_PER_SEC * elapsedTime);
ballVelocityY = ballVelocityY - (filteredAccelY * MAX_ACCEL_PER_SEC * elapsedTime);
```

While we're at it, we should also zero out the old filteredAccelX and filteredAccelY values when we reset the ball to the center of the view and set its velocity to 0.

Implementing Shake Detection

The other thing we said we were going to do with the filters was to detect a shake. For iPhone OS 3.0 users, we can let the system detect a shake and deliver an event to our MainView. That's handy, but we want to respond to the shake in the view controller, not the view. So, let's deliver the event to the controller by means of an informal protocol. Define a shakeDelegate property in MainView.h:

Accelerometer/BalanceBallFor3.0/Classes/MainView.h

```
@property (nonatomic, retain) id shakeDelegate;
```

Back in the MainViewController, you can set this delegate to be the view controller. While you're at it, you also need to make the view become the first responder in order to receive the UIEvents. So, let's perform both of these tasks in viewWillAppear:.

Accelerometer/BalanceBallFor3.0/Classes/MainViewController.m

```
- (void) viewDidAppear: (BOOL) animated {
    [super viewDidAppear: animated];
    [self.view becomeFirstResponder];
    ((MainView*)self.view).shakeDelegate = self;
}
```

Now we need the view to deliver the shake events to the controller. The view has to be able to become first responder and then to implement the motionBegan:withEvent: callback method. We'll just call a shakeMotionBegan: method on the delegate, if it has such a method:

Accelerometer/BalanceBallFor3.0/Classes/MainView.m

```
-(BOOL) canBecomeFirstResponder {
    return YES;
}
- (void)motionBegan:(UIEventSubtype)motion withEvent:(UIEvent *)event {
    [super motionBegan: motion withEvent: event];
    if ((motion == UIEventSubtypeMotionShake) &&
            [self.shakeDelegate respondsToSelector:
                    @selector (shakeMotionBegan:)]) {
        [self.shakeDelegate shakeMotionBegan: event];
    }
}
```

Back in the controller, we implement the callback by simply logging the event and resetting the ball's position.

> Accelerometer/BalanceBallFor3.0/Classes/MainViewController.m

```
-(void) shakeMotionBegan: (UIEvent*) event {
        NSLog (@"got system shake motion");
        [self resetBall];
}
```

But what if we can't assume the user has iPhone OS 3.0? To support shake-to-reset on iPhone 2.0, we can use the high-pass filter that we've already set up, which will mostly eliminate gravity and represent only significant user-applied force. So, as a simple implementation, we'll interpret any sharp movement on the x- or y-axis (in excess of SHAKE_THRESHOLD, which we defined as 2.0g) as a shake, and we'll respond by resetting the ball:

> Accelerometer/BalanceBallFor3.0/Classes/MainViewController.m

```
// detect shake
if ((fabsf (filteredAccelX) > SHAKE_THRESHOLD) ||
        (fabsf (filteredAccelY) > SHAKE_THRESHOLD)) {
        [self resetBall];
}
```

Notice how we use the floating-point absolute value function, fabsf(), so that whether the acceleration is positive or negative, we're concerned only with its magnitude and whether it's enough to count as a shake.

Try rolling the ball on your device with different kinds of filtering and see what you think. The difference between no filtering and the low-pass filter is subtle, since there's so much more gravitational force than that of your own tilting, which only goes to increase the pull toward the ground anyway. Still, there's a slight difference. On the other hand, the high-pass filter really doesn't make sense in this case, because you'd be hardpressed to apply more force than is going to be picked up by the small amount of gravity that doesn't get filtered out. Still, the payoff of the high-pass filter is in the shake detection. Flick the device right, left, up, or down, and you'll see the ball reset to the center.

Advanced Accelerometer Filtering

We said before that these filters are pretty simplistic, and they go only so far. For example, there's clearly some gravitational pull that gets past the high-pass filter, so you couldn't entirely depend on that for a game control.

Getting into more sophisticated filtering risks getting too deeply into digital signal processing, but we do have some ideas for further research.

If you're writing a tilt-controlled game, you might want to look at the accelerometer code in the Apple demo *Touch Fighter II*, which was demoed at WWDC 2008 and the late-2008 iPhone Tech Talk World Tour. This first-person space shooter sets up calibration values to represent the initial orientation and pull of gravity, resetting the calibration if the player's ship drifts all the way to one side of the screen, and uses these calibration values to average out incoming accelerometer data, trying to filter to just significant user movement. Unfortunately, as of this writing, the code is available only to attendees of those events.

The shake logic presented in this chapter accounts only for a single, strong jolt, strong enough to provide $2g$ of force on either the x- or y-axis. It might be a better interface to detect a back-and-forth movement, which wouldn't necessarily have to be as strong. A simple form of this might just note when a sufficient shake has been registered in one direction and then expect an equal shake in the opposite direction within a very short period of time (say, a few hundred milliseconds). A more sophisticated shake detector could look for the shake as a repeating signal along one axis. You could presumably do this with a Fourier Transform, which detects frequencies in functions. However, the iPhone doesn't have the Fast Fourier Transform (FFT) functions commonly found on desktops, so you'd have to provide your own implementation. Or, if you can count on your users having iPhone OS 3.0 installed, just depend on the system-delivered shake events from UIResponder.

One other thing you might want to do is to detect specific motions with the device: golf swings, pointing motions, sword slashes, and the like. This kind of thing is really popular on another accelerometer-based device, the Nintendo Wii video game console, so we asked some of our game developer friends where they learned the tricks of the trade. They said the best source for information is the Wii developer forums, which are available only to Nintendo licensees. Still, they did help us find some interesting academic research on the topic, including a paper out of the University of Oldenburg, "Gesture Recognition with a Wii Controller" [SPHB08], which uses a number of techniques such as modeling gestures onto points of a sphere. A Gamasutra article, "Where's the Wiimote? Using Kalman Filtering To Extract Accelerometer Data"

[Ras07], filters motion data by using time-dependent probability theory: given where the device has been going, where is it likely going next? Note that most of the published articles on accelerometer sensing are heavy on the math, rife with differential equations and matrix multiplication, which limits the number of developers who can understand and apply the material in code. We hope we'll see sophisticated accelerometer libraries, suitable for use by the everyday iPhone developer, emerge in the near future.

Chapter 23

Address Book

Your application can use phone numbers, addresses, or any other information a user might keep in their Address Book. You tie into the Address Book with view controllers in much the same way that you integrated with the photo library and camera. You'll also find that once again you'll end up using mostly C functions to work with the selected data. In this chapter, we will see how to use the two frameworks related to the Address Book to add that extra touch of personalization into our applications.

23.1 Address Book UI

The Address Book integration offers two view controllers that we can use to present existing contacts. The first is called ABPeoplePickerNavigationController, which allows us to choose any contact from the whole database. The second is called ABPersonViewController, which allows us to view and edit a single contact. The people picker is, as its name suggests, a navigation controller, and the person controller is a view controller.

An example application will help us understand these two controllers better. Let's create an application that will allow us to choose several of our contacts, list them alphabetically, and then choose individual contacts and see their phone numbers. Create a new navigation-based project called Contacts.

Since we will be interacting with Address Book classes, we need to add the AddressBook.framework and AddressBookUI.framework frameworks to our project. Right-click the Frameworks group, choose Add Existing

Framework, navigate to the iPhone SDK, choose the two frameworks, and click the Add button.

Integrating with the Address Book controllers is similar to the way we integrated with other controllers including the photo library in Chapter 20, *Drawing Images and Photos*, on page 397. The RootViewController will implement delegation protocols and then become the delegate of the Address Book controllers. Since it is the delegate, it will get callbacks from the Address Book controllers when appropriate stuff happens (like the user selecting a contact).

Specifically, we need to add the ABPeoplePickerNavigationControllerDelegate and ABPersonViewControllerDelegate protocols to the RootViewController and implement the required methods. We also need two collections: one to hold the names of the chosen contacts and the other to hold their identifiers. Open the .h file, and add the protocol declaration and the instance variable and property. The code should look like this:

AddressBook/Contacts/Classes/RootViewController.h
```
#import <UIKit/UIKit.h>
#import <AddressBookUI/AddressBookUI.h>

@interface RootViewController : UITableViewController
    <ABPeoplePickerNavigationControllerDelegate,
    ABPersonViewControllerDelegate> {
  NSMutableArray *contactNames;
  NSMutableArray *contactIDs;
}

@property(nonatomic, retain) NSMutableArray *contactNames;
@property(nonatomic, retain) NSMutableArray *contactIDs;

- (IBAction)makeNewEntry;

@end
```

23.2 People Picker Delegate

There are three methods to implement as part of this protocol. We are going to look at them one at a time. The peoplePickerNavigationControllerDidCancel: method is called when the user hits the Cancel button. Since the selection is canceled, all we really want to do in this method is clean up anything we set up to receive a selection and dismiss the people picker controller.

Here is the code for that method:

`AddressBook/Contacts/Classes/RootViewController.m`

```
- (void)peoplePickerNavigationControllerDidCancel:
(ABPeoplePickerNavigationController *)peoplePicker {
   [peoplePicker dismissModalViewControllerAnimated:YES];
   [peoplePicker autorelease];
}
```

In this code, all we are doing is dismissing the view controller and cleaning up the memory used. Since this is a cancel operation, we don't want to modify any data.

Next, we need to implement the method that is called when a user selects a contact. The peoplePickerNavigationController:shouldContinueAfterSelectingPerson: method is called when a contact is selected. If we want to continue to see the contact's detailed information, we should return YES; if we don't want to show the detailed information, we return NO. Most of the time if we return NO, we also want to dismiss the controller. For the example that we are building, we do not want to see the details, so we are going to return NO and perform the same dismiss that we did for the cancel case. We also want to grab the contact's name and identifier and place them both into their respective arrays. Here is the code:

`AddressBook/Contacts/Classes/RootViewController.m`

```
- (BOOL)peoplePickerNavigationController:
(ABPeoplePickerNavigationController *)peoplePicker
      shouldContinueAfterSelectingPerson:(ABRecordRef)person {
   NSString *name = (NSString *)ABRecordCopyCompositeName(person);
   [self.contactNames addObject:name];
   [self.contactIDs addObject:
    [NSNumber numberWithInt:ABRecordGetRecordID(person)]];
   [peoplePicker dismissModalViewControllerAnimated:YES];
   [peoplePicker autorelease];
   NSIndexPath *path = [NSIndexPath indexPathForRow:self.contactIDs.count - 1
                                    inSection:0];
   [self.tableView insertRowsAtIndexPaths:[NSArray arrayWithObject:path]
                  withRowAnimation:UITableViewRowAnimationRight];
   return NO;
}
```

We can cast the return value from ABRecordCopyCompositeName() to an NSString because behind the scenes the CFStringRef and NSString have the same memory layout. This concept is called *toll-free bridging* between Core Foundation and Foundation. You can read more about toll-free bridging of types in Apple's *Carbon-Cocoa Integration Guide* [App06b].

And finally, we need to implement peoplePickerNavigationController: shouldContinueAfterSelectingPerson:property:identifier:. For our example, we are not going to do anything with this method except return NO. Since we returned NO from peoplePickerNavigationController:shouldContinueAfterSelectingPerson: and then dismissed the controller, this method will not ever be called anyway. If you'd like to implement this method, you would do very similar things to what we did in the "should continue" method. Take the property, and copy its value into your own data structure for use in your application. The return value from this method tells the Address Book framework to either do the "default" action (like start composing an email or dialing a phone number) or nothing. If you return YES, the default action will take place; if you return NO, it won't.

There's one last method implementation before we talk about how to get our people picker controller to show up. Instead of setting up the arrays for the contact IDs and names in an init method, we implement the get methods to lazily create the arrays like this:

AddressBook/Contacts/Classes/RootViewController.m
```
-(NSMutableArray *)contactNames {
  if(nil == contactNames) {
    contactNames = [[NSMutableArray alloc] init];
  }
  return contactNames;
}
```

The sample code has both the contactNames and contactIDs methods implemented, but apart from the name of the instance variable, they are the same.

23.3 Creating and Configuring the People Picker

Now that we have the delegate methods implemented, it's time to make the person picker and animate it into view. To do that, let's start with adding a + button to the right side of the navigation bar (for more details about the navigation bar, see Chapter 6, *Navigation*, on page 103). We set the navigationItem's rightBarButtonItem to a new Bar Button Item. Here is the code:

AddressBook/Contacts/Classes/RootViewController.m
```
- (void)viewDidLoad {
  [super viewDidLoad];
  self.navigationItem.rightBarButtonItem = [[[UIBarButtonItem alloc]
       initWithBarButtonSystemItem:UIBarButtonSystemItemAdd
                                   target:self
                                   action:@selector(add)]
                                   autorelease];
```

```
    self.title = @"Selected Contacts";
}
```

Since we used the system item UIBarButtonSystemItemAdd, a + button will be placed in that top-right side of the nav bar. We also want to set the title of this view controller to Selected Contacts. And finally, the add should bring up the people picker. Here is the code for that method:

AddressBook/Contacts/Classes/RootViewController.m
```
- (void)add {
  ABPeoplePickerNavigationController *peoplePicker =
    [[ABPeoplePickerNavigationController alloc] init];
  peoplePicker.peoplePickerDelegate = self;
  [self presentModalViewController:peoplePicker animated:YES];
}
```

In this method, we are creating an instance of the people picker, setting its peoplePickerDelegate to self so that we get the callbacks, and then presenting the people picker as a modal view controller.

23.4 Person Controller

Next we want to make our application navigate to the details of a contact when the user selects a row from our table view. To do that, we are going to push an instance of ABPersonViewController onto the navigation controller's stack.

The usage pattern is, as you might expect, that we create the person view controller, set its delegate to self, and then push the person view controller onto the nav controller's stack. We will start with the implementation of the single method required by the ABPersonViewControllerDelegate protocol. Here is the code:

AddressBook/Contacts/Classes/RootViewController.m
```
- (BOOL)personViewController:(ABPersonViewController *)personViewController
          shouldPerformDefaultActionForPerson:(ABRecordRef)person
              property:(ABPropertyID)property
            identifier:(ABMultiValueIdentifier)valueID {
  ABPropertyType type = ABPersonGetTypeOfProperty(property);
  switch (type) {
    case kABStringPropertyType: {
      NSString *value = (NSString *)ABRecordCopyValue(person, property);
      NSLog(@"property value = %@", value);
      [value release];
      break;
    }
```

```
      case kABMultiStringPropertyType: {
        ABMutableMultiValueRef multi = ABRecordCopyValue(person, property);
        CFIndex index = ABMultiValueGetIndexForIdentifier(multi, valueID);
        CFStringRef value = ABMultiValueCopyValueAtIndex(multi, index);
        NSLog(@"value = %@", (NSString *)value);
        CFRelease(multi);
        CFRelease(value);
        break;
      }
      default:
        break;
  }
  [self.navigationController popViewControllerAnimated:YES];
  return NO;
}
```

This method is called when the user selects one of the attributes from the person view. The code demonstrates some of the API calls that we can use to use the Address Book types. First we get the type of the property that was selected, and then we switch on the type. We consider only two types here, but there are several others to choose from. We are logging only to the console in this method, but in your own code you'd want to do something more sophisticated.

Next we need to look at how to invoke the person view controller. When the user selects one of the contacts displayed in our table view, we want to have the person view controller navigate in. So, we need to implement the tableView:didSelectRowAtIndexPath: method to create and display the person view controller. Here is the code to do that:

AddressBook/Contacts/Classes/RootViewController.m

```
- (void)tableView:(UITableView *)tableView
didSelectRowAtIndexPath:(NSIndexPath *)indexPath {
  ABAddressBookRef addressBook = ABAddressBookCreate();
  ABRecordRef person = ABAddressBookGetPersonWithRecordID(addressBook,
              [[self.contactIDs objectAtIndex:indexPath.row] intValue]);
  ABPersonViewController *pvc = [[ABPersonViewController alloc] init];
  pvc.personViewDelegate = self;
  pvc.displayedPerson = person;
  [self.navigationController pushViewController:pvc animated:YES];
  CFRelease(addressBook);
  [pvc autorelease];
}
```

In this method, we are creating an ABAddressBookRef and then using that to find the person who our ID represents. We then pass that person into the person view controller's displayedPerson property, set self as the personViewDelegate, and then push the view controller onto the navigation stack. Since we did not copy the person object, we do not need to release it.

The Address Book object is conceptually a connection to the Address Book database. If you need to create new entries, you would start with a new instance of ABAddressBookRef and then use the other Address Book functions such as ABPersonCreate() and ABAddressBookAddRecord(). You need to make sure that access to an instance of ABAddressBookRef is always from one thread, or undefined behavior can result.

Now that we have our person view showing up, we can select a value. It will be logged to the console, and then the person view controller will animate out of view.

23.5 Adding New Contacts

As we mentioned earlier, the interface that we have to interact with the contact database is the ABAddressBookRef. Once we have this connection established, we can add, remove, and modify Address Book entries. For each thread in your application that needs access to the database, you need to make a new instance of ABAddressBookRef.

Let's add a method to our view controller that creates a single-person entry to the database. Our entry will be hard-coded to some data we will make up. In a more sophisticated application, you might be retrieving contact information from a web service or some other means. Once you have the data, though, the process and API you will use is the same.

In our example, we'll create a new person object, set the first and last names, add two phone numbers, and set the home address. Then we will add the person record to the database and clean up the memory we allocated. The process is straightforward, but if you've never done any C code, it might be a bit overwhelming looking at these API calls. Keep in mind two things: there is an introduction to working with Core Foundation in Section 17.1, *Using the Procedural-C APIs*, on page 358, and although this looks like a lot of code, it is doing only the five things outlined. Let's go through the code piece by piece. (To see all the code at once, grab the whole project and look at makeNewEntry in our RootView-Controller.) Here is the code to make the Address Book and the entry:

AddressBook/Contacts/Classes/RootViewController.m
```
ABAddressBookRef addressBook = ABAddressBookCreate();
ABRecordRef person = ABPersonCreate();
CFErrorRef error = NULL;
```

Despite the name, calling ABAddressBookCreate() does not create a new database; it merely opens a connection to the existing shared database. Calling ABPersonCreate() does create a new person object, and this is the object we are going to add all the data to and then pass off to our connection to the database to be stored. Next up we set the first and last names of our newly created person object:

AddressBook/Contacts/Classes/RootViewController.m
```
// set the first and last name properties
ABRecordSetValue(person, kABPersonFirstNameProperty,
                 CFSTR("Wonder"), &error);
ABRecordSetValue(person, kABPersonLastNameProperty,
                 CFSTR("Widget"), &error);
if(NULL != error) {
  NSLog(@"an error occurred");
}
```

Here we are using two Core Foundation string objects via the CFSTR() macro to set the first and last names of our person object. Remember that Core Foundation objects are toll-free bridged to their Foundation equivalents, so if you'd prefer to use NSStrings, you can do so via a cast like this:

```
ABRecordSetValue(person, kABPersonFirstNameProperty,
                 (CFStringRef)@"Wonder", &error);
```

Next we add two phone numbers to the person object. Phone numbers are multivalued properties in the Address Book schema, so the code is a bit more involved than setting the name. Here it is:

AddressBook/Contacts/Classes/RootViewController.m
```
// set a phone number
ABMutableMultiValueRef multi =
  ABMultiValueCreateMutable(kABMultiStringPropertyType);
ABMultiValueAddValueAndLabel(multi, CFSTR("(123) 456-7654"),
                             kABPersonPhoneMobileLabel, NULL);
ABMultiValueAddValueAndLabel(multi, CFSTR("(321) 543-7890"),
                             kABPersonPhoneWorkFAXLabel, NULL);

// add the phone numbers to the person record
ABRecordSetValue(person, kABPersonPhoneProperty, multi, &error);
CFRelease(multi);
if(NULL != error) {
  NSLog(@"an error occurred");
}
```

Instead of passing the phone numbers into the person object, we first create a multivalued object and add the phone numbers to that object. Then we pass the multivalue object to the person object. Notice that we are doing only minimal error checking here; in a more sophisticated application, you'd want to do more sophisticated error checking.

The phone number multivalue object contains strings, and the address multivalue uses a dictionary. Let's look at that next. Here is the code:

AddressBook/Contacts/Classes/RootViewController.m
```
// set the home address
CFStringRef keys[4] = {kABPersonAddressStreetKey,
                       kABPersonAddressCityKey,
                       kABPersonAddressStateKey,
                       kABPersonAddressZIPKey};
CFStringRef values[4] = {CFSTR("765 Four St."),
                         CFSTR("Fivesville"),
                         CFSTR("CO"),
                         CFSTR("80424")};
CFDictionaryRef data =
  CFDictionaryCreate(NULL,
                     (void *)keys,
                     (void *)values,
                     4,
                     &kCFCopyStringDictionaryKeyCallBacks,
                     &kCFTypeDictionaryValueCallBacks);
multi = ABMultiValueCreateMutable(kABDictionaryPropertyType);
ABMultiValueAddValueAndLabel(multi, data, kABHomeLabel, NULL);
CFRelease(data);

// add the address to the person record
ABRecordSetValue(person, kABPersonAddressProperty, multi, &error);
CFRelease(multi);
```

At first glance, this code looks much more complicated than setting up the phone number object. But when you dig into it, what is really happening is about the same. Instead of creating a string object with CFSTR(), we create a dictionary. It does take more code, but conceptually it is still just one object. After we create the dictionary, we add it to a new multivalued property and then add the address dictionary data to that multivalued property. After that is all set up, we add the multivalued property to our person object with the address property. It is more code, but it's basically the same thing that we saw with the phone number.

Finally, now that we have all the data in the person object, it's time to add the person to the Address Book, save the data, and clean up.

Here is the code:

AddressBook/Contacts/Classes/RootViewController.m
```
// add the person record, save and clean up
ABAddressBookAddRecord(addressBook, person, &error);
ABAddressBookSave(addressBook, &error);
CFRelease(person);
CFRelease(addressBook);
```

Many people get confused by moving to the C API of Core Foundation when they start dealing with the Address Book; don't worry if it is confusing at first. It takes some getting used to. But with experience you will master it. If you would like a primer on the C programming language, *A Book on C* [KP88] is a great resource.

Congratulations, you have successfully added the Address Book to an application. Using the knowledge you have gained here, you can add some really great personalization features to your application.

Chapter 24

iPhone Location API

Having location available on the phone changes your experience while traveling. You probably expect that when you're in a major metropolitan area, there is a great museum close to where you are. If instead you are in the middle of a remote part of the world on a hike, there probably isn't a museum nearby, but there might be an amazing spot to take photos.

The Core Location service is part of every iPhone, so our iPhone knows where we are and can therefore help us not miss the best museums or the most amazing sites. Instead of looking in a book that has places around "famous" spots, we have a device that can say "you are here" and then find great spots near where we are.

As with the other iPhone features, it's important that you stop to think about how best to take advantage of Core Location. Once you've decided what you want to do with location, you'll find that it's surprisingly easy to use the Core Location service on the iPhone. With just a handful of lines of code, we can fire up the location-oriented hardware and start getting updates about the location of the device. Let's get started.

24.1 Knowing Where

You'll use the CLLocationManager to turn the location services off and on and to specify the level of service you need. Let's build an example application that simply displays the location as reported by Core Location.

To get started, create a new view-based project called LocationDisplay. You will also need to add the Core Location framework to the project.

> **Location Manager's Delegate**
>
> In a typical application, you'd want the delegate object to be the controller that will be interacting with whatever model you have to keep track of the application's location data. In this example, we don't have a separate model, so we are simply going to have our view controller be the delegate. In a more complex application, you might want to do something like keep a running weighted average of the location data over time (as an example). The controller that manages the weighted average of the data would be a great candidate to be the delegate.

To do that, select the target for your project, and click ⌘-[i]. At the bottom of the General tab, click the + button, and then choose the Core Location framework.

In LocationDisplayViewController.h, declare a property named locationManager that is of type CLLocationManager. You also need to add the import for the Core Location header and the declaration that this class implements the CLLocationManagerDelegate protocol.

The next step is to create a new Core Location CLLocationManager object in LocationDisplayViewController's viewDidLoad method and set the delegate to the view controller. The delegate does not have to be a view controller; we could use any object as long as it implements the delegate protocol, but it is often convenient. After we create the manager and set its delegate, we call the startUpdatingLocation method to fire up all the location-based hardware and start looking for where the iPhone is. Here is the code required to get notified of location updates:

Location/LocationDisplay/Classes/LocationDisplayViewController.m
```
- (void)viewDidLoad {
  [super viewDidLoad];
  self.locationManager = [[[CLLocationManager alloc] init] autorelease];
  self.locationManager.delegate = self;
  [self.locationManager startUpdatingLocation];
}
```

Now that we have started the location service and set the delegate, we'll receive all the location updates when the delegate method locationManager:didUpdateToLocation:fromLocation: is called. We'll implement that method in a minute, but let's first build a UI that will show the

Figure 24.1: LOCATION COORDINATES

location that the hardware has found. We want two labels and two text fields, one for latitude and one for longitude. The UI is shown in Figure 24.1.

In the locationManager:didUpdateToLocation:fromLocation: method, we will need access to these two labels, so don't forget to add two IBOutlets to the view controller's header file and link them in IB. That way, we can set the text of the text fields as we receive location updates from the location manager.

Next we need to look at the delegate method that we implement to get notification of location updates. Here is the code:

Location/LocationDisplay/Classes/LocationDisplayViewController.m
```
- (void)locationManager:(CLLocationManager *)manager
   didUpdateToLocation:(CLLocation *)newLocation
          fromLocation:(CLLocation *)oldLocation {
  latitudeTextField.text = [NSString stringWithFormat:@"%3.5f",
                            newLocation.coordinate.latitude];
  longitudeTextField.text = [NSString stringWithFormat:@"%3.5f",
                             newLocation.coordinate.longitude];
}
```

This method is called by the CLLocationManager that we started in the viewDidLoad method. After we start the location manager, it sends this method to its delegate at least once (and typically several times; more on that in a moment). In this method, we are converting the two components of the coordinate to strings and placing those strings into the text fields that we placed on our UI. If you build and run the application in the simulator, the only location you will get is from Apple's campus in Cupertino, California. A typical Mac does not support location awareness, so the simulator can't either. If you'd like to see a more accurate representation of what your location is, push the application to your device and run it.

Now that we have seen the basic operation of the Core Location services, let's go back to the viewDidLoad method and adjust the service parameters on the location service to optimize for our application.

Core Location Service Parameters

As with every other aspect of iPhone software, the location service uses power. We need to be careful with how we configure the location service to make sure that we use as little power as possible and still meet the requirements of our application.

There are two parameters on the location manager that we can use to adjust how much power is used on our behalf. The first is called desiredAccuracy, which tells the location manager how accurate we want the results it sends to our code to be. There are five options ranging from kCLLocationAccuracyBest to kCLLocationAccuracyThreeKilometers. The more accurate we need the results to be, the more power will be consumed in getting that result. So, be deliberate in the way you set this property. If your application helps people find the closest coffee shop, then you need much better accuracy than if you are helping them find the nearest national park. Make sure that you set this to the largest value your application can reasonably work with.

The next property is called distanceFilter. This property gives us control over the frequency of updates that we get from the location service. If you are providing real-time walking directions around a city, then you will need to set this to a small number (say 5 or 10 meters). If you are providing more granular information, like how far your user rode her bike, then the location updates can be more like 100 meters. Keep in mind that just like the desiredAccuracy property, more updates require more power.

> ## Joe Asks...
> ### How Does Location Work Anyway?
>
> There are three components to the location service for an iPhone; for the iPod touch, there is one. Wifi-based location services are common between the iPhone and iPod touch. Wifi-based location works by querying a global database that maps wifi router MAC addresses to a geolocation. The iPhone has two additional means to determine the device's location. The first is via cell towers. Many towers (if not all) have their geolocation and broadcast it to interested devices. Via these locations, the iPhone can triangulate its position based on the location information it gets from the towers it can connect. Finally, almost every iPhone has an integrated GPS receiver that determines its position by triangulating based on the signals from three or four GPS satellites.
>
> Of the three options for location determination, GPS uses the most power, but as clients of the API, we have no way to ask that any one of these pieces of hardware be used or not. Our interface is only through the distanceFilter and desiredAccuracy properties.

Finally, on power consumption, the best thing you can do to limit the power consumed by the location service is to turn it off. If your application needs a fix on location only rarely, then make sure to send the stopUpdatingLocation when you have a location that is accurate enough for your application. This will allow the iPhone OS to turn off all the hardware related to getting your location. The most power efficient chip is the one that is turned off.

One last point about the location manager—the user can turn off location-based services in the Settings application. You should consider checking to see whether the location services have been disabled before calling startUpdatingLocation. The iPhone OS will prompt the user with an alert asking whether he wants to turn location services back on. If your application can work with the location services turned off, then this could be irritating to the user. If, on the other hand, your application won't work with the location service turned off, then the user might want the chance to turn it back on from your app.

As always, test the user experience in all cases to make sure the app flow is the best it can be for your users.

24.2 Location Updates

Now that we have an application that is updated by the location service, let's look at the information that we can get from these updates beyond the basic latitude and longitude.

When you implement the locatonManager:didUpdateToLocation:fromLocation:, you get not only the current best known location but also the previous known location. On the first call to this method, the oldLocation is **nil**, but the rest of the calls to this method get both old and new values. The old location can be useful if you want to do a simple form of averaging of the locations. If your app needs a very accurate location value, you probably want to do at least a little bit of averaging. The location objects that you get here are the subject of the rest of this section.

Accuracy

We have seen how to tell the Core Location service what accuracy we want, but just because we are requesting a particular accuracy does not mean that we will get it. Instead, with each delivery of a location update, the service also tells us the accuracy of that location. When the system is determining the accuracy of its location fix, it takes into account the actual means used to get the fix and many other factors. We should always look at the accuracy to make sure it is within the bounds needed for our applications.

There are two accuracy fields on CLLocation objects. The first is called horizontalAccuracy and tells us how sure the system is of the coordinate field on the location object (that is, the latitude and longitude). The second is verticalAccuracy, which tells how sure the system is of the reported altitude. Both properties will report a negative number when the system is not sure at all about the accuracy.

The location service is only as good as the data it is able to get. If the MAC address of the wifi networks that the device has access to are not in the database or if the cell towers in range don't provide altitude data, then the device can't know where it is. This is where we have to be very clever in the way we interpret the data. As we have seen, the verticalAccuracy property tells us whether the altitude number can be trusted, and the horizontalAccuracy value tells us whether the coordinate can be trusted. If either of these accuracy values is negative, then the corresponding reading should not be trusted. Even when the accuracy numbers are positive, on occasion they are potentially inaccurate. Thinking through how location determination works helps us understand why. When only one cell tower can be found (for the sake of the argument, ignore GPS and wifi for now), the device can't triangulate and knows only that it's in range of the tower it can access. Even when two towers are in range, the numbers can be off by quite a bit.

It is important to keep the accuracy numbers that the location service reports in mind as you use the location information. Averaging the numbers over time will help smooth out the aberrations. There are many different approaches you could take to do the averaging; the thing to do is experiment with various approaches to see which one works best with your application.

Distance

A location object can calculate the distance between itself and another location object with a call to getDistanceFrom:. This calculation can be used in a number of applications.

Let's build an example application that calculates the user's total distance and average speed and displays these numbers on the UI. The user interface with the values displayed in labels is shown in Figure 24.2, on the next page.

The code behind this sample is very similar to what we have seen before. In the viewDidLoad, we fire up a CLLocationManager and have it start updating our location. However, we are a bit more specific on the terms of service we'd like with this implementation than we were with the last. Here is the code:

Location/Distance/Classes/DistanceViewController.m

```
- (void)viewDidLoad {
  [super viewDidLoad];
  self.locationManager = [[[CLLocationManager alloc] init] autorelease];
  self.locationManager.distanceFilter = 10.0f;
  self.locationManager.delegate = self;
  self.locationManager.desiredAccuracy = kCLLocationAccuracyNearestTenMeters;
  [self.locationManager startUpdatingLocation];
  // 32 is a guess of a 'good' number
  self.locations = [NSMutableArray arrayWithCapacity:32];
}
```

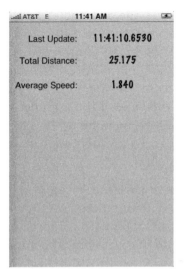

Figure 24.2: DISTANCE INTERFACE

In this method, we set the distanceFilter property to 10 meters because we want the notification to come in only for changes in distance greater than 10 meters. Then we set the desiredAccuracy to 10 meters with the kCLLocationAccuracyNearestTenMeters constant. This combination of service parameters will tell the hardware that we really only want to know to within 10 meters accuracy of where we are. We also allocate a mutable array to hold the location updates. Next up is the delegate method that processes the location updates. Here is the code:

Location/Distance/Classes/DistanceViewController.m

```
- (void)locationManager:(CLLocationManager *)manager
    didUpdateToLocation:(CLLocation *)newLocation
           fromLocation:(CLLocation *)oldLocation {
  // if we have a valid location, and its within 20 meters then stop
  // updating location, but turn it back on in 60 seconds
  if(newLocation.horizontalAccuracy > 0.0f &&
     newLocation.horizontalAccuracy < 20.0f) {
    if(self.locations.count > 3) {
      [self.locationManager stopUpdatingLocation];
      [self.locationManager performSelector:@selector(startUpdatingLocation)
                         withObject:nil
                         afterDelay:60.0f];
    }
```

```
        [self.locations addObject:newLocation];
        [self updateDisplay];
    }
}
```

In this method, we are looking for a valid accuracy (that is, the horizontalAccuracy is greater than zero) and that the accuracy is better than 20.0 meters. Keeping things in this range will ensure that we keep only valid data in our locations array. Next we make sure to capture at least three valid locations and then turn off the location manager. We then queue it up for checking again in sixty seconds. Grabbing three valid locations before turning off the service ensures that we have a valid set of data to do our calculations. Turning off the location manager for sixty seconds after we get a valid location will minimize our impact on power usage but still give us good enough results for our application. Finally, we update the user interface with the data so that our users can see how far they have moved. Here is the code for that method:

Location/Distance/Classes/DistanceViewController.m

```
- (void)updateDisplay {
    CLLocationDistance distance = [self totalDistanceTraveled];
    totalDistance.text = [NSString stringWithFormat:@"%5.3f", distance];

    NSTimeInterval time = [self timeDelta];
    // don't want to divide by zero
    if(time == 0.0f) {
        averageSpeed.text = @"0.000";
    } else {
        averageSpeed.text = [NSString stringWithFormat:@"%5.3f", distance / time];
    }
    NSDateFormatter *inputFormatter = [[[NSDateFormatter alloc] init] autorelease];
    [inputFormatter setDateFormat:@"HH:mm:ss.SSSS"];
    NSDate *date = [(CLLocation *)[self.locations lastObject] timestamp];
    lastUpdate.text = [inputFormatter stringFromDate:date];
}
```

In this method, we are just performing the calculations necessary to find the total distance traveled and the time taken to travel that distance and then using these two values to calculate the average speed. The timestamp label is also updated. Don't forget that in order to make all this work with the default view controller you get from the template, you need to add IBOutlets to the UILabels and configure the UI in Interface Builder.

Now let's look at how the total distance traveled is calculated. Here is the code:

Location/Distance/Classes/DistanceViewController.m
```
- (CLLocationDistance)totalDistanceTraveled {
  CGFloat totalDistanceTraveled = 0.0f;
  CLLocation *oldLocation = nil;
  for(CLLocation *location in self.locations) {
    if(nil == oldLocation) {
      oldLocation = location;
      continue;
    }
    totalDistanceTraveled += fabs([location getDistanceFrom:oldLocation]);
    oldLocation = location;
  }
  return totalDistanceTraveled;
}
```

In this method, we are summing the differences between the previous location and the new location. We take the absolute value because we count going forward and backward while still traveling. If we did not take the absolute value, we could show a zero amount traveled if the user went 1 kilometer north and then 1 kilometer south.

Finally, here is the calculation for the time delta:

Location/Distance/Classes/DistanceViewController.m
```
- (NSTimeInterval)timeDelta {
  NSDate *first = [((CLLocation *)[self.locations objectAtIndex:0] timestamp];
  NSDate *last = [((CLLocation *)[self.locations lastObject] timestamp];
  return [last timeIntervalSince1970] - [first timeIntervalSince1970];
}
```

This method simply calculates the time difference between the first location object and the last.

24.3 Compass

With iPhone OS 3.0, Apple added support for a magnetometer, which is basically a fancy word for a digital compass. The new features require one of the new iPhone 3Gs devices to work, but the great part is you really don't have a lot more to learn here. As the delegate of your location manager, you can implement the locationManager:didUpdate-Heading: method to receive updated heading information. All you need to do is tell your location manager to turn on the compass by sending the startUpdatingHeading.

Let's add the heading to our display of location manager information. Open the LocationDisplayViewController.xib file, and add a label and text field to the view. Change the label's text to Heading:. In the LocationDisplayViewController.h file, add an IBOutlet to point to the new text field, and call it headingTextField. Add the @synthesize statement to the implementation, and save your work. Now go back to Interface Builder, and connect the new outlet to the new text field. Now that we have the connections done, we can turn on the compass. Here is the code from the location display example with the additional code to turn on the compass on line 6:

Location/LocationDisplay02/Classes/LocationDisplayViewController.m

```
- (void)viewDidLoad {
    [super viewDidLoad];
    self.locationManager = [[[CLLocationManager alloc] init] autorelease];
    self.locationManager.delegate = self;
    [self.locationManager startUpdatingLocation];
    [self.locationManager startUpdatingHeading];
}
```

Simply calling the startUpdatingHeading turns on the compass and will cause the location manager to inform us with each heading update by calling the locationManager:didUpdateHeading: method. When we get this call, we are going to update the text field in the delegate method. Here is the code:

Location/LocationDisplay02/Classes/LocationDisplayViewController.m

```
- (void)locationManager:(CLLocationManager *)manager
        didUpdateHeading:(CLHeading *)newHeading {
    headingTextField.text = [NSString stringWithFormat:@"%3.2f",
                             newHeading.trueHeading];
}
```

In your own application, you will likely want to do something much more sophisticated with your user interface. For example, you might want to draw a view differently depending on which direction the user is facing.

The magnetometer occasionally needs to be calibrated. When the location manager discovers that a calibration step is necessary, it will send the delegate a locationManagerShouldDisplayHeadingCalibration:. If you return YES, a panel pops up and asks the user to move the device around in a figure-eight pattern. If you return NO, the panel is not displayed. Keep in mind that if you don't allow the calibration step to take place, all the heading updates will reflect uncalibrated results.

Another thing to keep in mind when using the compass feature is that only some iPhone OS devices support this feature. To discover whether the current device has a compass, you can get the headingAvailable property from your location manager. If the value is YES, the device has a compass and NO otherwise. You should check this property before you try to turn on heading updates.

As you can see, there are a lot of options in how to use the location manager and the Core Location service. It is vital that any application that we write with the location service is careful to be a good citizen and not consume more power than is necessary.

Chapter 25

Map Kit

Location services are fantastic and can add some cool features to our apps. Once your app knows where your user's device is, the next logical step is often to place that location on a map. That is where Map Kit comes in. With Map Kit, we can place locations onto a map and then present that map to our users. Map Kit provides the tiles (the images used in the map) and all the zooming and panning features.

Let's get started building an application to help us explore what is possible with the Map Kit.

25.1 Contact Mapper

Contact Mapper is an application to place selected contacts onto a map and show them relative to where you are. After starting up, the map shows your location with the nice blue button that is typical of the Maps application. You can then hit the Choose button to get to your contacts. After selecting a contact, a pin is placed at the contact's address, and then the map zooms to show both your location and the location of your contact. On the next page is a screenshot of the application running in the simulator with a single contact added.

When the user clicks the pin, the callout is shown (Kate Bell in the screenshot). Clicking the blue chevron brings the user to the contact's editing page. The blue dot shows the device's current location (in the simulator that defaults to Cupertino, California).

Map Kit provides a whole host of great features for this application. The user can pinch to zoom, move a finger around on the screen to move the map, and even flick to move a long way on the map. All the tiles from Google Maps are loaded automatically and zoomed/scaled as appropriate. Map Kit makes building this kind of application a breeze.

25.2 Showing a Map

Let's get started building ContactMapper with a new view-based application. We are not going to go step-by-step; there is too much Map Kit–specific stuff to cover. So, if you need a nudge to get the application working, grab the code from the book's code bundle under the MapKit folder.

In this first version of the application, we'll get the map to appear and to show the current location. To use Map Kit, we need to add the Map Kit.framework and CoreLocation.framework frameworks to the project.[1]

Open the ContactMapperViewController.xib file so we can work on the interface. If the view is not already open, double-click it. Open the library (⌘-⇧-L), and choose Map View. It's in the Data Views group. Here is a screenshot of the library with Map View selected:

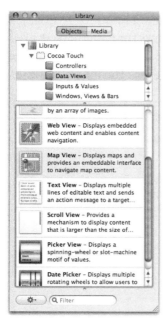

Drag the map view out onto the view. With the map view selected, open the Attributes inspector (⌘-1). Specify that the map should show the user's current location by turning on the switch box. We have the map type set to Map; the other choices are Satellite and Hybrid. In a typical app, you should provide a way for the user to switch between the three types unless it does not make sense for your user to switch. We are finished with IB for the moment, so save your work and hide (⌘-H) it.

In fact, we now have a map-based application. If you Build and Go, the application will appear and show the blue dot for where you are. Once loaded, the map can be zoomed in or out panned around—all the stuff you'd expect from a map on your phone.

---

1. To add frameworks, select the ContactMapper target (in the Targets group), right-click it, and choose Get Info. In the General tab, hit the + button at the bottom of the screen, choose the two frameworks, and click the Add button.

But we don't want to stop here. When the current location is found, we want the map to recenter over that location and zoom in so that the features/map is recognizable. The map view method setRegion:animated: does exactly what we want. The region parameter is a C structure, much like the CGRect we first discussed in the Chapter 19, *Drawing in Custom Views*, on page 385. The MKCoordinateRegion has two parts, the center and the span. The center is a CLLocationCoordinate2D with the latitude and longitude of the center point of the region. The span is an MKCoordinateSpan and specifies the change in longitude and latitude degrees that the region should encompass. Keep in mind that the actual area defined by the span will vary depending on how far you are from the equator; the further north you go, the closer together the longitude lines become.

When the user's current location is updated, we want to recenter the map over that location and change the span to about 0.15 degrees in each direction.

Recentering the map is a snap; we just create the new MKCoordinateRegion and tell the map view what its region is. Here is the code to do that:

MapKit/ContactMapper_01/Classes/ContactMapperViewController.m

```
- (void)setCurrentLocation:(CLLocation *)location {
  MKCoordinateRegion region = {{0.0f, 0.0f}, {0.0f, 0.0f}};
  region.center = location.coordinate;
  region.span.longitudeDelta = 0.15f;
  region.span.latitudeDelta = 0.15f;
  [self.mapView setRegion:region animated:YES];
}
```

We initialize the region to contain all zeros. Then we set the center and span of our region and tell the view to make that its region. Setting the animation flag to YES as you'd expect causes the map to animate to the new region. If you set it to NO, the view will jump to the new region. To make this work, we also need to have a pointer to the map view, so you guessed it—we need to add an IBOutlet to our interface and make the connection in IB.

The interesting part of this feature is knowing when the update should occur. Well, you already know at least one way to accomplish this. Our view controller can become the delegate of a location manager and recenter the map each time an update comes in; for more detail, see Chapter 24, *iPhone Location API*, on page 451. But it turns out that the map view already does that for us. When we turn on the Shows User

Location switch in IB or set the showsUserLocation property to YES, the map view will turn on the location manager and get updated with the current location. All we need to do is implement the proper map view delegate method. We will look at that in just a minute.

25.3 Map Annotations

There are two parts to each annotation on the map. The model piece of it is intended to be lightweight, so you can have many of them attached to a given map and not have to worry about the memory footprint. The other side is the view. Let's look at the model side first.

Map Kit defines the MKAnnotation protocol but no public implementations. So, in order to add an annotation to the map, we need to create our own implementation of this protocol. The protocol defines one property and two optional methods. The property is the location for the annotation; the methods are title and subtitle. The title is used as the text for the callout when the user clicks the annotation. The subtitle is displayed in smaller text under the title. Here is the header file for our annotation:

MapKit/ContactMapper_01/Classes/ContactAnnotation.h

```
#import <Foundation/Foundation.h>
#import <CoreLocation/CoreLocation.h>
#import <MapKit/MapKit.h>
#import <AddressBook/AddressBook.h>

@interface ContactAnnotation : NSObject <MKAnnotation> {
  CLLocationCoordinate2D _coordinate;
  NSString *_title;
  NSString *_subtitle;
  ABRecordRef _person;
}

+ (id)annotationWithCoordinate:(CLLocationCoordinate2D)coordinate;
- (id)initWithCoordinate:(CLLocationCoordinate2D)coordinate;

@property (nonatomic, assign) CLLocationCoordinate2D coordinate;
@property (nonatomic, assign) ABRecordRef person;
@property (nonatomic, copy) NSString *title;
@property (nonatomic, copy) NSString *subtitle;

@end
```

Since our annotation represents one of our contacts, we add a property named person for the contact. We also defined the title and subtitle properties instead of just defining the methods defined in the protocol. Finally, we have the initWithCoordinate: and annotationWithCoordinate: methods that do as you'd expect, initialize a ContactAnnotation, and create and return an autoreleased annotation, respectively. Here is the implementation for the ContactAnnotation:

MapKit/ContactMapper_01/Classes/ContactAnnotation.m

```
@implementation ContactAnnotation

@synthesize coordinate = _coordinate;
@synthesize title = _title;
@synthesize subtitle = _subtitle;
@synthesize person = _person;

+ (id)annotationWithCoordinate:(CLLocationCoordinate2D)coordinate {
  return [[[[self class] alloc] initWithCoordinate:coordinate] autorelease];
}

- (id)initWithCoordinate:(CLLocationCoordinate2D)coordinate {
  self = [super init];
  if(nil != self) {
    self.coordinate = coordinate;
  }
  return self;
}

@end
```

Now that we have our annotation, we need to look at how we are going to create and add the annotation to the map. Thinking back to the way we want the application to work, when the user clicks the Choose button, we want to bring up the people picker. After a contact's address is chosen, we want to add the annotation. To make that happen, we have to do a couple of steps:

1. Add the AddressBook and AddressBookUI frameworks to the project.

2. Add an action method to the view controller, and implement it to present the person chooser.

3. Add a toolbar to the bottom of the view, and resize the map view to fit.

4. Connect the Choose button to the action method.

5. Implement the people picker delegate protocol to get the address when it's clicked.

Let's look at the action method. Here is the code:

MapKit/ContactMapper_01/Classes/ContactMapperViewController.m

```
- (IBAction)choose {
  ABPeoplePickerNavigationController *picker =
  [[ABPeoplePickerNavigationController alloc] init];
  picker.peoplePickerDelegate = self;
  [self presentModalViewController:picker animated:YES];
  [picker release];
}
```

Now that we have the implementation, we need to add the declaration in the header file and then go back to IB to add the toolbar and make the connection from the button to the File's Owner.

For more detail on the people picker, see Chapter 23, *Address Book*, on page 441. Now that we have become the delegate of the people picker, we need to add that protocol to the header file and implement the methods. Of the three methods, we are going to look only at one, the peoplePickerNavigationController:shouldContinueAfterSelectingPerson:property: identifier: method. Here is the code:

MapKit/ContactMapper_01/Classes/ContactMapperViewController.m

```
Line 1  - (BOOL)peoplePickerNavigationController:
     -  (ABPeoplePickerNavigationController *)peoplePicker
     -      shouldContinueAfterSelectingPerson:(ABRecordRef)person
     -                      property:(ABPropertyID)property
     5                      identifier:(ABMultiValueIdentifier)identifier{
     -    if(kABPersonAddressProperty == property) {
     -      NSString *fullName = (NSString *)ABRecordCopyCompositeName(person);
     -      CLLocationCoordinate2D coordinate = {0.0f, 0.0f};
     -      self.newAnnotation = [ContactAnnotation annotationWithCoordinate:coordinate];
    10      self.newAnnotation.title = fullName;
     -      self.newAnnotation.person = person;
     -      [fullName release];
     -      ABMultiValueRef addresses =
     -              ABRecordCopyValue(person, kABPersonAddressProperty);
    15      CFIndex selectedAddressIndex =
     -              ABMultiValueGetIndexForIdentifier(addresses, identifier);
     -      CFDictionaryRef address =
     -              ABMultiValueCopyValueAtIndex(addresses, selectedAddressIndex);
     -      self.newAnnotation.coordinate = [AddressGeocoder locationOfAddress:address];
    20      [self dismissModalViewControllerAnimated:YES];
     -    }
     -    return NO;
     -  }
```

If the contact's address was chosen, then we grab the contact's full name, create an annotation, set the fullName to be the title, and set the annotation's person to be the selected contact. We also grab the address starting on line 13. We then geocode the address on line 19. We will look at the geocoding stuff in a moment (in Section 25.3, *TouchXML and the AddressGeocoder*, on page 472). After the user has chosen an address, we dismiss the person picker view.

Now our work is almost complete. Although we have created the new annotation, we have not added it to the map. We could do that in the people picker delegate method. However, we will get a much nicer animation if we place the code in the viewDidAppear: name. Here is the code:

MapKit/ContactMapper_01/Classes/ContactMapperViewController.m

```
- (void)viewDidAppear:(BOOL)animated {
  [super viewDidAppear:animated];
  if(nil != self.newAnnotation) {
    [self.mapView addAnnotation:self.newAnnotation];
    self.newAnnotation = nil;
  }
  if(self.mapView.annotations.count > 1) {
    [self recenterMap];
  }
}
```

First we check to see whether we have a new annotation to add, and if so, we add it and then set the newAnnotation to nil. If we have more than one annotation, we also recenter the map by calling the recenterMap method. Recentering the map is straightforward; we build a new region centered between the annotations with a span that covers all the annotations. Here is the code:

MapKit/ContactMapper_01/Classes/ContactMapperViewController.m

```
- (void)recenterMap {
  NSArray *coordinates = [self.mapView valueForKeyPath:@"annotations.coordinate"];
  CLLocationCoordinate2D maxCoord = {-90.0f, -180.0f};
  CLLocationCoordinate2D minCoord = {90.0f, 180.0f};
  for(NSValue *value in coordinates) {
    CLLocationCoordinate2D coord = {0.0f, 0.0f};
    [value getValue:&coord];
    if(coord.longitude > maxCoord.longitude) {
      maxCoord.longitude = coord.longitude;
    }
    if(coord.latitude > maxCoord.latitude) {
      maxCoord.latitude = coord.latitude;
    }
    if(coord.longitude < minCoord.longitude) {
      minCoord.longitude = coord.longitude;
    }
```

```
      if(coord.latitude < minCoord.latitude) {
        minCoord.latitude = coord.latitude;
      }
    }
    MKCoordinateRegion region = {{0.0f, 0.0f}, {0.0f, 0.0f}};
    region.center.longitude = (minCoord.longitude + maxCoord.longitude) / 2.0;
    region.center.latitude = (minCoord.latitude + maxCoord.latitude) / 2.0;
    region.span.longitudeDelta = maxCoord.longitude - minCoord.longitude;
    region.span.latitudeDelta = maxCoord.latitude - minCoord.latitude;
    [self.mapView setRegion:region animated:YES];
}
```

After identifying the minimum and maximum latitude and longitude, we create a new region based on this min and max and then tell the map view to set its region to this new region.

Now we have an application that does most of what we want. Build and Go, and choose one of your contacts to see the pin drop and the map resize to fit your current location and the location of the chosen contact.

However, our pin is red, and we really wanted it to be purple. The default color is red, and since we have not done anything special, that is what we get. Although that works, we want a purple pin. To get a purple pin, we need to implement the MKMapViewDelegate protocol.

There are several methods in the MKMapViewDelegate protocol; most of them are callbacks to let you know when interesting stuff is happening with the map view (that is, mapViewWillStartLoadingMap:). The one we are interested in now, though, is the mapView:viewForAnnotation: method. Like the table view, the map view keeps a set of annotation views that can be dequeued and reused. Our implementation of this method needs to take that into account and make sure to use the dequeued view when possible. Here is the code:

MapKit/ContactMapper_01/Classes/ContactMapperViewController.m

```
- (MKAnnotationView *)mapView:(MKMapView *)mapView
           viewForAnnotation:(id <MKAnnotation>)annotation {
  MKPinAnnotationView *view = nil;
  if(annotation != mapView.userLocation) {
    view = (MKPinAnnotationView *)
        [mapView dequeueReusableAnnotationViewWithIdentifier:@"identifier"];
    if(nil == view) {
      view = [[[MKPinAnnotationView alloc]
              initWithAnnotation:annotation reuseIdentifier:@"identifier"]
            autorelease];
    }
    [view setPinColor:MKPinAnnotationColorPurple];
```

```
      [view setCanShowCallout:YES];
      [view setAnimatesDrop:YES];
  } else {
    CLLocation *location = [[CLLocation alloc]
                            initWithLatitude:annotation.coordinate.latitude
                            longitude:annotation.coordinate.longitude];
    [self setCurrentLocation:location];
  }
  return view;
}
```

The first thing we do is check whether the current annotation is the current user location. We could return our own annotation view for the current user location, and it would work; however, if we return **nil**, then the default annotation view is used. The default annotation view for the current user's location is the blue dot that has the nice bouncing animation.

If this is not the user's current location, we dequeue an annotation view, and if there is not one to dequeue, we create one. Then we set up the annotation view the way we want it and then return the view. In our case, we are setting the pin color to purple, turning on callouts and specifying that the annotation should drop in with an animation.

If it is the user's current location, we call the setCurrentLocation:. This is where we get the Map Kit's integration with a location manager without having to implement the delegation methods ourselves.

TouchXML and the AddressGeocoder

Now let's talk about the AddressGeocoder. The geocoder uses a web service to encode the addresses. We pass in an address and out pops XML with the address and the latitude and longitude. As you saw back in the Chapter 12, *Connecting to the Internet*, on page 235, we can use the NSXMLParser to parse XML, and although that works, it can be quite tedious to parse the whole XML file when you really want only a couple of bits of information. To get at the data we want, we have a couple of options. The iPhone includes the libXML2 open source library, which works like a champ. But the API is entirely C-based, which makes it a bit of a cognitive disconnect when you'd rather be doing Objective-C. Enter the TouchXML open source library.

TouchXML provides an Objective-C wrapper over the libXML2 library. The project is located on the Google Code website at http://code.google.com/p/touchcode/wiki/TouchXML. Version 1.0.6 is included in the code for this chapter, so you don't have to download it. The driver behind

TouchXML is to provide the simplified XML APIs that are available on the Mac but not on the iPhone. So, if you want detailed documentation about one of the classes, you can replace the leading C with NS and look in the Mac documentation for details.

TouchXML provides us with XPath, which requires far less code to extract the information we need than doing the same with the parser. To get TouchXML working with your project, you need to do a couple of things:

- Get the code, either from the projects for this chapter or from the website.
- Add the source code for TouchXML and Tidy to your project.
- Add the libxml2 library to your linked libraries.
- Add the libxml2 header path to your header search path.

The TouchXML code includes the code for Tidy, so if you grabbed the code from TouchXML's website or from the code bundle for the book, you have the Tidy code. Let's look at adding the code to your project so you can use it instead of the parser.

First we need to add a group for TouchXML and then add a group under TouchXML for Tidy. Select the project, Ctrl+click, and choose Add > New Group. Name the group TouchXML. Select the group, right-click, and choose Get Info. On the General tab, click the Choose button near the Path item. Create a new folder called TouchXML, select it, and hit the Choose button. Repeat the steps for the Tidy group, select the TouchXML group, right-click, choose Add > New Group, name it Tidy, and specify its path is the Tidy folder. Here is a screenshot of the TouchXML folder in our project:

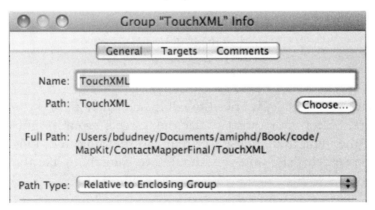

Depending on where you got the code, the directory structure will be different. If you are grabbing the code from the book's source bundle, then everything is under the TouchXML directory and the Tidy directory under that. If you are grabbing the source from the TouchXML project site, the TouchXML code is under Common/Source and Common/Source/Creation. The Tidy source is under Externals/tidy/src and Externals/tidy/include. You need all the files from both TouchXML and Tidy.

In the Finder, select all the files for TouchXML, and drag them into the TouchXML group in Xcode. When prompted, make sure to select the "copy files" checkbox. Do the same for the Tidy files, dragging them into the Tidy group, and make sure to copy them as well. When you are done, Xcode should look roughly like this:

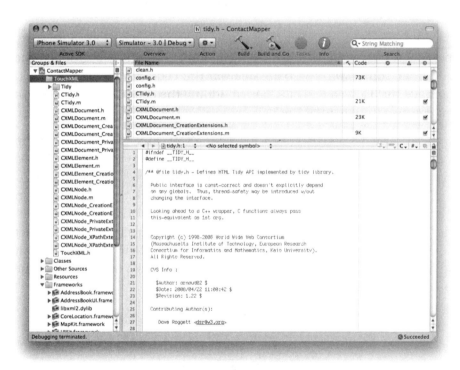

Now that we have all the code, we need to tell Xcode where to find the libXML2 headers. Select your target under the Targets group. Right-click or ^-click the item, and choose Get Info. On the Build tab, type header s into the search field, and you should see something like the next screenshot.

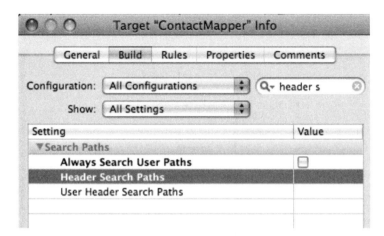

Double-click the line that reads Header Search Paths. In the sheet that pops up, click the + button, and type /usr/include/libxml2 into the Path column. When you are done, it should look like this:

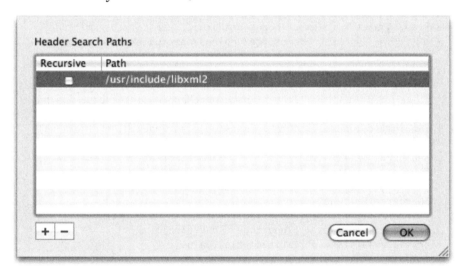

Click the OK button, and you should be able to build with TouchXML and Tidy as part of your project.

There are other ways to approach this; we could instead make a static library from the TouchXML and Tidy source, but that is beyond what we will cover.

Now that we have TouchXML set up and running, let's take a look at the AddressGeocoder class and how it uses TouchXML to get an address back from one of the geocoding services.

Here is part of the implementation:

MapKit/ContactMapper_02/Classes/AddressGeocoder.m

```
Line 1  + (BOOL)geocodeStreetAddress:(NSString *)street
                              city:(NSString *)city
                             state:(NSString *)state
                               zip:(NSString *)zip
     5                     country:(NSString *)country
                      intoLocation:(CLLocationCoordinate2D *)location {
            BOOL success = NO;
            NSURL *url = [self urlForAddress:street city:city state:state
                                         zip:zip country:country];
    10      NSError *error = nil;
            CXMLDocument *doc = [[CXMLDocument alloc]
                                 initWithContentsOfURL:url
                                 options:CXMLDocumentTidyXML error:&error];
            CXMLElement *element = [doc rootElement];
    15      NSDictionary *namespaceMappings =
                [NSDictionary dictionaryWithObject:@"http://earth.google.com/kml/2.0"
                                            forKey:@"kml"];
            NSArray *status = [element nodesForXPath:@"//kml:Status/kml:code"
                                    namespaceMappings:namespaceMappings
    20                                          error:&error];
            if([@"200" isEqualToString:[[status objectAtIndex:0] stringValue]]) {
                NSArray *coordElements = [element nodesForXPath:@"//kml:coordinates"
                                               namespaceMappings:namespaceMappings
                                                           error:&error];
    25          NSString *coords = [[coordElements objectAtIndex:0] stringValue];
                NSArray *components = [coords componentsSeparatedByString:@","];
                NSNumberFormatter *formatter = [[NSNumberFormatter alloc] init];
                NSNumber *longitude =
                    [formatter numberFromString:[components objectAtIndex:0]];
    30          NSNumber *latitude =
                    [formatter numberFromString:[components objectAtIndex:1]];
                location->longitude = [longitude floatValue];
                location->latitude = [latitude floatValue];
                [formatter release];
    35          success = YES;
            }
            return success;
        }
```

On line 9, we turn the address into a URL suitable for the geocoding service we are using. This URL will of course be service specific. Next we grab the XML from the Web and turn it into a CXMLDocument on line 13. Behind the scenes, TouchXML takes the URL, downloads the whole XML document, parses it, and turns it into a DOM tree for us to use. Downloading the XML on the main thread might not be the best option for your application, so evaluate carefully if you need to place this into a background thread. For this example, we are not going to go into placing our XML parsing and downloading into a background thread. Next we set up the name spaces used by the XML document. We then get and check the status to make sure that we got a valid response from our server on line 20. Then if everything has gone as planned, we have a valid response, and we turn that into a latitude and longitude starting on line 24.

This is not all the code involved in the AddressGeocoder; for the complete class, make sure to download the code bundle and add the class to the project.

Our application is nearly finished. We can select contacts from the people picker and have them placed on a map. The last feature we need to implement is seeing the details about a contact when we click one of the pins. Let's look at that now.

25.4 Selecting an Annotation

When the user clicks one of the annotations on the map, the map view displays the callout view. The callout can have two accessory views. Although we can use any subclass of UIView as the accessory view for our annotation, it is typical to use a UIControl. If you use a control, the map view will take care of handling the events for you. If you write your own view, then you will have to deal with the events yourself.

In the Maps application, the accessory view used is a UIButton with its type set to UIButtonTypeDetailDisclosure. Since users are so used to that paradigm, we are going to do the same in our application. Here is the modified code that builds the annotation view:

MapKit/ContactMapper_02/Classes/ContactMapperViewController.m

```
- (MKAnnotationView *)mapView:(MKMapView *)mapView
           viewForAnnotation:(id <MKAnnotation>)annotation {
   MKPinAnnotationView *view = nil;
   if(annotation != mapView.userLocation) {
     view = (MKPinAnnotationView *)
        [mapView dequeueReusableAnnotationViewWithIdentifier:@"identifier"];
     if(nil == view) {
       view = [[[MKPinAnnotationView alloc]
                initWithAnnotation:annotation reuseIdentifier:@"identifier"]
               autorelease];
       view.rightCalloutAccessoryView =
       [UIButton buttonWithType:UIButtonTypeDetailDisclosure];
     }
```

```
             [view setPinColor:MKPinAnnotationColorPurple];
15           [view setCanShowCallout:YES];
             [view setAnimatesDrop:YES];
         } else {
             CLLocation *location = [[CLLocation alloc]
                                     initWithLatitude:annotation.coordinate.latitude
20                                   longitude:annotation.coordinate.longitude];
             [self setCurrentLocation:location];
             [location release];
         }
         return view;
25   }
```

On line 11, we set the right callout accessory view to be a button. Notice that we don't have to set up the target/action stuff for the button. The map view takes care of it for us and calls the mapView:annotationView: calloutAccessoryControlTapped: method when the accessory button is clicked.

This callback is similar in many ways to the method that is invoked by the table view when a row is clicked. The map view calls us when the user taps on the control to let us know so we can do whatever the user expects. Here is the implementation:

MapKit/ContactMapper_02/Classes/ContactMapperViewController.m
```
- (void)mapView:(MKMapView *)mapView
  annotationView:(MKAnnotationView *)view
calloutAccessoryControlTapped:(UIControl *)control {
    ContactAnnotation *ann = (ContactAnnotation *)view.annotation;
    ABPersonViewController *personVC =
    [[[ABPersonViewController alloc] init] autorelease];
    UINavigationController *nav =
    [[[UINavigationController alloc] initWithRootViewController:personVC]
     autorelease];

    personVC.navigationItem.leftBarButtonItem =
    [[[UIBarButtonItem alloc]
      initWithBarButtonSystemItem:UIBarButtonSystemItemStop
                           target:self
                           action:@selector(stopEditingPerson)]
     autorelease];

    personVC.displayedPerson = ann.person;
    personVC.personViewDelegate = self;
    [self presentModalViewController:nav animated:YES];
}
```

We create the person view controller, set it up in a nav controller so we can add buttons to the nav bar, and then set the person that we want displayed. Just before displaying the person view, we set our selves as the delegate. In our implementation of the person viewer delegate method, we simply return NO. But in your application, it might make sense to allow the user to dial a number or send an email from this view.

The Map Kit has some great functionality that gives our applications something far more interesting than what we could achieve with the location manager alone.

Chapter 26

Application Integration

Integrating your application with others on the system adds a level of sophistication that your users are sure to appreciate. For example, if your application displays phone numbers, your users will be very grateful if you allow them to click the phone number and have the phone application launched.

In this chapter, we will see how to get our applications to launch the others on the system. The actual code to do so is minimal; the interesting parts of the integration come down to what you want the user experience to be.

We'll also show you how to allow other applications to launch your application in the same way that Apple has made it possible for us to launch Mail and Safari. We will see how in the last half of this chapter. Let's start with an application that will launch Safari and search for a term with Google.

26.1 Launching Other Applications

Applications that support URLs can be launched with a call to the UIApplication's openURL:. As long as the URL is well formed and the application is properly registered, the iPhone OS will take care of the rest. Let's look at how to launch Safari with this mechanism.

Let's create a new project, add the UI to let us enter a search term, and add the code that will launch Safari and run a search at Google. Create a new view-based project called Searcher. Open the SearcherViewController.xib file, and add a text field to the view. Set its placeholder text to "query term" in the Attributes inspector (⌘-1). Then set the text field's delegate to be the file's owner (Ctrl+drag from the text field to the File's Owner object). We are finished with the UI, so you can quit Interface Builder if you'd like.

Now we need to implement the UITextFieldDelegate protocol on our SearcherViewController class so that we can capture the search term from the text field. (For more information on the delegate protocol, see Section 4.7, *Configuring the Text Field*, on page 70.) Open the SearcherViewController.h, and add the protocol declaration; then open the SearcherViewController.m file, and add an implementation of the textFieldShouldReturn: that looks like this:

AppIntegration/Searcher/Classes/SearcherViewController.m
```
- (BOOL)textFieldShouldReturn:(UITextField *)field {
  [field resignFirstResponder];
      return YES;
}
```

As we saw in Section 4.7, *Configuring the Text Field*, on page 70, the textFieldShouldReturn: is a great place to resign first responder status so the keyboard animates out. The next method that we implement is textFieldDidEndEditing:. Here is the code for that method:

AppIntegration/Searcher/Classes/SearcherViewController.m
```
- (void)textFieldDidEndEditing:(UITextField *)field {
  NSString *query = @"iPhone";
  if(field.text != nil && [field.text length]) {
    query = field.text;
  }
  NSString *urlString = [NSString stringWithFormat:@"http://google.com?q=%@",
                        query];
  NSURL *url = [NSURL URLWithString:urlString];
  [[UIApplication sharedApplication] openURL:url];
}
```

If there is a query term, we grab it; otherwise, we default to searching for iPhone. Once we have the term, we construct a URL with it. Then we get our application object via the sharedApplication method and ask our application to open the URL with openURL:.

When the user taps the return button after typing in a term, Safari will be launched and will open the URL and thus display search results.

Although it is great to be able to reuse these other applications via this mechanism, it does cause your application to quit, so you should consider the implications of that closely as you take advantage of this feature. If the launch to a new app is a logical stopping point in your application and the user is unlikely to want to return to your app unless they are starting a new cycle, then this type of integration makes great sense. If the user is going to want to come back to your application after a short time and continue where they left off, you will need to take that into account with the way your application launches[1] to make sure the experience is what your users want and expect.

In this example, we have seen HTTP links. However, Cocoa Touch supports several other schemes. Phone uses the tel: scheme, and SMS uses the sms:. Unlike the other applications, Maps, YouTube, and the AppStore use specially constructed http: URLs.

All the applications that can be launched from a URL will be launched in the same way. The only difference is in the construction of the URL. You can read the details about the various URL schemes that are supported in the iPhone docs by doing a full-text search for *URL Scheme Reference*.

Now that we have seen how to launch an application with a URL, let's go see how we can make one of our applications launchable.

26.2 Becoming Integration Ready

In this section, we are going to see what needs to be done to make your application launch with a URL. There are just two steps:

1. Add an entry to your Info.plist file.

2. Implement the application:handleOpenURL: method.

And that is it. As you can see, getting your application integration ready is straightforward. Let's look at an example that can be launched via a URL and uses the resource specifier of the URL to update a label on the UI.

Start another view-based project called Integrated. Open the IntegratedViewController.h and add two instance variables, one NSString named message and an IBOutlet called messageLabel of type UILabel. Also add a property for message so we can set it from the app delegate (which we will update shortly).

---

1. For example, you could save the state of where your application is when the other app is launched and then restore that location when the user returns.

Our header looks like this:

`AppIntegration/Integrated/Classes/IntegratedViewController.h`

```
@interface IntegratedViewController : UIViewController {
  NSString *message;
  IBOutlet UILabel *messageLabel;
}

@property(nonatomic, retain) NSString *message;

@end
```

Now let's go update the UI and make our connections. Open the IntegratedViewController.xib file. Add a UILabel to the view, and set its text to default. Connect the IntegratedViewController's messageLabel outlet to the newly added UILabel (the IntegratedViewController is the File's Owner). We are done with the UI, so you can quit Interface Builder if you'd like.

Next we need to implement the setMessage: method in the IntegratedViewController so that whenever the message is set, the label also gets updated. Here is the code:

`AppIntegration/Integrated/Classes/IntegratedViewController.m`

```
- (void)setMessage:(NSString *)newMessage {
  [message release];
  message = [newMessage retain];
  messageLabel.text = message;
}
```

In this method, we are making sure to honor the retain attribute we set in the property declaration by releasing the old message and retaining the new message. Then, after doing the housekeeping, we update the messageLabel. That is all we need to do here.

Next up we need to implement the application:handleOpenURL: method. The implementation of this method can be very simple or very complex depending of course on the needs of your application. However, there are some things that you should be very deliberate in thinking through:

- The URL is a public API; it's difficult to change once others start to use it.

- What malicious use could be made of the URL? You must make sure to guard against the attacks you have thought of and the ones you have not.

- Limiting the complexity (and thus flexibility) of the URL will make it easier for others to integrate.

For our example, we are going to do only minimal work on checking the URL to make sure it is not malicious (not that the app needs that; it's just for illustration). And since this is just an illustrative application, there is no real complexity to consider, so our URL scheme/ implementation is very simple. We choose integrated:// as our scheme and will take the resource specifier (the part after :// in the URL) as the message. Here is the implementation of application:handleOpenURL::

AppIntegration/Integrated/Classes/IntegratedAppDelegate.m
```
- (BOOL)application:(UIApplication *)application handleOpenURL:(NSURL *)url {
  BOOL flag = NO;
  if([[url resourceSpecifier] length] < 25) {
    viewController.message = [[url resourceSpecifier] substringFromIndex:2];
    flag = YES;
  }
  // if the message is longer than 25 it might be malicious
  return flag;
}
```

In this code, we are first checking to see whether the resource specifier is less than 25 characters, and if so, we use it; otherwise, we assume it might be malicious and ignore it. If the URL meets our approval, then we grab the resource specifier and pass that along to the viewController's message property, which updates the UI. We are almost there; there's one last thing to take care of, and that is letting the iPhone OS know that our application understands integrated URLs.

To register our application's ability to open integrated URLs, we need to add a CFBundleURLName array key to our Info.plist. The plist file format is well documented in the iPhone docs, so we are not going to go into any of the detail here. This is the code you'd add to the bottom of the file:

```
<key>CFBundleURLTypes</key>
<array>
        <dict>
                <key>CFBundleURLName</key>
                <string>com.pragprog.amiphd.integrated</string>
                <key>CFBundleURLSchemes</key>
                <array>
                        <string>integrated</string>
                </array>
        </dict>
</array>
```

We are declaring here that our application understands URLs with the scheme integrated. To launch our application, we need to have a URL that starts with integrated://, so build and install the application in the simulator (with the Build and Go button). Once the application is launched, stop it by hitting the Home button. Then launch Safari and type in the URL integrated://message. When you hit the Go button, Safari will quit, and our application will come to the front with the default message changed to message.

Very cool! You have an application that can be launched via URLs and is open for integration with other applications on the iPhone.

Registering your new scheme does not make your URLs recognized by the text system. That means that sending an email between your users, or an SMS message won't automatically recognize the URL. However, if you enclose the URL in angle brackets (that is, < and >), the system will understand that it's a link. When your users click it, the proper application will be launched. In our example here, adding this <integrated://worked> to an email will allow the user to click the link and have the integrated app launch and display "worked."

Of course, you can also embed your links into proper HTML anchor tags like this: <a href="integrated://worked">say worked</a>. How you want your users to see links to your app will determine how you end up putting links into text. If you want your users to know that the link launches your app, then the first option is better. If your users don't need (or want) to know that your app is behind the link, then the second option is probably better.

In this chapter, you have seen how to launch other applications with URLs and not just launch them but also pass information along in the URL. You have also learned how to make your application have these same features so that other applications can launch your app.

Chapter 27
Debugging

As you've worked through this book and developed your own code, we're sure that, like us, you've had no problems along the way. Your code has built the first time, has run correctly without any hitches, and has done so with blazing efficiency. After all, this stuff is easy, right?

Right?

No, of course not. And it didn't work for us the first time either. Every code example in this book has been banged off the compiler more times than any of us would like to count.

Developing a working application, particularly on a platform that's both new to you and to the world at large, is a process of constant refinement, which implicitly means it's one of repeated failure. It will probably go something like this:

1. First your code won't compile.
2. Then, it will compile, but it will crash.
3. Then, it won't crash, but it won't work right.
4. Then it will work right, but it will be slow.
5. Then it will be fast.
6. *And then it will be freaking awesome.*

In this chapter, we'll look at the first three of these problems—how you'll use the tools provided by the iPhone SDK to make your code stable and correct. Once that's done, you can move on to Chapter 28, *Performance Tuning*, on page 511, in which you'll make your code run as fast as possible on the device.

27.1 Understanding and Fixing Build Errors

The first problem you're likely to encounter is the compiler error. It's as total a failure as can be, because regardless of the merits of most of your code, the parts that are kicking up build-time errors are so broken that you can't even run your application until they're resolved. But as severe as the problem is, the cause is often trivial; a lot of times, all you have to do to clear a compiler error is fix a typo.

In fact, let's deliberately blunder into that kind of an error right now. The sample project TrivialCompilerErrors features a one-view GUI whose only contents are a single UITextView, wired up to the variable textView in the TrivalCompilerErrorsViewController. The only work we'll have this class do is try to set the text when the view comes up.

Debugging/TrivialCompilerErrors/Classes/TrivialCompilerErrorsViewController.m
```
- (void)loadView {
    [super loadView];
    textView.txt = @"Hi mom";
}
```

As you can see, all we want the application to do is set the UITextView's text property to a static string. But here, we've made a typo by using the property name txt, when the correct property name is in fact text. One thing that might tip you off is that when you type a known property or function name of a class, Xcode's syntax highlighting turns it purple; the fact that txt remains black is a subtle warning that this code is broken.

What happens when you build this code? You get a build error. You'll see a little status indicator in the bottom right of the project window, and every source window open in this project looks like this:

The total number of errors in this status bar are shown with the white "x" in a red circle; if you had any build warnings, they would appear as a white "!" inside a yellow triangle. Each of the warning/error counts works like a hyperlink; you can click the number to view the Build Results window, shown in Figure 27.1, on the facing page. Within this window, you can click each error or warning in the top pane, and you'll be shown the offending line in the bottom pane, with a little pop-up that indicates the error.

UNDERSTANDING AND FIXING BUILD ERRORS ◀ 489

Figure 27.1: XCODE BUILD RESULTS WINDOW

In this case, the error is "error: request for member 'txt' in something not a structure or union".[1] You can fix the error anywhere it's visible: in the Build Results window, in the project window (after you select the source file from the list), in the source file's own editing window, and so on. Just change txt to text, build again, and you'll be all set.

While we're in such a simple project, let's tweak a few other parts of the code to create some common errors that you'll see in your everyday work. One of the most common mistakes for new Objective-C programmers is to forget to declare your objects as pointers. Let's see what that looks like. Go to the TrivialCompilerErrorsViewController.h file, and find the declaration of the text view:

IBOutlet UITextView* textView;

---

[1]. This is arguably either a terrible error message or an argument against the decision to reuse the dot operator, originally defined in C to denote members of a struct, for Objective-C properties, which is what text actually is. The error has nothing to do with structures or unions, but you wouldn't know that if you took the message at face value.

We'll mess that up by removing the * character, meaning it's no longer a Objective-C object reference:

`IBOutlet UITextView textView;`

When we do this, we get three errors. Two of the errors (one each for TrivialCompilerErrorsAppDelegate.m and TrivialCompilerErrorsViewController.m, because both import TrivialCompilerErrorsViewController.h) complain of the *statically allocated instance of Objective-C class 'UITextView'*, while the third reports that *'struct UITextView' has no member named 'text'*, which is really a side effect of not having an actual UITextView instance and therefore not knowing what its properties are.

Notice, by the way, that the files that Xcode reports as having errors aren't actually the source of the problem. The actual error is in the TrivialCompilerErrorsViewController.h header file, but the compiler reports errors in the implementation .m files that *import* that header. When you get errors, it's important to consider the nonobvious nature of some of the messages. Fortunately, in this case, if you click the errors in the Build Results window, Xcode will highlight the offending line from the header file.

27.2 Understanding and Fixing Importing/Linking Errors

Another common mistake is to have code that is correct but doesn't build because the information about building the code is incorrect. These are two common errors of this variety:

- Not importing needed header files
- Not linking needed frameworks

The first will be familiar to anyone who's worked with any flavor of C. When you fail to #import or #include a needed header file, every reference you make to an object, method, or procedural function defined by that header file coughs up an error in your build.

For example, in Chapter 10, *The SQLite Database*, on page 185, we imported the SQLite3 database's header file in order to be able to use its functionality. If we had forgotten to include the header file—#include <sqlite3.h>—then we would have seen errors or warnings for every attempted call to SQLite, claiming either that the functions were undefined or that we were implicitly declaring them, as shown in Figure 27.2, on the facing page.

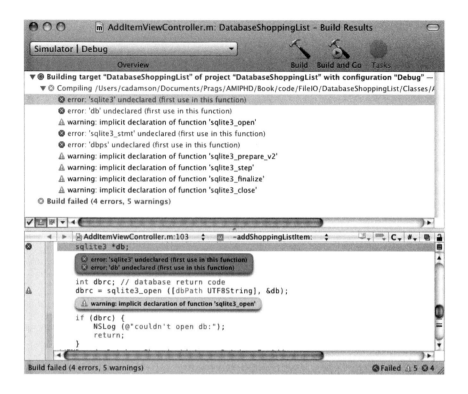

Figure 27.2: BUILD RESULTS WITH MISSING #INCLUDE

A similar problem results from failing to add a framework or library to the Xcode project. Although using the #include, #import, or @class declaration will satisfy the compiler by providing it with class, method, and/or definitions, the linker still needs to have a path to the .framework bundle or .dylib file that actually contains the framework or library code. If not, you'll get a "Symbol(s) not found" linking error, like the one shown in Figure 27.3, on the next page.

When you see this error and realize you haven't added the library or framework to your project, it's a simple matter to right-click the Frameworks folder in the Groups & Files section of your Xcode project, choose Add Existing Framework, and navigate to the needed framework bundle or library file (taking care to ensure the path is within the SDK directories, like /Developer/Platforms/iPhoneSimulator.platform or /Developer/Platforms/iPhoneOS.platform, and not in the Mac OS X /System/Library/Frameworks directory).

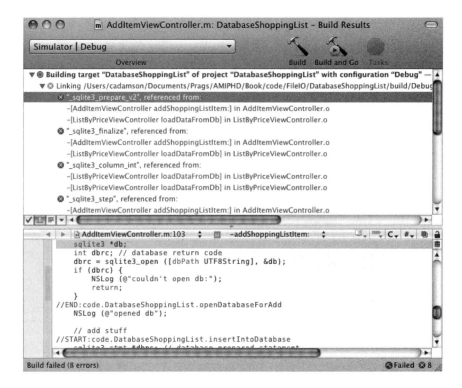

Figure 27.3: BUILD RESULTS WITH MISSING LIBRARY

For the new developer, both of these errors are fairly shocking because the sheer number of errors can be overwhelming—nobody is happy when a build breaks with more than 100 errors and warnings. But in time, you'll find that both of these common mistakes are almost obvious by the extent of their build results, and the fix to each is quite simple.

27.3 Using iPhone SDK Documentation

Thinking back to the first example that broke our build, the typo in which we tried to access the nonexistent txt property of a UITextView, consider this question: what if you honestly thought the property was called txt? Or what if you didn't think it was a property at all and instead tried to use a method setText: that turns out not to exist?

In such a case, it would be a good time to go looking at the documentation. After all, perhaps the best way to avoid bugs is to *know what you're doing*.

As mentioned earlier, the syntax coloring can be a tip. Once you type a valid method, function, or property, Xcode will color it according to your editor preferences. If you're stuck, you might also try using Xcode's code completion features. When you're a few characters into a class name, method name, function name, or property name, Xcode will attempt to autocomplete it if you stop typing briefly. If the proposed completion isn't what you want, you can bring up the next one with ^. or see a list of completions with ⌥ESC.

When you need more information than a code completion offers, Xcode provides documentation for all the iPhone SDK in a Documentation Viewer window, which you can bring up from the Help menu or with the keyboard combination ⌘⌥?. As shown in Figure 27.4, on the following page, the Documentation Viewer provides documentation for your installed SDKs—typically Mac OS X, Java, and one or more versions of the iPhone SDK—as well as for Xcode itself. These "doc sets" are shown in the narrow left pane, and the first time you bring up the viewer, you may see a Get or Subscribe button next to a given documentation set. Assuming you have the bandwidth, subscribing is probably the best option, so you'll always have local copies of the latest documentation. The alternative is that if you don't have the current doc sets locally installed, you'll sometimes find that when you look something up in the viewer, clicking the search result will open up your default web browser and load a documentation page from Apple's developer website (http://developer.apple.com/).

In the case of Figure 27.4, on the next page, we've looked up documentation for the UITextView class by typing its name into the search field at the upper right of the toolbar. Below this, you'll notice a strip that affects the behavior of the search. The leftmost segment determines whether to look for the term in APIs (class, method, or function names, constants, and so on), in filenames, or in the full text of the documentation. The next segment determines whether to apply the search to all doc sets or the doc set selected on the left panel. After that, another pair of buttons determines whether the search applies to all languages or just Objective-C, C++, JavaScript, and C. Finally, the last modifier specifies whether the query should find documents starting with your search term, containing it, or exactly matching it.

The table in the upper portion of the window shows the search results. So, a search for APIs containing the search term UITextView matches the UITextView class, the UITextViewDelegate class, and several constants

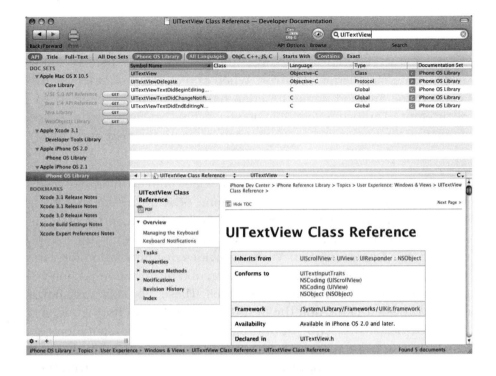

Figure 27.4: XCODE DOCUMENTATION VIEWER WINDOW

used for notifications. The top result will automatically be selected from this table, and the corresponding help document will be displayed in the HTML viewer below. So, in our rhetorical case of not knowing how to set the text of a UITextView, we'd just scroll through the description of the class down to the Tasks section, where we find a link to the text property, which when clicked goes to a description and definition of the property further down in the file.

There are several shortcuts you can use to quickly access the documentation from your code. You can Option+double-click any text in your source to bring up the Documentation Viewer, with the term you clicked prepopulated in the search box (you don't have to select the text first—Xcode will use the surrounding whitespace to figure out what term you clicked). This is great for looking up documentation for unfamiliar methods and classes. But what if you have an object reference and you don't remember exactly what class it's an instance of? In this case, you can right-click (or Ctrl+click) the reference to bring up a contextual

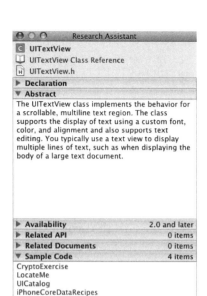

Figure 27.5: XCODE RESEARCH ASSISTANT

menu with, among other things, a variety of search options. Perhaps the most helpful is Jump To Definition. This will take you to wherever the object is defined, presumably a header file, whether in your code or in a framework you've imported. The definition will declare the object's class, at which point you could just Option+double-click the class name to search for its documentation in the Documentation Viewer.

Another documentation option is to use the Research Assistant, a floating window that is brought up with the Help menu's Show Research Assistant item (^ ⌘ ?). This window (shown in Figure 27.5) searches for whatever text is selected and—if it is a class, method, C function, or other documented item—presents an overview of documentation related to the selection, including its definition, an abstract (typically the first paragraph of its documentation), SDK availability, sample code that uses the item, and more. Blue items in the window are hyperlinks that take you to documentation, sample code projects, or definitions in header files. Rather than opening the Research Assistant on an as-needed basis, try leaving it open as you move through your code, and watch as it updates its contents each time you cursor over a term it recognizes.

27.4 Understanding and Fixing Interface Builder Errors

Thus far, we have focused exclusively on problems with code, which is a natural tendency for programmers. However, when developing in Cocoa, it's highly possible to make mistakes that don't expose themselves in code... because the mistakes are actually in Interface Builder.

Perhaps the simplest IB problem is making changes to your GUI (or other content loaded from a nib file) without remembering to save the nib file in IB. When you build and run against the old version of the nib, you could be missing GUI widgets, you could fail to get notifications from UI events, or you might not see changes to the GUI that you expect your code to cause, such as setting the contents of text views. In cases like these, you need to remember to just quickly go to IB and look at the nib's Close button; if there's a dot in the middle, then you have unsaved changes that need to be saved before you try again.

Similar IB problems can be caused by failing to correctly wire up your components in the first place. Let's consider two highly common examples. The first is when you add a view and forget to make its connections. If you expect to refer to this view from code—for example, to set the text of a UITextView as we did before—you'll typically declare a reference to it in one of your header files, preceding the declaration with the IBOutlet keyword. Then in IB, you'll change the class of File's Owner to match the class that owns the reference. By performing these two steps, you'll be able to wire a connection from the view to the outlet via IB. However, if you forget to actually wire up the connection, the calls from your code to interact with the view won't do anything, because there's nothing connecting the reference in your code to the view when it's loaded from the nib. What makes this trickier is that it's a silent failure. Trying to call methods on or set properties of the uninitialized reference to the view won't kick up an error or crash. It just won't do anything.

A similar problem comes from when you forget to connect a view's delegate connection to the class that implements the delegate methods, or a UITableView's dataSource to the class that implements UITableViewDataSource. You'll see this manifested as a table that doesn't show any data or respond to input, even though you're sure you implemented all the delegate and table source methods.

Another common IB mistake, fortunately, does kick up an easy-to-spot error message. Let's say you've developed a custom view controller and a custom view, and you make them visible in code, perhaps by pushing the view controller onto the stack of a UINavigationController. To do this, you'd typically allocate the view controller and initialize it with something like initWithNibName:bundle:. One very common error will actually crash your application when you first try to load the view.

How do you find it? Bring up Xcode's Debugger Console window from the Run menu (or with the keyboard shortcut ⌘ ⇧ R), and notice the following message right before the gdb debugger dumps a bunch of call stack addresses:

```
2008-10-10 15:53:22.149 FilesystemExplorer[7959:20b] *** Terminating app due to
uncaught exception 'NSInternalInconsistencyException', reason: '-[UIViewController
loadView] loaded the "FileOverviewView" nib but no view was set.'
```

What has happened here? Simple—you've created the custom view in IB and perhaps set the File's Owner class, but you've forgotten to connect the File's Owner's view to the custom view in the nib. The view controller needs to have a valid view in order to appear in the GUI, and without this connection, it doesn't have one. The crucial step of making this connection is shown in Figure 27.6, on the next page.[2] If you don't do it before you try to run your app, you'll remember to do so after you crash.

27.5 Debugging

Even if you don't crash, you'll often find yourself wondering why your application is behaving incorrectly. The code may compile, the connections in IB may be correct, but the application logic is somehow wrong. At this point, you need to take a detailed look at your code as it runs and determine why it behaves the way it does.

Logging to Standard Output

The program crash in the previous section logged its error to the system's "standard output" (also known as *stdout*), which we viewed in Xcode's Console window. Since these log messages go to the system's standard output, they can also be viewed in the Console.app application (located in /Applications/Utilities), where they'll appear alongside the

---

2. As you can tell from the error message and the figure, we have deliberately broken and now fixed one of the custom views from the FilesystemExplorer application in Chapter 8, *File I/O*, on page 129 in order to generate the error message and the figure in this chapter.

Figure 27.6: CONNECTING A VIEW CONTROLLER TO ITS VIEW

log messages from every other application and system process. Fortunately, Console.app offers a filter at the upper right, so you can restrict your view to just the current application (by typing its name into the filter box, because all log messages are preceded by their process name).

As an experienced developer, chances are you're familiar with writing text to standard out by using something like C's printf(), Python's print(), or Java's System.out.println(). And chances are, you'll want to do it with your iPhone applications. Debugging misbehaving code by leaving yourself a bunch of "I'm doing X" messages may not please the purists, but it's a bread-and-butter technique for many developers.

In Objective-C, you'll use the function[3] NSLog() to send messages to stdout. NSLog() takes an NSString reference, followed by zero to many additional variables. In its simplest form, you can log a single string:

```
NSLog (@"Hello iPhone");
```

---

3. Notice this is a procedural *function* and not an object-oriented *method*.

This sends the String Hello iPhone to stdout, where you see it in the Console window.

The more sophisticated, and valuable, use of NSLog() is to use the string as a *format string*. In this case, you use special character sequences to indicate where variables' values are to be inserted into the string before it's sent to stdout. For example, if you were in a method that had an NSString variable called userName, you could log its value with the following statement:

NSLog (@"User name is %@", userName);

The %@ sequence is used to provide a String representation of any Objective-C value. Technically, it retrieves the value of the NSObject method descriptionWithLocale: (or, if that method is undefined, description:), which Cocoa classes typically implement to provide a string representation that would be useful at debugging time. Of course, an NSString just returns its own value, which we use here to insert the userName string into the format string.

A complete list of the substitution sequences is in the "String Format Specifiers" section of Apple's *String Programming Guide for Cocoa*, which you can find by just looking up the documentation for Foundation's NSLogv() function (the low-level function called by NSLog()). Some of the most commonly used format strings are summarized here:

Sequence	Meaning
%@	String representation of an object, the result of calling descriptionWithLocale: or, failing that, description:
%d, %D, %i	Signed 32-bit integer (a long)
%u, %U	Unsigned 32-bit integer
%x	Unsigned 32-bit integer expressed as hexidecimal
%f	64-bit floating point number (a float)
%s	A string derived by interpreting the value as the address of a null-terminated string of 8-bit characters (a typical "C string," in the system encoding)
%S	A string derived by interpreting the value as the address of a null-terminated string of 16-bit Unicode characters

Let's look at how this could be useful by offering up another deliberately broken application. The PathologicalPrimeCounter is an application that simply allows the user to enter a maximum value and find all the

Figure 27.7: PRIME NUMBER COUNTER APPLICATION

prime integers up to that value. A working version of the application is shown in Figure 27.7. This is a single-view application with a UITextField connected to the view controller as countToField, a UITextView connected as primesView, and a UIButton that calls handleGoTapped: when tapped.

To see how NSLog() can help us, let's use it to debug a logic error. The sample application is set up to call countPrimesToSelectedValue when either the button is tapped or the user presses Enter after inputting a maximum value. Here's an initial, *incorrect*, implementation of that method:

Debugging/Primes/Classes/PathologicalPrimeCounterViewController.m
```
- (void) countPrimesToSelectedValue {
        int maxPrime = [countToField.text intValue];
        if (maxPrime < 1)
                return;
```

```
        // for each int up to maxPrime, check every divisor up to
        // maxPrime/2+1.  there are more efficient prime counters,
        // but this is easy to understand
        for (int i=2; i<maxPrime; i++) {
                int maxDivisor = i/2 + 1;
                BOOL isPrime = YES;
                for (int j=1; j<maxDivisor; j++) {
                        if (i%j == 0) {
                                isPrime = NO;
                                break;
                        }
                }
                if (isPrime) {
                        NSLog (@"%d is prime", i);
                }
        }
}
```

In this version, we haven't worried about populating the text view yet... we're just writing out all the primes to standard output for now. Each discovered prime will be written out with the statement:

`NSLog (@"%d is prime", i);`

In this case, there's one character substitution sequence, %d, whose value is provided by the additional argument, i.

So, bring up the Console window with Command+Shift+R, do a Build and Go, enter a maximum number, press Go, and...

Nothing happens.

So, what's going on? Is the loop not running? Are primes being rejected improperly? Let's test the second hypothesis by adding a log statement (right before isPrime = NO) to the block that rejects a potential prime, i, because it's divisible by some smaller integer, j. We'll use a format string that prints out both of these values:

`NSLog (@"%d is not prime, divisible by %d", i, j);`

As you can see, the format string reserves space for two signed integers, which are then provided as the next two values to the NSLog() statement.

Run this, and the output makes the bug obvious.[4]

---

4. In the actual output, each line is preceded by a timestamp. We've removed that so that each line of output fits on one printed line in the book.

```
PathologicalPrimeCounter[13308:20b] 2 is not prime, divisible by 1
PathologicalPrimeCounter[13308:20b] 3 is not prime, divisible by 1
PathologicalPrimeCounter[13308:20b] 4 is not prime, divisible by 1
PathologicalPrimeCounter[13308:20b] 5 is not prime, divisible by 1
PathologicalPrimeCounter[13308:20b] 6 is not prime, divisible by 1
PathologicalPrimeCounter[13308:20b] 7 is not prime, divisible by 1
PathologicalPrimeCounter[13308:20b] 8 is not prime, divisible by 1
PathologicalPrimeCounter[13308:20b] 9 is not prime, divisible by 1
PathologicalPrimeCounter[13308:20b] 10 is not prime, divisible by 1
```

The bug is that for every potential prime, we start our loop of potential divisors, j, at 1, and since every integer divides evenly by one, we wrongly conclude that every value of i isn't a prime. The fix is to start the j loop at 2.

Using the Debugger

NSLog()-based debugging may be a bread-and-butter technique for a lot of us, but it has its limits. It worked in the previous case only because we had some inkling of what was going on, so we had some idea what to actually put in our NSLog() statement. So, what do you do if you truly have *no idea* what your code is doing?

Let's push this example a little harder. Instead of fixing the inner loop by changing the initial value of j to 2, let's make things even worse by changing it to 0. Run this, enter a value for the maximum prime, tap Go…and you won't even see your log messages in the Console window. Instead, you'll see a message from gdb, the GNU debugger, which looks something like this:

```
Loading program into debugger...
GNU gdb 6.3.50-20050815 (Apple version gdb-962) (Sat Jul 26 08:14:40 UTC 2008)
Copyright 2004 Free Software Foundation, Inc.
GDB is free software, covered by the GNU General Public License, and you are
welcome to change it and/or distribute copies of it under certain conditions.
Type "show copying" to see the conditions.
There is absolutely no warranty for GDB.  Type "show warranty" for details.
```

What has happened here? Actually, your application has *crashed*, and the debugger has taken control (assuming you launched from Xcode; if you ran directly on the simulator or device, the app goes black and returns to the home screen). What you're left with is the Console window ready to take command-line input to gdb. If you're familiar with gdb, you're welcome to issue commands here, but most of us will be more productive working with Xcode's debugger GUI.

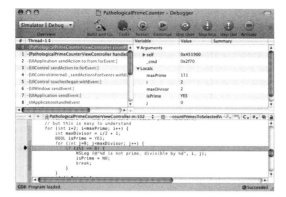

Figure 27.8: XCODE'S DEBUGGER

In Xcode, bring up the debugger from the Run menu or with the keyboard combination ⌘ ⇧ Y. The contents of this window are shown in Figure 27.8. At the top left, you see the current call stack for the active thread. Scroll down, and you can see the stack of calls that got all the way from the main() that Xcode created for you in main.m up to the method call you were in when you crashed. Method and function calls for which source is available—generally meaning your code and not the various system libraries—are shown in black text, with the rest in gray. Click one of the black lines of text, as we've done here, to show the source in the bottom pane (if you click a gray line, you'll see the assembly code for that call). In the source pane, the red arrow at the left indicates where the application execution has stopped.

At the upper right, the debugger shows a table of variables and their current values. Take a look at the offending line of code:

`if (i%j == 0)`

Then take a look at all the variables on the right (you can also examine the value of the variables by mousing over them in source, as shown in Figure 27.9, on the next page). There are only two variables that matter on this line: i has a value of 1, and j has a value of 0. So, what's wrong with 1 % 0? The problem is that it's division by zero, which is always a crasher.

In fact, since the debugger has captured the state of the application immediately prior to the crash, you can confirm this for yourself by letting the application continue. Click the Continue button from the

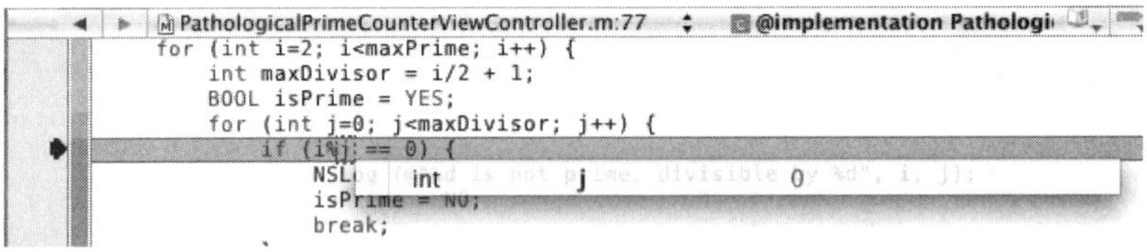

Figure 27.9: VIEWING VARIABLE VALUES IN SOURCE CODE

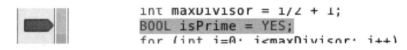

Figure 27.10: ADDING A BREAKPOINT

debugger's toolbar, and you'll see the crash message at the bottom of the debugger: Program received signal: "EXC_ARITHMETIC".

There's a lot more you can do with the debugger than just watch your application crash. In fact, you carefully step through the code one line at a time and even reset the values of variables as you go. Let's try that. Stop the application if it's running, and go to the PathologicalPrimeCounterViewController.m source. In the wide gutter at the far left, next to the BOOL isPrime = YES; statement, click the mouse to create a breakpoint, as shown in Figure 27.10.

Build and Go to run the application again. With the breakpoint set, the application will stop as soon as execution reaches the breakpoint. This will give you an opportunity to carefully examine the app's state right up to the crash. With the application stopped, the list of variables shows i equals 2, isPrime is YES, and j is out of scope.

At this point, you could click the Continue button to plow ahead and crash, or you can use the three Step buttons to inch toward the crash. Each of these allows you to proceed ahead through a certain number of method or function calls. Step Into executes the next statement, and if that statement is itself a method call, function call, or other complex statement, execution will stop within the statement (with the call stack and source display updated, if necessary). In a series of deep calls, this can show you who's calling whom, but it can also be burdensome to dig several layers deep. Step Out is the corresponding command: if you're in a method call, it allows execution to continue until the method returns. And if you don't care about the interior of the method call at all, you can use Step Over to execute the method call, however many calls deep that ends up being, and stop again when the top-level call returns.

In this case, click Step Into once to advance past the breakpoint. This will bring you into the for statement. j is now in scope but is not initialized until your next step, so it has some random value. Step Into again, and its value will be set to 0.

You are now on the brink of the crash. Step again, and you'll move on to the modulo statement that will divide by zero and crash the application. But you know what? *Let's not crash.* In the variables table, double-click j's value, and notice that it's editable. Go ahead and set it to 2, and then step into or over again. By changing the value at runtime, you're able to continue without crashing!

Of course, your next time through the loop, you'll crash, but you now see what you have to change in your code to make it work. You've proven that avoiding 0 as a value for j eliminates the crash, and the fix is just one Build and Go away.[5]

Moreover, in the big picture, it's quite powerful to be able to stop your code at breakpoints, inspect the values of variables, and even edit those values while the app is running to see what kinds of values will make your app work correctly.

27.6 Finding Over-Released "Zombie" Objects

In the previous section, we saw how you can use the debugger to find and fix a divide-by-zero crashing bug. Most crashes will not be that obvious, however. By far, the most common crashing bug is **EXC_BAD_ACCESS**, which always results from some sort of pointer error, such as failing to initialize a pointer, using a referenced numeric value (rather than the pointer) as an address, and so on.

---

[5]. When developing Mac OS X applications, you can sometimes use Xcode's Fix command to insert fixed code into a running application, but this seems not to work on the iPhone Simulator, at least when we tried it.

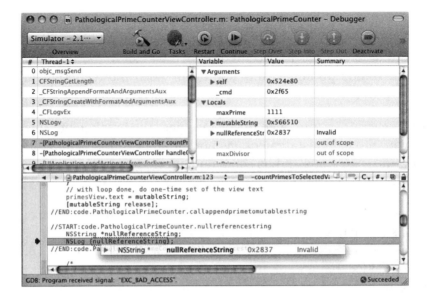

Figure 27.11: An **EXC_BAD_ACCESS** crash caused by a broken pointer

For a simple example, consider a simple failure to assign a pointer:

Debugging/Primes/Classes/PathologicalPrimeCounterViewController.m

```
NSString *nullReferenceString;
NSLog (nullReferenceString);
```

Since nullReferenceString is never allocated and assigned, the reference is bogus, and the first attempt to reference the nonexistent object will crash with **EXC_BAD_ACCESS**. Fortunately, this is extremely easy to spot with the debugger, as shown in Figure 27.11.

As you can see, the debugger is stopped on the offending NSLog() call, and mousing over the nullReferenceString shows us that it is "invalid," which you can also see in the variable list in the top-right pane.

So, that was easy enough to debug, but consider this attempt to log a message at the bottom of the countPrimesToSelectedValue method:

Debugging/Primes/Classes/PathologicalPrimeCounterViewController.m

```
NSString *headerString = [NSString stringWithFormat:
        @"Counted primes to %@", countToField.text];
NSLog (headerString);
[headerString release];
```

This might look perfectly harmless at first glance, but it crashes every time you run it. Worse yet, investigating the crash with the debugger doesn't yield an obvious problem. None of our application code is implicated in the crashed thread's stack, because the crash occurs when NSPopAutoReleasePool attempts to send a message to an object... and, unfortunately, we can't even tell what object or method call is to blame.

We'll reveal the cause first and then show you a tool for tracking this kind of crash. The problem is an *over-release* of an object. The class method stringWithFormat: creates an autoreleased NSString, so the subsequent [headerString release] is not appropriate. What happens is that our release reduces the reference count to zero, freeing the object, but then the autorelease pool tries to release the object again later, sending another release message to the object that's already gone and whose reference is now bogus. The fix is either to eliminate the release or to replace stringWithFormat: with the nonautoreleasing alloc and initWithFormat: pair.

Thing is, even if you suspect an over-release, it can be hard to find just *which* object is being over-released.

This is a job for *zombies*, specifically, NSZombies. In Cocoa, *zombies* are objects that, while dead, are still walking around and causing trouble. Fortunately, there are tools available in Xcode to find and hunt zombies. What we can do is to enable a build setting that, instead of freeing objects when their reference count gets to 0, converts them into NSZombie objects. This class's purpose is to log any call to an instance of it, since that means the code is trying to call a method on a dead object.

In your project window, under Groups & Files, expand the Executables item, and select the PathologicalPrimeCounter application. Bring up its Inspector with Get Info (⌘I or the Info toolbar button), and choose the Arguments tab. In the bottom half, "Variables to be set in the environment," use the + button at the bottom to add two variables: NSZombieEnabled and MallocStackLogging, setting both to the value YES, as shown in Figure 27.12, on the following page

Bring up the console (⌘R), and run again. You'll see some messages about where "malloc stack logs" are being written, usually into your /tmp directory:[6]

⇒   `launchd(2018) malloc: stack logs being written into`
     `/tmp/stack-logs.2018.launchd.odBbpt`

---

6. We've wrapped the output to fit the book's layout.

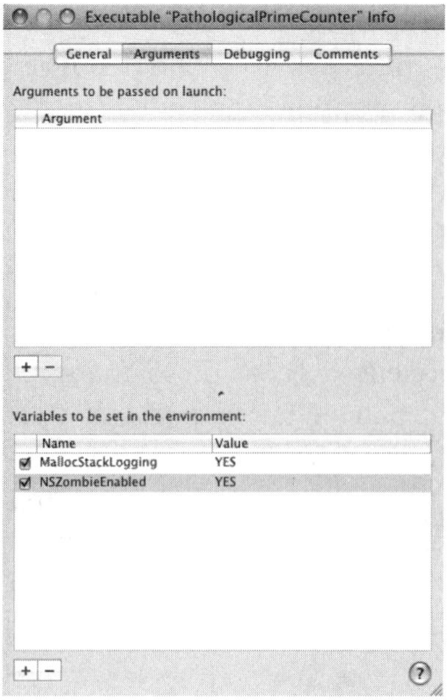

Figure 27.12: ENABLING NSZOMBIEENABLED IN THE EXECUTABLE'S ENVIRONMENT VARIABLES

Go ahead and count some primes to cause the crash. This time, right before the crash, you'll get a message from the zombie:

```
2009-03-10 16:56:01.128 PathologicalPrimeCounter[2018:20b] ***
  -[CFString release]: message sent to deallocated instance 0x591f40
```

Now we have something. We now know the message being sent to the over-released object (another release), the class of the over-released object (CFString, the Core Foundation equivalent of NSString), and even the address of the zombie. That tells us the end of the story, but what about the beginning; how do we trace back to where things went wrong? That's what the malloc tracing is for. Using the zombie's address, we can find when it was allocated.

In the Console window, after the (gdb), type this:[7]

```
shell malloc_history pid address
```

---

7. The shell command sends the rest of the command to the Darwin shell, so you could also just issue the malloc_history command directly to the command line in Terminal or xterm.

where *pid* is the process ID shown in parentheses in the malloc: log messages, and *address* is the address of the zombie as reported by the "deallocated instance" message. Given the previous log messages, here's what your interaction with the console will look like:

```
(gdb) shell malloc_history 2018 0x591f40

Call [2] [arg=32]: thread_a024c720 |0x1 | start | main | UIApplicationMain |
 -[UIApplication _run] | GSEventRun | GSEventRunModal | CFRunLoopRunInMode |
 CFRunLoopRunSpecific | PurpleEventTimerCallBack | SendEvent |
 _UIApplicationHandleEvent | -[UIApplication sendEvent:] | -[UIWindow sendEvent:] |
 -[UIControl touchesBegan:withEvent:] |
 -[UIControl(Internal) _sendActionsForEvents:withEvent:] |
 -[UIControl sendAction:to:forEvent:] | -[UIApplication sendAction:to:from:forEvent:] |
 -[PathologicalPrimeCounterViewController handleGoTapped:] |
 -[PathologicalPrimeCounterViewController countPrimesToSelectedValue] |
 +[NSString stringWithFormat:] |
 -[NSPlaceholderString initWithFormat:locale:arguments:] |
 _CFStringCreateWithFormatAndArgumentsAux | CFStringCreateCopy |
 __CFStringCreateImmutableFunnel3 | _CFRuntimeCreateInstance | malloc_zone_malloc
```

You may get multiple Call[]s here, some of which may not even be from your application (that is, they're logged from other applications). But if you look for your own classes, you can find the call stack that represents the creation of the object that eventually became a zombie. In the previous output, you can see that this object was created via the NSString class method stringWithFormat:, called from inside countPrimesToSelectedValue. With that, you should be able to look at the source, figure out which object is the troublemaker, and ultimately figure out how you've managed to over-release it.

Once you've figured out your over-release, it's important to remove the NSZombieEnabled and MallocStackLogging environment variables (or set them to **NO**) before you ship. After all, NSZombieEnabled works by converting objects that should be freed into NSZombie objects instead, meaning that *no objects are ever freed* when you have NSZombieEnabled set.

Armed with these tools to track and kill zombies, you should be able to work through over-release crashes.[8] Combine them with the debugger techniques described earlier, and you'll be able to analyze and understand any pointer problems that lead to **EXC_BAD_ACCESS** crashes.

---

8. And, using the techniques introduced in Section 28.3, *Investigating Performance with the Clang Static Analyzer*, on page 523, you may be able to avoid creating over-release bugs in the first place.

Oh, and did we mention that everything the debugger does—setting breakpoints, stepping through code, and viewing and altering variables—works when your application is running on the device, not just on the simulator? The same goes for viewing your logging output. Xcode gives you tremendous tools for developing iPhone applications, and the more you use them, the more you'll find them invaluable.

Chapter 28
Performance Tuning

When we introduced the prime counter application in Section 27.5, *Debugging*, on page 497, we said we were going to put the resulting primes in a UITextView. Now that the logic to find primes is working, let's do that. As a simple way to do this, let's append the text in the view every time we find a new prime. Here's a simple method to do that:

Debugging/Primes/Classes/PathologicalPrimeCounterViewController.m
```
- (void) appendPrimeToPrimeArea: (int) aPrime {
    NSString *primeString = [[NSString alloc] initWithFormat: @"%d ", aPrime];
    primesView.text = [primesView.text stringByAppendingString: primeString];
}
```

Each time this method is called with a new prime number, it creates an NSString consisting of the number with a trailing space, gets the UITextView's current string, and replaces it with a new value, created by appending the new string to the old value. Call it in the **if** block that reports found primes (while you're there, comment out the no-longer-needed NSLog() statement):

Debugging/Primes/Classes/PathologicalPrimeCounterViewController.m
```
if (isPrime) {
    // NSLog (@"%d is prime", i);
    [self appendPrimeToPrimeArea: i];
}
```

Run this, and you'll see output more or less like the working app shown in Figure 27.7, on page 500. Ask for primes up to 100, and they pop right up. Enter 1,000, and they're displayed after a slight pause. Ask for primes to 10,000, and you'll have to wait a few seconds. Ask for 100,000, and... well... it looks like our application is hanging pretty badly.

Figure 28.1: SELECTING A PROCESS TO PROFILE WITH SHARK

Yet, we're not doing anything *that* expensive. Sure, we have a nested loop that we iterate more than 100,000 times, but half of those will break out on the first trip through the inner loop, since all even numbers are divisible by 2, the first value for j. The smart programmer knows that *something* is rotten. The question is... what?

Some programmers will get ahead of themselves at this point and begin to throw various optimizations at the problem—hacks with loops and side effects meant to cut down the number of instructions executed. The smart programmer asks him or herself, "Where's the slowdown, and what can I do about it?"

28.1 Investigating Performance with Shark

Xcode includes a number of performance analysis tools. One that has been around for a number of years is called *Shark*, so named because of the relentless, predatory instinct required for hunting performance problems. Shark is located in /Developer/Applications/Performance Tools, and although it was originally intended for use with Mac applications, it proves enormously helpful at analyzing apps running in the iPhone as well.

To use Shark, begin by running the application in the simulator with a typical Build and Go. Once it's running, switch over to Shark, where you'll be presented with its main window, shown in Figure 28.1. In this window, one pop-up offers a choice of profiling types (we'll use the default, Time Profile) and a kind of target (either a specific process or the entire system). Set Process for the target, and then look up the PathologicalPrimeCounter in the final pop-up. Note that the name may be truncated, as it is in the figure.

With the target selected, click Start to begin profiling. Now you can return to the iPhone Simulator and continue using the application, knowing that Shark is watching where the app spends its time. Enter 10000 for the maximum prime value, and tap Go. When the text area is

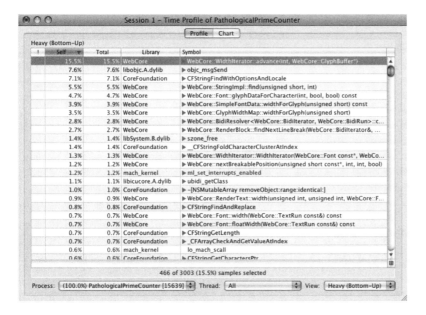

Figure 28.2: SHARK TIME PROFILE RESULTS, "HEAVY" VIEW

finally populated, go back to Shark, and click Stop to end the profiling. At this point, Shark analyzes the data it has collected and produces a Time Profile window, as shown in Figure 28.2.

The Time Profile window has two basic viewing modes, controlled by a pop-up at the bottom right. In heavy (bottom-up) mode, shown in Figure 28.2, you can see which method and function calls account for the most time spent by the application. The table is sortable by its various columns: Self and Total (to be explained shortly), Library, and Symbol. Just taking an uninformed first look at the table, something should stand out: the application is spending nearly all of its time in the WebCore library. Calls to WebCore methods account for seven of the ten most called methods, and these methods alone represent nearly half of the time our application is spending.

Now, take a closer look at what these methods are: WidthIterator::advance(), Font::glyphDataForCharacter(), SimpleFontData::widthForGlyph()... even without knowing the first thing about what any of the methods are, you may start to hypothesize that our application is spending an extraordinary amount of time performing graphic layout, presumably for the text we're putting in the UITextView.

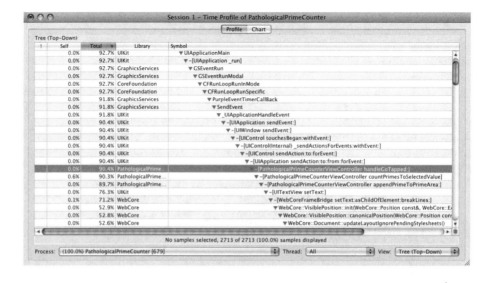

Figure 28.3: SHARK TIME PROFILE RESULTS, TREE VIEW

We can verify this by switching view modes from heavy (bottom-up) to tree (top-down). Now instead of seeing where we spend our time by method, we get a stronger idea of who's calling those methods. The table starts with the lines start(), main(), and UIApplicationMain(), which is the top of the call stack to start an iPhone application. Further down, you'll see the application's main run loop (CFRunLoopRunInMode()), which is sending the tap event ((UIApplication sendAction:to:from:forEvent:)). Further down, you'll finally see the methods from our application, PathologicalPrimeCounter, as shown in Figure 28.3.

This is where the Self and Total columns come into play. The Self method represents the percentage of the application's time spent in this method itself, while Total includes the time spent in everything called by this method. As you can see, of the three PathologicalPrimeCounter methods on the stack, our handling of the tap event consumes a total of 90.6 percent of the application's time. However, effectively zero time is spent in handleGoTapped:, and only 0.6 percent of the application's time is actually spent in the loops of countPrimesToSelectedValue. Nearly all the time spent by our application, 89.7 percent to be exact, is spent in the GUI methods called by appendPrimeToPrimeArea:.

By profiling in this way, we know where we should, and shouldn't, look for performance optimizations. Whether or not our prime counting algorithm is efficient, it accounts for less than 1 percent of our run time, so all the "loop unrolling" optimizations in the world won't make a difference there. The problem is in the expense of putting the contents into the UITextView. If we think about it further, the fact that these layout operations seem to be expensive should lead us to an alternative approach. Instead of resetting the view's contents every time we find a prime, what if we set the view's contents only once, at the bottom of the loop?

Let's try that. Instead of calling appendPrimeToPrimeArea: each time we find a prime, we'll instead append the prime to an NSMutableString and then use that to set the view's contents only when we've found all our primes. So, write a simple method to use a string formatter to add a prime number and a trailing space to a mutable string:

`Debugging/Primes/Classes/PathologicalPrimeCounterViewController.m`

```
- (void) appendPrime: (int) aPrime
            toMutableString: (NSMutableString*) mutableString {
        NSString *primeString = [[NSString alloc] initWithFormat: @"%d ", aPrime];
        [mutableString appendString: primeString];
}
```

And rewrite countPrimesToSelectedValue to call the String appender. Here is the rewritten loops and the post-loop call to set the view's text.

`Debugging/Primes/Classes/PathologicalPrimeCounterViewController.m`

```
NSMutableString *mutableString =
        [[NSMutableString alloc] initWithCapacity: maxPrime * 5];
for (int i=2; i<maxPrime; i++) {
        int maxDivisor = i/2 + 1;
        BOOL isPrime = YES;
        for (int j=2; j<maxDivisor; j++) {
                if (i%j == 0) {
                        // NSLog (@"%d is not prime, divisible by %d", i, j);
                        isPrime = NO;
                        break;
                }
        }
        if (isPrime) {
                // NSLog (@"%d is prime", i);
                [self appendPrime:i toMutableString: mutableString];
        }
}
// with loop done, do one-time set of the view text
primesView.text = mutableString;
[mutableString release];
```

Try it and behold the results. Running the original version in the simulator on an eight-core Mac Pro—we're not about to put such inefficient code on the iPhone hardware itself—counting primes up to 10,000 took 2.5 seconds to display and up to 50,000 took 66.8 seconds. Switching to the version that builds up an NSMutableString and sets the view's text once cut those times to 0.02 seconds and 0.25 seconds, respectively. With a fairly simple change, we've gone from over a minute down to a quarter second, which is exactly what you'd expect when we can eliminate the repeated calls that were accounting for nearly all of our execution time.

And that's the power of profiling: you can determine exactly where your application is spending time to find hotspots where you can attempt either to write more efficient code or to find alternative approaches that eliminate the hotspot entirely.

Using Shark to Profile Applications on the Device

Although it's not immediately obvious, Shark can also be attached to iPhone OS applications running on the device.

Assuming you have already configured your environment to code-sign your application for running on an iPhone OS device connected to your Mac, run your application (either via Xcode's Build and Go or by simply running it from the device's home screen), and then start up Shark if it's not already running. In Shark, choose the menu item Sampling > Network/iPhone Profiling (⇧ ⌘ N). This causes a new pane to slide out the bottom of the Shark window, as shown in Figure 28.4, on the facing page.

This pane exposes Shark's remote debugging options, controlled by the radio buttons at the top of the pane. Selecting "Profile this computer" will collapse the pane and return to local profiling mode. What we want is the last option: "Control network profiling of shared computers." Select the checkbox next to your device's name (Cloud in this example), and use the pop-ups to configure the kind of sampling (the default Time Profile is what we've used thus far) and the target process. If your app is running, you should see its short name, like PathologicalPrim, in the Target pop-up list. Then just click Start at the top of the window. From this point, Shark works just like before: run your application through its hotspots, stop the sampling in Shark, and examine the profile results.

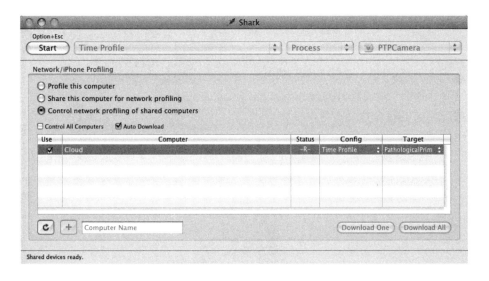

Figure 28.4: PROFILING PROCESSES ON THE DEVICE WITH SHARK

28.2 Investigating Performance with Instruments

While Shark is powerful and straightforward at finding extraordinary hotspots, its approach is limited in a key way: it treats all activity within its sampling period as more or less the same. It can't catch short segments of time, usually the result of some external event, that may account for most of the activity that you're interested in. In the previous example, nearly all the work was done after tapping the Go button, but Shark shows all activity within its sampling period equally, whether it happened before or after we started grinding primes.

This is where the Instruments application comes in handy. Instruments keeps a time-based log of activity in your program and shows it on a timeline, so you can see how your application consumes CPU, memory, and other resources in response to activity within the application's lifetime.

Let's take a quick look at how Instruments would diagnose the same problem with the PathologicalPrimeCounter. Revert to the inefficient version of countPrimesToSelectedValue; if you've downloaded the sample code, all the various versions of this method are commented out, so just comment in the one that says it calls appendPrimeToPrimeArea:. To run with Instruments, build the app again and then go to the Run

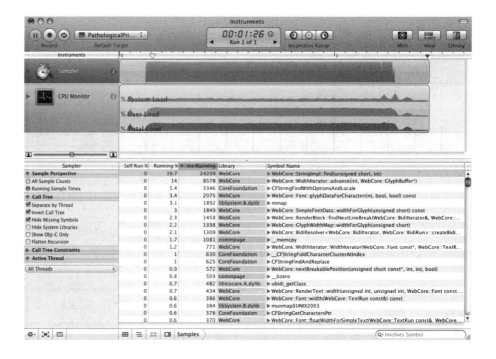

Figure 28.5: INSTRUMENTS WINDOW WITH SAMPLER INSTRUMENT TIMELINE

menu and select Start with Performance Tool, selecting CPU Sampler from the submenu. The Instruments application will start up and show its main window. Switch to the iPhone Simulator, and use the application as before, entering some nice high maximum prime value (we used 50,000 again). When you're finished, switch back to Instruments, and click the Stop button to stop sampling. Now let's take a look at the Instruments window, shown in Figure 28.5

The top portion of the Instruments window shows a timeline, as you might be used to from media editing applications like GarageBand or Final Cut Pro. This timeline represents the time that your app was running, from when you started recording (that is, when Instruments started up, since launching from Xcode causes Instruments to begin recording immediately) to when you clicked stop. This timeline shows two "tracks" in the figure: one for the Sampler instrument and one for the CPU Monitor. Above the timeline, there's a small white triangle; this is the *playhead*, which you can drag back and forth to pick out

> ### Joe Asks...
>
> #### If Instruments Does CPU Sampling and More, Why Bother with Shark?
>
> Instruments has big advantages over Shark: it combines a bunch of tools in one place (CPU sampling, leak detection, and so on) and has a timeline concept that lets you see these metrics over time. Apple has made it clear that Instruments is its tool of the future.
>
> Still, for a quick look at where your cycles are going, the simplicity of Shark can sometimes trump the comprehensive approach of Instruments. It's pretty easy to get an overall sense of where your cycles are going in Shark and to drill down to CPU hogs with Shark's two straightforward views. In Instruments, there's a lot of mental correlating between the timeline and the two detail views.
>
> If you want to learn only one tool, go with Instruments. But we still like Shark too.

specific moments in the application's execution for further analysis. Slide it back and forth, and you'll see pop-ups along the various tracks showing their status at that time, such as the CPU usage (as a percent) in the Sampler track.

As you can see from the Sampler track, there's an extraordinary burst in activity about 14 seconds into the application's lifetime, indicated by the purple bar graph in the track. As you might have guessed, this represents the time after the user clicked the Go button, and the application was busy generating primes and doing lots of what turned out to be unnecessary layout within the UITextView. With the Sampler track selected, you can also see an overview of function and method calls in the bottom of the window, called the *detail pane*. By setting some of the options over to the left, you can screen out system calls that aren't relevant to your analysis (Objective-C message dispatching, for example) and focus on where your application is spending its time. Although the layout and presentation is different from Shark, you can see once again that the application is spending the overwhelming majority of its time in WebCore layout methods.

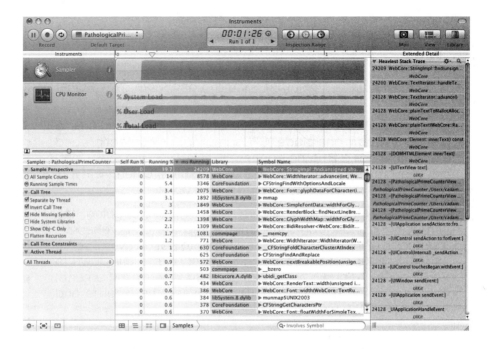

Figure 28.6: INSTRUMENTS WINDOW WITH THE EXTENDED DETAIL PANE

The Instruments window also has an *extended detail* pane on the right, though it's hidden by default. To show it, select Extended Detail from the View menu, use the keyboard combo ⌘ E, or click the half-filled rectangle button at the bottom of the Instruments window, below the detail pane. With both panes visible, you can click an item in the detail pane to show more information about the selected item in the extended detail pane. For example, with the Sampler track selected, if you click one of the method names in the detail pane, the extended detail pane will show the "heaviest stack trace" that results in that method call. In Figure 28.6, we can see a stack trace in its extended detail pane that implicates our PathologicalPrimeCounter in the call stack for the WebCore call that accounts for more than 39 percent of the runtime activity.

When you launch Instruments from Xcode, you do so by indicating one instrument you want to use, like the CPU Sampler we launched with earlier. But once Instruments is running, you can add more instruments as tracks, allowing you to investigate how your application uses different system resources. To see available instruments, bring up the Library window with its menu item (in the Window menu) or via the keyboard combo ⌘ L. The Library shows all the instruments that you can drop into a Mac or Xcode Instruments session. Of course, some of the instruments don't apply to iPhone, like the Core Data or Java tools; when you select a tool, the description pane at the bottom of the Library window will indicate whether the tool is for use with Mac, iPhone, or both.

So, although we've eliminated our waste of CPU power, let's make sure we're not wasting memory either. Select the ObjectAlloc instrument, and drag it to the list of instruments at the left side of the window. Click Record to run the app again. Now that we know we can crunch a lot of primes in very little time, go ahead and count all the primes up to 50,000. When the text view fills in, switch back to Instruments, and take a look at the timeline. You can see that at the same time that there's a small burst of activity in the CPU Monitor, there's also a bump in the ObjectAlloc graph. That means that when we go into our prime-counting routine, we're creating a lot of objects. Moreover, careful examination of the graph shows that the object count isn't going down. You can confirm this by looking at the table of categories[1] in the detail pane and clicking those categories that interest you. In Figure 28.7, on the next page, we've clicked CFString (which is the Core Foundation equivalent to NSString), which causes a red bar graph to be drawn into the purple bar graph of All Allocations. In fact, you can deselect All Allocations with its checkbox, leaving only CFString selected; the resulting graph shows the count of objects going straight up and not coming down.

With a mass of objects being created and never released, it's highly likely we're leaking memory.

To investigate, open the Extended Detail window, and click the arrow that appears next to CFString when you mouse over it. This will replace the table of categories with a list of all the CFString instances, when they were created, and what call created them. Since our memory leak seems to occur later in the timeline, we can sort by time and scroll down in the table to find instances created later in the application's lifetime. What we'll find is thousands of instances created by -[NSPlaceholderString initWithFormat:locale:arguments:]. Click one, and we'll see the call stack in the extended detail pane and, smack in the middle, a set of calls

---

1. Categories, in this usage, is understood to be Cocoa classes and Foundation opaque types, which are not classes per se.

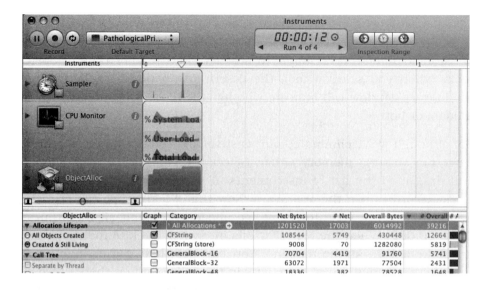

Figure 28.7: EXAMINING A POSSIBLE MEMORY LEAK WITH THE OBJECTALLOC INSTRUMENT

from PathologicalPrimeCounter. Double-click the last of these, and it will take you to the exact line that created the leaked string. It's in appendPrime:toMutableString:. Let's look at that method again:

Debugging/Primes/Classes/PathologicalPrimeCounterViewController.m
```
- (void) appendPrime: (int) aPrime
           toMutableString: (NSMutableString*) mutableString {
    NSString *primeString = [[NSString alloc] initWithFormat: @"%d ", aPrime];
    [mutableString appendString: primeString];
}
```

Do you see the leak? We create an NSString, append it to the NSMutableString...*and never release it.* Once we've appended it to the NSMutableString, we're no longer interested in it and can release it immediately. Add one line to the bottom of this method:

```
[primeString release];
```

Build again, switch to Instruments, and record. Set the ObjectAlloc instrument to only graph CFStrings, and you should see the much better result illustrated in Figure 28.8, on the facing page. There's a spike in the object count right around the time that the CPU gets busy counting primes and filling the UITextView, but then it drops right off, since we've cleaned up all the temporary string objects.

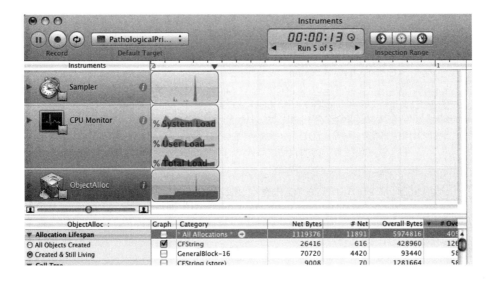

Figure 28.8: EXAMINING A FIXED MEMORY LEAK WITH THE OBJECTALLOC INSTRUMENT

28.3 Investigating Performance with the Clang Static Analyzer

Profiling your code with Shark or Instruments is exceptionally useful at identifying your code as it runs under real-world conditions. Still, you won't catch *everything* through profiling. Bugs lurking in branches of code that you don't typically take in your testing won't show up; and it'll only be when an end user tries something you hadn't anticipated that those issues will manifest themselves. And as we saw in the Instruments section, your application generates thousands of objects as a normal part of its execution, meaning it's easier to find a bug that leaks a lot of objects at once than continues to leak objects slowly over a long period of time.

Static analysis offers a radically different approach to finding bugs. Rather than examining your code as it runs, static analyzers model the code through their own techniques and attempt to find readily identifiable mistakes. In just the last few months, an Apple-sponsored open source project has released the *LLVM/Clang Static Analyzer*. It works with Xcode projects, including iPhone applications.

> **Clanging on the Bleeding Edge**
>
> If you are running Xcode 3.2 or newer on Snow Leopard, the Clang Static Analyzer is installed and integrated into Xcode. If you are developing on Leopard, you will need to install the command-line version of the analyzer yourself. You'll find instructions for setting up the command-line version of the Static Analyzer in a PDF included with the book's downloadable code.
>
> Although the interface to the Static Analyzer is changing, the tests it performs and the reports it puts out are likely to remain pretty stable, which is what we focus on in this section.

The analyzer is an offshoot of the *Clang* project, which provides a front end to the LLVM compiler[2] for C, C++, Objective-C, and Objective-C++.

Assuming that you have the Analyzer installed on your Mac (see the the sidebar on the current page sidebar for a link to our install docs for the command-line version), let's test it on some real code. Be sure that your project builds without errors with its default options. If it doesn't, you may see a bewildering error message saying that scan-build can't find a needed version of gcc. In practice, we've seen this when a project defaults to building for the device instead of the simulator, and the current environment isn't provisioned for code-signing. This may be the case if you download someone else's code, the project is set to build for the device, or you haven't joined the paid iPhone developer program and gotten an app-signing certificate yet. You can get around this by changing the project's properties to build for the simulator by default, rather than the device. In your Xcode project, select the project icon from the very top of the Groups & Files tree, and then click Get Info in the toolbar or the File menu. As shown in Figure 28.9, on the facing page, set Base SDK to Simulator, rather than Device.

---

2. The Low Level Virtual Machine (LLVM), hosted at http://www.llvm.org/, is a multipronged approach to improving application performance by employing a virtual instruction set that can be intelligently adapted to its host environment by offering the ability to optimize its native bytecode at runtime and even post-installation.

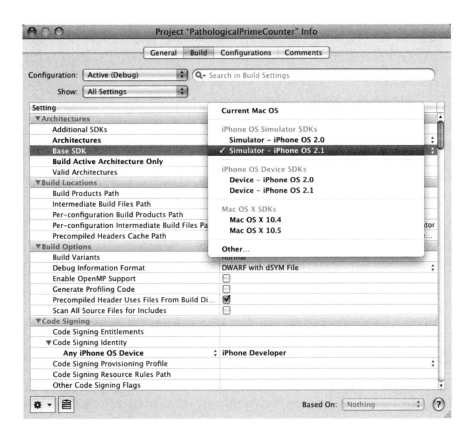

Figure 28.9: SETTING THE BASE SDK TO THE SIMULATOR FOR USE WITH THE CLANG STATIC ANALYZER

With all this business worked out, try kicking off a scan from the project directory. If you're using the command line, type scan-build xcodebuild. If you can run a scan from Xcode, choose Build & Analyze.

Assuming all goes well,[3] you'll be notified that a bug has been found.

In the web view, each bug that scan-build finds is listed in the Reports area, with links to view a report of the bug. Click the View Report link to see a step-by-step assessment of the bug. The web page generated from the command line looks like Figure 28.10, on the next page. If you are

---

3. We're sure you'll let us know on the book's forum (http://forums.pragprog.com/forums/83) if it doesn't. Also, one of our readers pointed out that doing a clean build as part of running the Static Analyzer can help while iteratively fixing bugs.

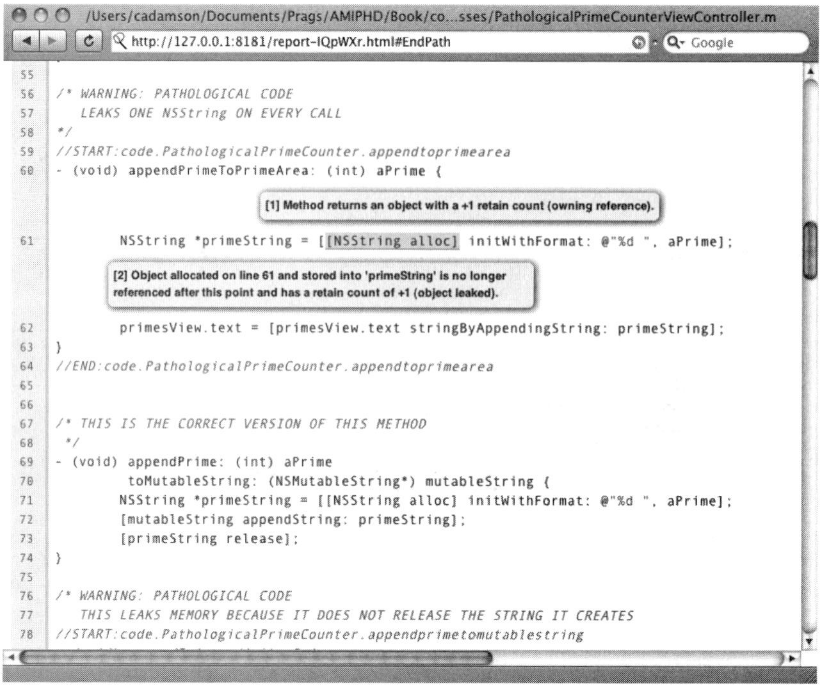

Figure 28.10: VIEWING THE CLANG STATIC ANALYZER REPORT OF AN OBJECTIVE-C MEMORY LEAK

running the Static Analyzer from within Xcode, you will see the bugs reported much the same way that errors and warnings are. Despite the differences in the interface, the resulting bug report contains this information:

- "Method returns an object with a +1 retain count (owning reference)."

- "Object allocated on line 61 and stored into 'primeString' is no longer referenced after this point and has a retain count of +1 (object leaked)."

In this case, the analysis has found another case where we allocate an NSString instance and never release it.

So, outside of finding memory leaks, what does the tool do? Here's a short list of some of the most useful checks it performs:

- Proper use of retain and release
- Unused instance variables
- Uninitialized variables
- Unreachable code paths
- Referencing null pointers
- Division by zero
- Dead stores (assigning a value to a variable that is never used)
- Objective-C method signatures with type incompatibilities
- Missing dealloc
- Unused instance variables

What's interesting about this list is that some of these tests are things you would never catch with profiling. We found the leaking objects in Section 28.2, *Investigating Performance with Instruments*, on page 517 largely because we were leaking so many of them, so fast. With so many objects naturally created as part of an application's life cycle, it might be a lot harder to find a bug that steadily but slowly leaked objects over a long time. However, a reference-counting bug like that is something that scan-build could catch. Trying to write to a variable that's already been freed is potentially a crashing bug, and catching it early with static analysis could save you, your testers, and your users some pain. Finally, it's worth remembering that profiling won't catch bugs in branches of your code that you don't take in your testing, but the Static Analyzer looks at every possible path through your code, again finding bugs that you might otherwise miss.

The Clang Static Analyzer is still very early and incomplete, but it does enough right that it could be a useful addition to your testing toolkit.

What's Next?

In this chapter and the previous one, we've looked at a number of tools provided by Xcode to help you ensure that your application runs correctly and efficiently. We started in the previous chapter by looking at how to deal with common compilation errors and how to find documentation and moved on to more sophisticated debugging techniques for diagnosing and resolving more serious problems. In this chapter, we moved into performance by investigating CPU activity, first with Shark and then with Instruments, the latter tool also giving us the ability to track object allocations that leaked memory. We also stepped onto the cutting edge by trying out the Clang Static Analysis tool to find memory leaks and other bugs at build time.

There are a few Instruments that we haven't specifically tried out in this chapter, which will be of use to some iPhone application developers. There are instruments to measure the graphics performance and CPU usage of Core Animation and OpenGL ES calls, as well as a System Usage instrument to measure I/O calls such as opens, closes, reads, writes, and so on.

There are also a number of Mac-specific Instruments, including a highly desirable UI Recorder, which allows you to record a set of interactions with your GUI and then "replay" them in order to find and fix performance issues specific to *how* the application is being used. Although these tools may not be available for the iPhone SDK now, the fact that there are Mac equivalents provides some hope that they might be ported over in the future.

Although it's not a performance tool per se, iPhone SDK 3.0 also adds integrated support for *unit testing*, which allows you to write code that tests the logic and runtime behavior of your code, thereby allowing you to automate some of your testing. You do this by selecting the project's Targets and adding new Unit Test Bundle targets. You then create new classes in your project that target the unit test bundle target, rather than the application target. The process is fairly involved, particularly for runtime tests, so if unit testing is an important part of your development process, you'll want to study the "Unit Testing Applications" chapter of the *iPhone Development Guide* [App08d].

Chapter 29

Before and After

So, here we are. We have come a long way since the opening paragraph encouraged you to build something great. We're sure after reading the book your head is teeming with ideas. In this chapter, we'll take a step back and talk about the stuff around the technology that goes into a successful application.

We kick off our discussion with starting correctly. Before you can write code, you have to decide what you are going to write in the first place. Of course, this can be a huge part of the battle to get something shipping. It's hard to narrow down what you are going to build into a defined subset of all the great ideas flying around. It is, however, one of the most important things you can do to make sure you are able to ship. If you don't plan, you can end up with a kitchen-sink application. The kitchen-sink apps can be fun to build but can often be hard to sell. We will talk about some of the strategies you can use to nail down the exact feature set you want to build.

Once the major functionality of your application is in place and working, it is important to go through your application looking for places that you can polish. Users appreciate a polished app and are more likely to talk about it and share it with their friends. This is one of the areas that is so often ignored but is so important to the success of your application.

After you have built your application, it is important to test it. Getting your app in testers' hands can be a daunting task. The next section talks you through the basics of what you need to do to set up a successful beta test via ad hoc distribution. We will also cover some things to do to polish your application for your testers so that their initial impression is as positive as possible.

Now that your application is polished and tested, it's ready to go the next step: promotion. In the final section of this chapter, we talk through some ideas of what you can do to generate buzz about your app. There is a lot to generating buzz, and much of what you do will depend on the community your app is aimed at. We will talk about several general ideas that can help you get attention.

Let's get started at the beginning.

29.1 Starting Right

When we have a great idea, we're so tempted to fire up Xcode and start code slinging. Long, hard nights of refactoring have taught us that this is the wrong approach. Although we can make great progress on fleshing out our ideas with Xcode and Interface Builder, the cost of change goes up dramatically when an idea is committed to code.

Before you are ready for Xcode, you should do at least two things: define an application definition statement and draw out a paper prototype. Instead of coding up a prototype to further understand your idea, flesh it out in your mind and on paper. Write down what your application is going to do and what the intended audience is. Use a pencil and paper, really.

The application definition statement, ADS for short, is probably one of the most important things you can do for your application. This statement, one sentence long, helps gel what your application is all about. It helps you make decisions on features and pieces of your UI that will make your application much more focused and clean than it would be otherwise.

Writing an ADS is not hard, but it will force you to articulate what your application is all about. There are three parts to any good ADS:

- The differentiator
- The solution
- The audience

The differentiator is what makes your application special. Is your application going to be "easy to use"? Then that can be your differentiator. This part of your ADS sets your application apart from other applications in the space. Instead of easy to use, you might want the application to be "graphically enticing." Whatever it is that you want your application to stand out for should be the differentiator.

The solution defines what problem your application is going to solve. Are you going to provide some information, organize information, or allow your users to share photos? The solution you are going to provide defines the problem your app is going to solve for your user. From this part of the statement you will get a list of features, and then you will delete all that do not apply to your audience.

The audience defines who the app is written for. Are you building a tool for serious photographers or an entertaining application for children? Keeping focused on your audience helps you cull your feature list to those that are important for your audience. If you are setting out to make an entertainment app for kids, your set of features and the user interface you choose to present them will be vastly different than if your audience is salespeople on the road.

To some, this sounds crazy. "I could spend a whole week writing down an ADS and drawing out my UI and have no code to show for it!" Although you won't have any code to show for your efforts, you will, however, have a much better idea of how your idea translates into a real application.

All too often we have sat down with Xcode and started coding only to realize hundreds of lines of code later that the metaphor we had in mind does not work. At that point we're faced with the daunting task of rewriting the application or trying to tweak the metaphor. The cost of a rewrite can be huge, so we end up tweaking. What we end up with is not what we wanted the application to be.

Practically, using pencil and paper to draw out your application translates into some broad ideas:

- Really use pencil and paper, not OmniGraffle. Although it's tempting to use software to plot out your UI, that often leads to tweaking the drawing rather than fleshing out the idea. The point is to get your ideas written down, not to learn the intricacies of getting a pixel-perfect drawing.
- Draw your ideas on index cards or screen-sized sheets of paper. Then you can simulate the real UI look by sliding in the next screen when a button is clicked. This makes for a great way to

show your users what the app will really look like without having to write any code.
- As you find parts of your interface that don't work, throw out those pieces of paper and quickly create another sketch. Paper drawings feel more dispensable, and that's what you want at this point—cheap ideas that can be thrown away if they aren't working.
- Spend time on "the stuff that matters." It is not always easy to tell "what matters," but think about what makes your application unique, what will be on each screen, how will it be presented—that kind of stuff. Don't spend a lot of time on stuff you will simply be reusing, like a nav bar or picker list. In other words, it's important to draw out what your application looks like, not to get the drawing of a particular picker list exactly right.
- Share and review the paper with someone else. If you have a team, make sure they are involved in reviewing (and even coming up with the paper in the first place). If you don't have someone who can review the paper, make sure to share it with someone who is a "representative" user of your application. The more feedback you get at this point, the easier building the perfect app will be.
- Since everything is on paper at this point, it's cheap to add both layouts to your application (landscape and portrait) and very cheap to throw out one that does not work. Users love getting a different layout when they rotate the device, but if that layout does not make sense, it's better to find that out on paper than after many hours of coding.

Now that you have a good idea of what your application will look like, it's time to start thinking outside the box. Drawing a UI on paper gives us insights into what our application will look like and even how it will work. But, there is a lot of cool technology that comes with the iPhone, and now that you know what your app will do and how it will look, it is time to think about the "wow factor." What can you add to your application that will make your users want to talk about it with their friends? Is there some way to add a "shake gesture" that makes sense? Can you use the "always on" Internet connection to add a social aspect to your app?

These kind of "thinking outside of the box" features are what will add that extra something to your app. These kinds of features are another reason to favor paper prototypes rather than code. It's easy to add location awareness to paper. Just draw it out, and if it does not work for the application, it is just as easy to pitch it.

Keep in mind that when putting together the "wow factor" features, it's easy to get carried away. And if you are carried away in code, the emotional attachment is much greater. Paper is easy to crumple up and put into the waste bin, so it's much easier to get rid of paper "wow factor" features that don't work than it is to delete them from code.

As always, the important thing to think about in any feature you add to your application is the user. If the great feature or gesture only adds clutter to the app, then it's much better to leave it out. Even if it's cool looking, in the end it detracts from the core of your application.

Consider a "shake gesture" again for a moment. It is kind of cool that you can shake an iPhone and get something to happen. But does the shake make sense to your users? Ask yourself these kinds of questions. Will my users know that they can shake the device to get this function or feature? If not, is there another way for them to invoke the function? Questions like this are very important and can be easily missed in the excitement of adding the "wow factor" features.

Another great way to find these cool features is to show your prototype to users and get their feedback. Often your potential users will have some great ideas for cool features that you had not even thought of.

It is doubly hard when writing the feature into code. Who wants to think through whether the "wow factor" is actually useful when in the midst of coding? We hate to keep repeating this, but it's another great reason to put your idea on paper first. It's lots easier to pitch the idea if you have invested only some graphite and paper to it.

It is often these seemingly little things, choosing what to pitch, what small tweak to add, what cool gesture to incorporate. These things make people want to show off the application to their friends. But if one of these features or gestures is clumsy or just gets in the way, it can have the opposite effect that you were hoping for. Ultimately, the reason to build any app in the first place is to make someone's life better, not more complex. Always keep your user at the center of your decisions, and you will find it hard to go wrong.

Now that we have talked about the front end of the process, let's fast-forward to the other end and talk about polishing up your application before letting it loose in the wild for some testing.

29.2 Polish

Once the application is working, it's often tempting to send it out for testing right away to get all your hard work in front of some real users and get their feedback. Although early feedback and testing is important, we think it is best to take a step back after the heavy lifting is done and look for areas to polish.

Polish can mean many different things, but the general idea is to make your application seamless and simple to use. Questions like these are important to consider:

- How does my app handle interruptions, such as incoming calls?
- Are the major features/functions easy to get to and use?
- What is the perceived performance profile? Does the application start quickly enough?
- Do any areas in the application that use more power than it should /could?

Although none of these questions has a universal answer, it is important that you ask them before calling your application done. Users really appreciate the attention to detail that is paid in these areas even if they don't send you an email saying "great job" on handling incoming phone calls.

Another area to consider when polishing up your application is the startup image. On the device it can take a bit of time for your application to start up, especially if it's doing some network access or other intensive process before becoming active.

One trick of the trade to improve perceived performance is to add a Default.png file to your application. When your application is starting up, this image is loaded and displayed to the user right away. So, what many applications do is take a typical screenshot of what the application might look like at startup and make that the Default.png file. This gives the users the impression that the application is going before it really is. In addition to the Default.png, you can add one file for each URL scheme that your application can load. For example, in Chapter 26, *Application Integration*, on page 481, we built an application that was opened via URLs with the integrated scheme. We could add a Default-integrated.png image to that application, and each time it is launched with a URL, that image would be used instead of the Default.png image.

Handling interruptions gracefully is equally important. Know your users and how they expect to use your application. If your application has very little context, then restarting where the user left off is not very important. But as the amount of context built up increases, it becomes more important to be able to restore the previous state of your application with each launch. In our applications, we use SQLite to store the context so that we can restart at the same spot each time the application is launched (for more info on SQLite see Chapter 10, *The SQLite Database*, on page 185). Every application will be different; just make sure to think through how your users expect the application to work, and make it work that way.

One last bit of polish to take care of before you email the beta version of your application out to testers is the icons. You need two, one for the iTunes icon that will be used to represent your application there and the other for your application on the iPhone home screen.

For your application icon, you need a couple of things. The icon for the app should be 57x57 pixels in PNG format. Add it to the Resources section in Xcode. You also need to update Info.plist to specify the file. The code to do that is listed here:

```
<key>CFBundleIconFile</key>
<string>icon.png</string>
```

Just replace the icon.png value with the name of your icon, and the iPhone will display that instead of the generic white blob.

For iTunes you need to add an image file called iTunesArtwork (note the lack of extension) to the Resources group in Xcode for your project. This file is used by iTunes to represent your application under the Applications tab. The file should be a 512x512 PNG or JPEG image and ideally be the same as the icon for your application on the device.

Now that your application is polished, it's time to send it to a group of real-world users for beta testing.

29.3 Other Features

Another thing to consider is using some of the web-based features that Apple introduced with iPhone OS 3.0, namely, in-application purchase and push notifications.

In-application purchase allows your users to purchase additional add-ons to your application after they have purchased your application. Is it possible for you to offer additional features or components of your application to your users after the fact? If so, it might make sense to integrate with the in-app purchase features. No only do you need to write the pieces in your application, but you also need a web server on the Internet that can interact with your application and Apple's servers. Look at the developer documentation for more details about this feature.

Another cool feature of the 3.0 release is push notification. Does it make sense for your application to inform your users of new content? If so, you should consider integrating with the push notification service. This feature also requires you to have a server that is available on the Internet and set up to communicate with Apple's push notification servers.

29.4 Beta Testing

Testing is one of the most important stages in your application's life cycle. When your application makes it way out into the hands of a real-world user is when the rubber finally meets the road. It is vitally important to your application that you don't scrimp on this step. Allow adequate time for your users to use the application and to give you feedback.

Listen to your testers. We know it might seem silly to say that, but why bother getting people to test your application if you don't listen to them? The problem is that sometimes testers' feedback is going to be negative. It is often hard to take negative feedback, but it is often some of the most helpful. It is so much easier for your application to address the negatives in this early stage than it is to dig out of a bunch of negative reviews on the App Store. That brings us to another really important point.

Choose your beta testing group carefully. You want to have people who will take the time to test and give you honest feedback. If you get only positive feedback, it's important to go fishing for what people did not like. Push hard for what did not work or what people did not like. Although it's nice to get positive feedback, it rarely helps the application get better.

The details of the process to get your application into beta testers' hands is constantly evolving, but the basic idea is that you need to get an "ad hoc distribution profile" for the tester's device. In order to do that, you need the device ID from each device that will be used in the test. To get that, you can connect the device to a computer with Xcode installed, open the Organize window, choose the device, and grab the ID from the summary page. Or if you don't have access to the device, have the user connect their device, launch iTunes, and select the device. In the Summary tab of iTunes, clicking the Serial Number field switches to the Identifier field. After switching to Identifier, the user can hit Copy (or Command+C) to copy the number into the pasteboard. They can then launch mail, paste the ID into a new mail message, and send it to you.

Once you have your testers' device IDs, you can take them to the iPhone developer portal and get an ad hoc distribution profile for each device. Follow the instructions from Apple exactly because any deviation can lead to a failure to deploy on your users' devices.

Now that you have your application in real users' hands, you need to provide great support. When bugs are reported, make sure to send the tester a note letting them know you got the bug report and that you are looking into it. People are motivated by positive feedback, and it's really important to keep your testers motivated.

Finally, a note on finishing well—you should be careful to address as many bugs as you can during the beta phase, but be careful of scope creep disguised as a bug report. Keep focused on the "main thing" that your application addresses. Many great ideas will come in from your testers. If it's not a bug, put it aside for later implementation. Too many applications have languished under feature creep. Shipping software beats perfect software every time.

29.5 Getting into the Store

There has been a lot of press about Apple's application approval process, and some of it can be a bit discouraging. Don't let that get to you, though. As with so many things in life, bad news makes headlines, while good news goes unnoticed. Chances are very high that you will get approved, but here are a few ideas to help to ensure a first time success.

First things first, make sure to get your legal stuff settled well before you want to release an app. That means establish a bank account for your business, get a taxpayer identifier, and set up all the other local requirements in place to be a real company. Make sure you get the contracts signed with Apple as well as getting this information to them.

It takes time for the machinery to grind through all this and actually verify you are who you say you are, so don't expect to be ready on day one of submission to be able to ship an app. Also keep in mind that there are thousands of developers with many new ones coming online every day. Apple has a lot of people doing this work, but it still takes time to process each new developer.

When the docs tell you not to use private APIs, they really mean it. If your application links to and uses private frameworks, you will be rejected. Apple documents this kind of stuff fairly extensively and provides many warnings about it, but many rejections still happen because people ignore these restrictions. It might be tempting to use text to speech, but if you do, you will be rejected.

If you do get rejected and the reason seems bogus to you, make sure to take a deep breath, step away from it for a day or so, and then look at it again with fresh eyes. Most of the app rejections are valid, and changing the application really is the right thing to do.

If after a bit of time away from the issue you decide the rejection is bogus, try resubmitting. The reviewers are human, after all, and a fresh set of eyes might look at your application differently. Also keep in mind that the reviewers have to review hundreds of applications every day, and mistakes happen, so don't take it personally. Just resubmit, and see whether a new reviewer has a different perspective. Check with colleagues to see whether there's a restriction you missed or whether Apple might be making a mistake. If your app gets rejected again, refer to the previous paragraph.

29.6 Promoting Your Application

Marketing and promoting your application can be a full-time job. If you are an independent developer, it is easy to get sucked into doing this full-time. If you are not an independent, you might have folks to take care of all this for you.

If you are not that person with a team to handle your marketing and promotion, then read on. If you are forced to wear multiple hats and are looking for the biggest bang for the buck, this section is for you.

Getting the word out can be tough, but the real secret is to build a network of people that know and trust your work so that they can help you. Get to know people via online social networks like Twitter, Facebook, or LinkedIn. Joining local developer meetups is another way to build a network. Then once your product is ready, make sure to ping your network with the facts. Don't spam people, but politely let them know your app is ready and available.

Another often underutilized group of people is your testers. If someone has had the chance to use your application and liked it, make sure to ask them for a review on iTunes as well as asking them to spread the word.

Press releases can be effective at getting the word out. A comprehensive list of sites that review iPhone applications would be impossible to maintain because of the very dynamic nature of the marketplace. But spending thirty minutes googling around for sites can be quite productive and help you find several avenues to get people exposed to your application.

Please make sure to customize your press releases for the site. If you write a generic press release, you will get a generic response (all too typically the spam filter). If you have a game and are writing to a site that specializes in games, focus on the "gameness" of your app. If you are writing to a site that reviews all kinds of apps, write about the fun or challenge of your game. Use your common sense, and write a specific press release to a specific person or site. They will appreciate it and will be much more likely to publish you.

Finally, a word about support—when someone sends you an email asking questions about your application or asking for a feature, and so on, make sure to get back to them quickly and politely. We know it sounds crazy to have to tell people to be polite to their customers, but sometimes user requests can seem a bit crazy; always tell them thank you and be polite. In the end, a polite response to a support email might make the difference between an angry customer who talks bad about your application and a customer who recommends your stuff to their friends.

This chapter is far from comprehensive of all the things that you can and will have to do to have a successful application. What we covered is what has worked (or not) for us as independent iPhone application developers. It is our hope that the information contained in this chapter will help you attain whatever your goals are with your application. If you have other ideas, please share them on the book's forum. Good luck!

Appendix A

Bibliography

[Apa09]   Apache Software Foundation. Authentication, Authorization and Access Control. http://httpd.apache.org/docs/2.0/howto/auth.html, 2009.

[App06a] Apple, Inc. *Cocoa Fundamentals.* http://developer.apple.com/documentation/Cocoa/Conceptual/CocoaFundamentals/CocoaDesignPatterns/CocoaDesignPatterns.html#//apple_ref/doc/uid/TP40002974-CH6-SW1, 2006.

[App06b] Apple, Inc. *Toll Free Bridging.* http://developer.apple.com/documentation/Cocoa/Conceptual/CarbonCocoaDoc/Articles/InterchangeableDataTypes.html, 2006.

[App07a] Apple, Inc. *Carbon-Cocoa Integration Guide.* http://developer.apple.com/documentation/Cocoa/Conceptual/CarbonCocoaDoc/index.html, 2007.

[App07b] Apple, Inc. *Core Audio Overview.* http://developer.apple.com/documentation/MusicAudio/Conceptual/CoreAudioOverview/, 2007.

[App07c] Apple, Inc. *Introduction to Quartz 2D Programming Guide.* http://developer.apple.com/documentation/GraphicsImaging/Conceptual/drawingwithquartz2d/dq_intro/chapter_1_section_1.html, 2007.

[App08a] Apple, Inc. *Archives and Serializations Programming Guide for Cocoa.* http://developer.apple.com/DOCUMENTATION/Cocoa/Conceptual/Archiving, 2008.

[App08b] Apple, Inc. *Event-Driven XML Programming Guide for Cocoa.* http://developer.apple.com/documentation/Cocoa/Conceptual/XMLParsing/, 2008.

[App08c] Apple, Inc. *iPhone Developer Program.* http://developer.apple.com/iphone/program/, 2008.

[App08d] Apple, Inc. *iPhone Development Guide.* http://developer.apple.com/IPhone/library/documentation/Xcode/Conceptual/iphone_development/, 2008.

[App08e] Apple, Inc. *Key-Value Coding Programming Guide.* http://developer.apple.com/documentation/Cocoa/Conceptual/KeyValueCoding, 2008.

[App08f] Apple, Inc. *Obtaining your iPhone Development Certificate.* http://developer.apple.com/iphone/manage/certificates/team/howto.action, 2008.

[App08g] Apple, Inc. *Predicate Programming Guide.* http://developer.apple.com/documentation/Cocoa/Conceptual/Predicates/predicates.html, 2008.

[App09a] Apple, Inc. *HTTP Live Streaming Overview.* http://developer.apple.com/iphone/library/documentation/NetworkingInternet/Conceptual/StreamingMediaGuide/, 2009.

[App09b] Apple, Inc. *Introduction to Core Data Programming Guide.* http://developer.apple.com/documentation/Cocoa/Conceptual/CoreData/cdProgrammingGuide.html, 2009.

[App09c] Apple, Inc. *iPhone Application Programming Guide.* http://developer.apple.com/iphone/library/documentation/iPhone/Conceptual/iPhoneOSProgrammingGuide, 2009.

[App09d] Apple, Inc. *Technical Note TN2188: Exporting Movies for iPod, Apple TV and iPhone.* http://developer.apple.com/technotes/tn2007/tn2188.html, 2009.

[App09e] Apple, Inc. *The Objective C Programming Language.* http://developer.apple.com/documentation/Cocoa/Conceptual/ObjectiveC, 2009.

[Dud08] Bill Dudney. *Core Animation for OS X: Creating Dynamic Compelling User Interfaces.* The Pragmatic Programmers, LLC, Raleigh, NC, and Dallas, TX, 2008.

[GL06]   David Gelphman and Bunny Laden. *Programming with Quartz, 2D and PDF Graphics in Mac OS X*. Morgan Kaufman, San Francisco, 2006.

[Hip09]   Inc. Hipp, Wyrick & Company. Sql features that sqlite does not implement. http://www.sqlite.org/omitted.html, 2009.

[KP88]   Al Kelley and Ira Pohl. *A Book on C*. Addison-Wesley Professional, New York, 1988.

[KR98]   Brian W. Kernighan and Dennis Ritchie. *The C Programming Language*. Prentice Hall PTR, Englewood Cliffs, NJ, second edition, 1998.

[Pan09]   R. Pantos, Ed. Http live streaming (internet-draft). http://www.ietf.org/internet-drafts/draft-pantos-http-live-streaming-01.txt, 2009.

[PT08]   Maija Palmer and Paul Taylor. Google homes in on revenues from phones. *Financial Times*, February 13, 2008.

[Ras07]   Bertis Rasco. Where's the wiimote? using kalman filtering to extract accelerometer data. *Gamasutra*, June 20, 2007.

[RSWLH07] Jr. Richard S. Wright, Benjamin Lipchak, and Nicholas Haemel. *OpenGL SuperBible*. Addison Wesley Longman, Reading, MA, fourth edition, 2007.

[SPHB08]   Thomas SchlÃűmer, Benjamin Poppinga, Niels Henze, and Susanne Boll. Gesture recognition with a wii controller. In *Proceedings of the 2nd international conference on Tangible and embedded interaction*, pages 11–14, New York, 2008. ACM Press.

[Ste09]   Daniel H Steinberg. *Cocoa Programming: A Quick-Start Guide for Developers*. The Pragmatic Programmers, LLC, Raleigh, NC, and Dallas, TX, 2009.

[Sto08]   Brad Stone. iphone users love that mobile web. *The New York Times*, June 19, 2008.

[Uni04]   Inc. Unicode. Unicode technical standard 35: Locale data markup language (ldml). Unicode Technical Reports, 2004.

[Zar09]   Marcus Zarra. *Core Data: Apple's API for Persisting Data under Mac OS X*. The Pragmatic Programmers, LLC, Raleigh, NC, and Dallas, TX, 2009.

Index

Symbols
#pragma mark, 262
* character, 48, 490
+ character, 31
- character, 31
%@ format specifier, 36, 499
_ characters, 272
:// resource specifier, 485

A
a-law, 336
AAC, 335
Accelerometer, 423–439
   axes of, 424f
   device orientation, 424–425
   filtering data, 434f, 432–439
   overview, 423
   raw data access, 426–431, 432f
   shake detection, 425–426, 434, 436
   simulator and, 423
Accessory type, 108
Actions, 30, 34, 237
   adding, 152
   adding to view controller, 58–60
   button, 51–55
   implementing, 36–37
   target/action paradigm, 55
Add button, 91
Address Book, 441–450
   adding contacts to, 447–450
   people picker configuration, 444–445
   people picker delegation, 443–444
   person controller, 445–447
   user interface, 441–442
AddressGeocoder, 472, 472 – –477
ADPCM, 336
ALAC, 336
Alignment, 32
alloc, 359

Allocator, 359, 360
Angle brackets, 65
Animation, 409–422
   customizing, 413
   delegation, 414
   layers, 411f, 418f, 417–419
   OpenGL and, 419–422
   overview, 409–410
   UIView, 410–416
Animation blocks, 410, 412
Apache, Bonjour-enabled websites, 270, 271f
App Store, 2, 537–538
Event-Driven XML Programming Guide for Cocoa (Apple, Inc.), 263
Application anatomy, 38–40
Application definition statement (ADS), 530
Application delegate, 38, 41
Application ID, 138
Application memory, managing, 44–45
Application music player, 306
Applications, 529
   approval for, 537–538
   beta testing, 536–537
   crashing, 502
   flippable views, 164f, 163–165
   integration of, 481–486
      launching other applications, 481–483
      overview, 481
      preparation for, 483–486
   navigation-based, 105–108
   photos and, 397
   polishing, 534–535
   promoting, 538–539
   starting and planning for, 530–533
   web-based features, 535–536
Arcs, 388

Arrays, 80, 122
  PreferenceSpecifiers, 175
Aspect Fit, 400f, 401
Asynchronous file reading, 148–152
Attributes inspector, 32, 39, 98
  tab bar controller, 119
Audience, 530, 531
Audio category, 353
Audio Conversion Services, 369
Audio file formats, 330
Audio File Services, 368
Audio File Stream Services, 368
Audio formats, 300
Audio Queue Services, 327, 368
Audio session services, 366–367
Audio sessions, 353–355
Audio Unit Graph, 370
Audio Units, 369–370
Audio/visual formats, 300
Authentication, Authorization and Access Control (Apache), 248
AuthenticationChallengeViewController, 251
Autorelease pool, 38
Autoreleasing, 44
AVAudioPlayer, 350, 352
AVAudioRecorder, 329f, 327–331
  delegate callbacks, 341
  methods for, 341
  recording levels, 347f, 343–348
  tab-based application, 328
  using, 339–343
AVAudioSession, 353–355
Average power level, 343
AVFoundation
  AVAudioRecorder, delegate callbacks, 341
  AVAudioRecorder, methods for, 341
  AVAudioRecorder, using, 339–343
  encoded formats, 337f, 335–339
  key-value pairs, 335
  playing audio, 348, 349f, 348 − −352
  properties, 333
  recording levels, monitoring, 347f, 343–348
  sample application setup, 328–331
  settings, 333
  tab layout, 329f
  uncompressed audio formats, 333f, 331–335
Axes, accelerometer, 424f

B

BalanceBall example, 426, 432f
Bar graph drawing, 390f
Beta testing, 536–537
Bezier curves, 388
Bluetooth, *see* Game Kit
Bonjour
  enabling Apache, 270, 271f
  network services, 269–271
  philosophy of, 269
  service discovery, 273f, 271–277
  web viewing, 276f
A Book On C: Programming in C (Kelley & Pohl), 450
Boolean preference, 178f
Breakpoint, 504f
Browsers, *see* Web browsers
Bug, in iPhone OS SDK, 405
Bugs, *see* Debugging
Build errors, 489f, 488–490
Bundle, settings, 175, 176f
Bunny, Laden, 393
Button action, 51–55
Buttons, 91
  callback method, 153
  done, 230
  text for, 33

C

cabinetController, 109
CAEAGLLayer, 419
.caf files, 330
CALayer, 416
Callback method, 153
Callbacks, 243
Carbon-Cocoa Integration Guide (Apple, Inc.), 443
Cell identifier, 85
Cell reuse, 84
Cell styles, custom, 96f, 94–98, 99f
Cells, table, 83f, 81–85
center, 412
Certificate, developer, 19
CFNetwork, 244
CFRelease, 360
CGContextDrawImage(), 401
CGRect, 84
Checkmark, 108
Circles, 388
Clang Static Analyzer, 525f, 526f, 523–528

Classes, 24, 26–28
    collection, 27
    described, 28
CLLocationManager, 451, 454
Clock application, 118
Clock, constant updates for, 172–173
Cocoa Fundamentals Guide: Cocoa Design Patterns (Apple, Inc.), 51
Cocoa Programming: A Quick-Start Guide for Developers (Steinberg), 24
Cocoa Touch, 24
    classes, 26–28
    Core Foundation references and, 359
    delegation, 40–43
    development process, 58
    drawing, 385
    fundamentals, summary of, 48–49
    reference counting system, 44
    schemes and, 483
    service resolution, 274
    support, 483
    URL Loading System, 240
    see also Objective-C
Code navigation, 18
"Coding in Objective-C 2.0" (screencasts), 23
Collection classes, 27
Command objects, 374
Compass, 460–462
Compiler errors, 489f, 488–490
Completion process, 363
Conference application, *see* CoreData/Conference project
Connections inspector, 39, 60, 70
Contact Mapper
    AddressGeocoder, 472–477
    annotations, 467–477
    annotations, selecting, 477–479
    described, 463–464
    recentering a map, 470
    showing a map, 464–467
ContactAnnotation, 467
Container formats, 299
Content view, 95
contentView, 95
Coordinates, 453f
Core Animation, 409–422
    customizing, 413
    delegation, 414
    layers, 411f, 418f, 417–419

    OpenGL ES, 419–422
    overview, 409–410
    UIView, 410–416
Core Animation for OS X: Creating Dynamic Compelling User Interfaces (Dudney), 414, 416, 419
Core Audio, 357–371
    Audio Conversion Services, 369
    Audio File Services, 368
    Audio File Stream Services, 368
    Audio Queue Services, 368
    audio session services, 366–367
    Audio Units, 369–370
    four-character codes, 340
    OpenAL, 370
    overview, 357
    Procedural C APIs, 358–360
    settings, 333
    system sounds, 360–366
Core Data, 203–234
    CoreData/Conference project
        changing tracks, 225–228
        Core Data stack, 211–212
        creating, 207–210
        description of, 204–207
        fetching tracks, 224
        modeling classes, 213–219
        navigation, 229–234
        table views, 220–224
    overview of, 203
    persistent object stores for, 209
    vs. SQLite, 186, 203
Core Data Documentation (Apple, Inc.), 223
Core Data stack
    building, 211–212
    components of, 210
    modeling classes, 213–219
    objects, 208
CoreData/Conference project, 204–207
    adding and deleting tracks, 205
    changing tracks, 225–228
    Core Data stack, 211–212
    editing tracks, 205
    fetching tracks, 224
    modeling classes for, 213–219
    navigation, 229–234
    selecting as session, 206
    table views, 220–224

Core Data: Apple's API for Persisting
   Data under Mac OS X (Apple, Inc.),
   222, 223
Core Foundation
   Cocoa and, 359
   memory management, 359
   rules for, 360
   string objects, 448
Core Graphics, 385–395
   bar graphs, 390f
   drawing model, 386f, 385–386
   filled view, 389f
   graphics context, 392–393
   paths, 387–392
   redrawing, 394
   vector drawing, 386–387
Core Location, 451–462
   accuracy and, 456
   compass, 460–462
   components of, 455
   coordinates, 453f, 451–456
   distance, 457, 458f
   location updates, 456–460
   service parameters, 454
   see also Map Kit
CoreData/Conference project
   creating, 207–210
create table, 187
CreateDirectoryViewController, 131
CreateFileViewController, 131, 157
createNewDirectory, 155, 156f
createNewFile, 154
Ctrl+drag, 60, 70
Currency formatter, 68
Curves, 388
customizableViewControllers property, 126

D

Data storage, see Filesystem
Databases, 185–202
   adding to device, 191–193
   Core Data vs. SQLite, 186, 203
   creating, 186–188
   creating on desktop, 193
   function calls for, 194
   inserting values, 195–197
   reading values, 198–201
   sample application, 189–191
   SQLite overview, 185
Dead reckoning, 285
dealloc, 58

Debugging, 487–510
   breakpoint, 504f
   compiler errors, 489f, 488–490
   debugger, 503f, 497–505
   importing/linking errors, 491f, 492f, 490–492
   Interface Builder errors, 496–497, 498f
   objects and, 506f, 508f, 505–510
   overview, 487
   remotely, with Shark, 516
   using documentation, 494f, 495f, 492–495
   see also Performance
Decibels, 345
Declarations, 48, 59
Delegate protocols, 41, 49, 149
Delegation, 40–43
   Address Book, 443–444
   animation, 414
   images, 403
   location manager, 452
   table view, 77
deleteObject:, 226
Deleting, table data, 85, 86, 88f
Detail Disclosure Button, 108
Detail pane, 519
Developer certificates, 19
Development tools, see Instruments; Interface Builder; Xcode
Device orientation, accelerometer, 424–425
Dictionary
   data and, 113, 114
   preferences, 176
Differentiator, 530
Directories, 131, 135–139
   creating and deleting, 153f, 156f, 158f, 152–158
DirectoryViewController, 131, 138
Disclosure Indicator, 108
displayName, 283
Distance, 457, 458f
Dot, 377
Documentation, 10, 492
   Research Assistant, 495f
   shortcuts to, 494
Documentation Viewer, 494f
   Research Assistant, 494f
Documents directory, 131
done button, 230

done, 170
Dot operator, 36, 47
Double taps, 378
Drawing, 385–395
   bar graphs, 390f
   filled view, 389f
   graphics context, 392–393
   model for, 386f, 385–386
   paths, 387–392
   planning, before coding, 532
   redrawing, 394
   vectors, 386–387
drawRect:, 390, 394
Dudney, Bill, 414, 416, 419
DVDCase app, 105–108

E

editButtonItem, 88f
Editing table, 85–87, 88f
Editor pane, 13
Email, 265f, 264–267
Encoded formats, audio, 337f, 335–339
Entities, 210, 214, 215
Errors, *see* Debugging
Event model, 374f, 373–375
Event-driven parser, 256
EXC_BAD_ACCESS, 505, 506f
Explicit animation, 410
Extended Audio File Services, 369
Extended detail, Instruments app., 520, 520f

F

Fast Fourier transform, 438
Fetched results controller (FRC), 220, 224
File's Owner, 34, 35, 97, 107, 169, 237, 497
File/resource list, 13
FileContentsViewController, 131
FileOverviewViewController, 131
Filesystem, 129–162
   adding actions, 152
   alternatives to, 129
   asynchronous file reading, 148–152
   creating and deleting directories, 153f, 156f, 158f, 152–158
   directories, 130–132
   directories in, 135–139
   FilesystemExplorer application, 132f

FilesystemExplorer application, creating, 133–139
file attributes, 143f, 139–144
files, reading data from, 148f, 144–148
navigation flow of, 130f
overview of, 129
property lists, 161–162
refactoring code, 133, 134f
writing to files, 158–161
FillingView, 386
Filter predicates, 319
Filters, 432
Flippable view, 164f, 163–165
   changing and updating, 169–172
   preferences, 166–169
FlippableClock, 165
FlipsideViewController, 170
Format specifier (%@), 36
Format strings, 499
Forward declarations, 59, 72
Four-character codes (4cc), 340
Fourier transform, 438
Fragmentation, 1
Frameworks, 29, 490

G

Game alerts, 291f
Game Kit
   data mode, 285
   GKSession communication, 286–290, 291f
   networked game, 279f, 279–280
   networking game logic, 284–285
   overview, 277–278
   peer picker delegate, 281–284
   peer picker setup, 280–281, 282f
   peer-to-peer chat, 290–292
   receiving data, 289
   state changes, 287
   support for, 278
Game logic, 284–285
Gelphman, David, 393
General media items property keys, 307
"Gesture Recognition with a Wii Controller" (Schlömer), 439
GKPeerPickerController, 282f
GKSession, 283, 286–290, 291f
GPS receiver, 455
Graphic user interface, *see* Interface Builder (IB); User interface

Graphics context, 392–393
Grouped style, tables, 75, 76f, 83f

H

Header files, 24, 48, 57
    ConferenceAppDelegate, 208
    errors with, 490
    forward declarations, 59
    instance variables, 258
    Session class, 218
    SQLite API, 192
    Track class, 218
Heads-up-display (HUD), 34
Hello iPhone project, 12f, 11–13
    anatomy of, 38–40
    parts of, 28
    returning the greeting, 29
Hello World project, 19–21
High-pass filter, 432, 433, 435
htpasswd, 249
HTTP authentication, 247f, 252f, 246–253
HTTP LIVE Streaming (Apple, Inc.), 297
Human Interface Guidelines (HIG), 14

I

IB, *see* Interface Builder
IBAction, 30–31
IBOutlet, 30–31, 47
Icons, 535
Identifier, 19
IMA4, 336
Images, 397–407
    display, customizing, 402f, 400–402
    drawing, 399f, 400f, 398–400
    formats of, 398
    image picker, 404f, 402–406
    overview, 397
    startup, 534
    video, 406–407
Implementation files, 24
Implicit animation, 410
#import, 490
Importing/linking errors, 491f, 492f, 490–492
#include, 490, 491f
indexPath, 80
init method, custom, 57
initWithSessionID, 282
insertTrack(), 225
Installation, developer tools, 10

Instance variables, 47, 48, 65, 80, 113, 258
    editing tables, 90
    flippable views, 168
    writing data to files, 159
Instruments, 10
Instruments application, 518f, 520f, 522f, 517–522, 523f
integrated://, 485
Integration, 481–486
    launching applications, 481–483
    overview, 481
    preparation for, 483–486
Interactive application, 29
Interface, *see* User interface
Interface Builder (IB), 10, 13–18
    audio encoding settings, 337f
    browser, building, 236, 237f
    code navigation, 18
    connecting outlets and actions, 68
    custom cell styles, 96f, 94–98, 99f
    debugging, 496–497, 498f
    editing view controller in, 73–74
    file types for, 16
    File's Owner, 97
    GUI layout in, 31–35
    IBAction, 31
    images in, 399
    Movie class, 60
    navigation objects, 92
    table views, 76f
    target/action paradigm, 55
    user interface, adjusting, 16–18
    see also Table views; UIWebView; User interface
Internet, 235–267
    building a browser, 235–240, 241f
    HTTP authentication, 247f, 252f, 246–253
    overview of, 235
    reading network data, 242f, 241–246
    sending mail, 265f, 264–267
    XML parsing, 255f, 257f, 263f, 253–264
    see also Peer-to-peer networking
Internet Low-Bitrate Codec, 336
Introduction to Quartz 2D Programming Guide (Apple, Inc.), 393
iPhone
    audio formats, 330
    audio recording capabilities, 354

components, 76
filesystem in, 129
fundamentals, summary of, 48–49
Hello World application, 12f, 11–13
image formats supported by, 398
impact of, 1–2
Interface Builder, 13–18
   user interface, adjusting, 16–18
Internet use, 235
keyboard, 42, 66
media application names, 304
Multi-Touch feature, 70, 373
rotation, 16
running apps on, 19–21
SDK documentation, 494f, 495f, 492–495
software stack, 24
terminology, for this book, 10
tools for, 10
versions, 354, 405
versions of, 166, 278, 426
Xcode and, 9–11
see also App Store
iPhone Application Programming Guide (Apple, Inc.), 300
iPhone Developer Program (Apple, Inc.), 10, 19
iPhone Development Book (Apple, Inc.), 528
iPod, 303–326
   application name, 304
   library for, 314–324
   media library search, 317
   media queues, 318f
   playback, 303–312
   playback control, 312–314
   playback status, 306–308
   player notifications, 310–312
   time properties, 310f, 309–310
   time, setting, 324
   UISearchDisplayController, 316f
   user interface, 305f, 304–306

K

Key path, 223
Key-Value Coding (Apple, Inc.), 219, 223
Key-Value Coding, 101
Key-value pairs, 176, 335
Keyboard, 42, 66, 70
Keyframe animations, 414

Keywords, *see* IBAction; IBOutlets

L

Labels, 15
Launching applications, integration and, 481
Layers, 409, 418f, 417–419
   CAEAGLLayer, 419
   CALayer, 416
   OpenGL and, 419–422
layoutSubviews, 421
Level meters, 343, 345
Library, 14, 132
   iPod, 314–324
   libXML2, 472
libsqlite3.dylib, 191
libXML2 library, 472
Linear PCM, 331–333, 333f, 335
Lines, 388
Linking errors, 491f, 492f, 490–492
Location, 451–462
   accuracy, 456
   compass, 460–462
   coordinates, 453f, 451–456
   delegation, 452
   distance, 457, 458f
   iPhone components of, 455
   location updates, 456–460
   service parameters for iPhone, 454
   see also Map Kit
Lossy, 335
Low-pass filter, 432, 433, 435

M

Magnetometer, 460–462
Mail, navigation through, 103–104
main(), 40
make-table-script, 188
malloc:, 509
Managed object context (MOC), 205, 210, 212
Managed object model (MOM), 210, 212–219
Map Kit, 463–479
   Contact Mapper
      AddressGeocoder, 472–477
      annotations, 467–477
      annotations, selecting, 477–479
      overview, 463–464
      recentering a map, 470
      showing a map, 464–467

overview, 463
Map View, 465
Markers, 113
Media formats, 299–301
Media Player framework, 314, 315
Memory, 521, 522, 523f
    autorelease of, 38
    Core Foundation, 359
    managing, 44–45, 359
    in Movie class, 57
    properties and, 46
    releasing, 37, 49
    see also Performance
Messages, 26
Method calls, 26
Method declaration syntax, 31
Method declarations, 48
Method signatures, 43
Methods, map annotations, 467
MKAnnotation, 467–477
MKMapViewDelegate, 471
modalTransitionStyle, 166
Models, building, 56–58
Models, for tables, 79
Movie class example, 56–58
    sorting, 101, 102f
    table data, modeling, 79
MovieEditorViewController, 72, 88–89
MPMediaItem, 323f
MPMoviePlayerController, 293–297
    movie selection, 296–297
    notifications, 298–299
    supported media formats, 299–301
    user interface, 295–296
Multi-Touch feature, 70, 373
    event model and, 375
    gestures, 384f, 379–384
    multiple touch, 380f
    tapping, 378–379
    tracking touches, 376f, 376–377
    zoom, 382f
Multivalue objects, 448

N

Navigation, 18, 87–94, 103–115
    applications based on, 105–108
    controller for, 104–105
    mail, 103–104
    navigation bar, customizing, 110–115
    of Track, 229–234
    view controllers and, 109–110
    view controllers, popping, 115
    see also Tab bar controllers
Navigation bar, 78
Navigation controller, 78, 90, 104–105
Navigation items, 78
Navigation objects, 78f, 92
Navigation-based application template, 208
Network keys, 284
Networked games, 279f, 279–280
New projects, 11
nextResponder, 374
.nib files, 16
    adding, 68
    contents of, 39
    owner of, 33, 97
    refactoring, 135f
    tab bar controllers, 124
    view controllers and, 62
Non-FTP protocol, 244
Non-HTTP protocol, 244
Notifications center, 311
NSDateFormatter, 168
NSFileManager, 137
NSLog(), 498, 500, 501, 506
NSNetServiceBrowser, 272, 273f
NSOutputStream, 158
NSStream, 158
NSString, 161
NSTimer, 172
NSZombie, 507, 508f

O

ObjectAlloc, Instrument, 521, 522f
objectForKey, 123
Objective-C, 23
    classes in, 24
    coding in, 24–26
    method calls in, 26
    method declaration syntax, 31
    necessity of, 25
    parameters, 26
    vs. Procedural C, 358
    properties, 45–48
The Objective-C 2.0 Programming Language (Apple, Inc.), 23
Objective-C protocol, 41
Objects, 48
    autoreleasing, 44
    command objects, 374

Core Data stack, 208, 210
debugging, 506f, 508f, 505–510
delegation, 40–43
memory and, 49
multivalue objects, 448
navigation items, 78, 78f
opaque types and, 358, 359
over-release of, 507
as pointers, 489
processing objects, 374
table view data source, 76
Observers, 311
Obtaining your iPhone Development Certificate (Apple, Inc.), 19, 21
1718449215, 340
Opaque, 420
Opaque types, 358, 359
OpenAL, 370
OpenGL ES, 419–422
OpenGL SuperBible (Wright, Lipchak & Haemel), 420, 421
openURL:, 482
Organizing data, *see* Navigation; Tab bar controllers; User interface
Outlets, 30, 34, 237
 adding to view controller, 58–60
Over-released objects, 507

P

P2PTapWar game, 279f, 279–280
Parameters, 26, 359
 GKSession, 282
Parsing XML, 255f, 257f, 263f, 253–264
Password-protected web site setup, 246–249
Paths, 389f, 390f, 387–392
pause, 341
PCM, *see* Pulse code modulation
Peak power level, 343
Peer picker delegate, 281–284
Peer picker setup, 280–281, 282f
Peer-to-peer chat, 290–292
Peer-to-peer networking, 269–292
 Bonjour services, 273f, 276f, 271–277
 Bonjour-enabled, 271f, 269–271
 Game Kit overview, 277–278
 Game Kit, game logic, 284–285
 Game Kit, GKSession, 286–290, 291f
 Game Kit, networked game setup, 279f, 279–280

Game Kit, peer picker delegate, 281–284
Game Kit, peer picker setup, 280–281, 282f
Game Kit, peer-to-peer chat, 290–292
overview, 269
see also Internet
Performance, 511–528
 cell reuse and, 84
 Clang Static Analyzer, 525f, 526f, 523–528
 Instruments application, 518f, 520f, 522f, 517–522, 523f
 Instruments vs. Shark, 519
 overview, 511
 profiling, 515, 516
 Shark and, 512f, 513f, 514f, 512–516, 517f
Persistent object store (POS), 209, 211
Persistent store coordinator (PSC), 210
Photos, uses for, 397
 see also Images
Picker, 169, 170, 190, 349, 403, 443–444, 469
Pin color, 471
Pinch gesture, 382
Placeholder text, 71
Plain style, tables, 75, 76f, 83f
Playback status, 306–308
Playhead, 518
Podcast item property keys, 307
Pointers, 505
Polishing applications, 534–535
Population data, ordering, 121
Power levels, 343
Predicate Programming Guide (Apple, Inc.), 234
Preferences, 132, 163–184
 boolean, 178f
 changing and updating, 169–172
 constant updating (clock), 172–173
 flippable view, 164f, 163–165
 flips vs. bundles, 177
 loading, 181–184
 managing, 166–169
 overview of, 163
 Settings application, 175f, 176f, 178f, 179f, 173–180, 181f
PreferenceSpecifiers, 175

Prime number counter application, 500f
Primitives, 27
Procedural C APIs, 358–360
Processing objects, 374
Products, 29
Profiling, 515, 516, 523
Programming with Quartz (Gelphman & Bunny), 393
Promoting applications, 538–539
Properties, 49
    adding, 57
    dot operator, 36
    layers, 409
    support for, 47
    variables as, 45–48
Properties Inspector, 332
Property lists, 161–162
Protocols, 41, 49
Pulse code modulation (PCM), 331, 333, 333f, 335
Push notifications, 535–536

Q

Queue, 315, 318f, 323f, 324

R

Readability, 46
record, 341
Recording levels, monitoring, 347f, 343–348
Redrawing, 394
Reference counting system, 44
removeItemAtPath:error:, 156
Research Assistant, 495, 495f
Resolution, 274
Resources, 29
Responder chain, 374
Reusing cells, 84
Root.plist, 175, 176f, 179f
RootViewController, 78–81, 93f
RootViewController, 133, 134f
Rotation, of device and text, 16–18
Rows, 75
Run loop, 149

S

Safari WebKit, *see* UIWebView
Sampling, 331
Sandbox approach to security, 130

save:, 226, 230
scan-build, 525, 527
Scrubbing, 304, 313
Search result tables, 321
selectedTimeZone, 171
Sending mail, 265f, 264–267
Service record, 274
Session entity, 215
Session, 232
sessionMode, 283
setMessage:, 484
Settings application, 175f, 176f, 178f, 179f, 181f, 173–184
Settings.bundle, 175
Shake detection, accelerometer, 425–426
sharedApplication, 482
Shark, 512f, 513f, 514f, 512–516, 517f, 519
Shopping list database
    application details, 189–191
    creating, 186–188
    inserting values, 195–197
    reading values, 198–201
Signatures, 43
Simulator, 423
Single taps, 378
Software stack, 24
Solution, 530, 531
Sort descriptors, 100, 122
Sorting table data, 102f, 98–102
Sound, measuring, 345
Source files, 28
Springs, 16
SQLite database, 185–202
    adding to device, 191–193
    vs. Core Data, 186, 203
    creating, 186–188
    creating on desktop, 193
    function calls for, 194
    home page, 185
    inserting values, 195–197
    overview of, 185
    reading values, 198–201
    sample application, 189–191
SQLite's C API, 191
sqlite3, 186, 187
sqlite3_open(), 195
Startup image, 534
States example (tabs), 118
    configuration for, 127

customizing with view controllers, 121–125
tab bar template, 119–120
Static analysis, 523
Steinberg, Daniel, 23
stop, 341
Streaming support, 297
Streams, 146, 148, 151
String programming Guide for Cocoa (Apple, Inc.), 499
String representation, 36
String-appender, 515
Strings
   cell identifiers, 85
   format specifier for, 36, 499
   format strings, 499
Struts and Springs configuration, 16
Symbols list, 19
Synthesize statements, 57
Synthesizing, 46
System Settings application, 175f, 176f, 178f, 179f, 181f, 173–184
System sounds, 360–366
SystemSoundsDemo, 361–366

T

Tab bar controllers, 117–127
   creating, 119–120
   multiple, working with, 125–127
   use of, 117–118
   view controllers for, 121–125
   see also Navigation
Table view data source, 76
Table view delegates, 77
Table views, 75–102
   adding items to, 91, 92
   basic view, 82f
   cells, 83f, 81–85
   custom cell styles, 96f, 94–98, 99f
   custom cells, assigning values to, 98
   custom cells, loading and using, 96
   data modeling, 78–81
   data, sorting, 102f, 98–102
   editing, 85–87, 88f
   media queue, 321
   navigating with, 87–94
   navigation, 77–78
   parts of, 76f, 75–77
   tab bar controllers and, 122
   UITableView, 77–79
   visual styles for, 75

Tags, 98
Tapping, 378–379
Target/action paradigm, 55
Technical Note TN2188: Exporting Movies for iPod, Apple TV, and iPhone (Apple, Inc.), 301
Templates, 11, 119
   MediaPlayer, 294
   navigation-based, 208
   Utility Application, 166
   utility application project, 316
   view-based, 236
Temporary (tmp) directory, 132
Testing, 536–537
Text fields, naming, 32
textFieldDidEndEditing:, 67, 482
textFieldShouldReturn:, 482
Thread handling, 46
Tidy, 473, 475
Time Profile window, Shark, 513
timeLabel, 168
Timeline, Instruments application, 518
Timing curves, 413
Toll-free bridged, 359
Toolbar, 12
Touch Fighter II, 438
TouchXML, 472, 473, 475
Track changing, 225–228
Track fetching, 224
Track table views, 220–224
Tracks, 204, 214, 218
TrackEditingViewController, 229
transform, 417
TrivialCompilerErrors, 488
twentyFourHourSwitch, 169
Twitter, 253
   loading data, 254
   parsing data, 256
   public timeline, 263f

U

u-law, 336
UIAccelerometerDelegate protocol, 427
UIActionSheet, 153f
UIBarButtonItem, 93f
UIButton, 477
UIImage object, 398, 400, 401
UIImagePickerController, 404f, 402–406
UIImageView, 397, 398
UIKit, 373
   animation layers and, 416

UIKit framework, 27
UIPickerView, 348
UIResponder, 425–426
UISearchDisplayController, 316f, 318
UITabBarDelegate protocol, 127
UITableView, 77–79, 82, 83f, 84
UITableViewCell, 95
UIView, 410–416
UIWebView, 239f, 235–240, 241f
Uncompressed audio formats, 333f, 331–335
Underscore characters, 272
updateTimeView, 168
URL Loading System, 241, 244
URL, launching application with, 483
User interface
   Address Book, 441–442
   AudioRecorderPlayer, 333
   browsers, 236, 237f
   images and, 399
   in Interface Builder, 31–35
   iPod, 305f, 304–306
   media queues, 318f
   Movie class model, 60
   .nib files, adding, 68–72
   planning for, 532
   rotation, adjusting for, 16
   tab bar controllers, 117–118
   video playback and, 295–296
   see also Drawing; Interface Builder; Preferences; Table views
User-defined property keys, 307
uses24Hour, 171
Utility application project template, 316
Utility Application template, 165, 166

V

Variables
   adding, 57
   outlets as, 30
   as properties, 45–48
   in source code, 504f
Vector drawing, 386–387
Video, 293–301, 406–407
   MPMoviePlayerController, 293–297
      movie selection, 296–297
      notifications, 298–299
      supported media formats, 299–301
      user interface, 295–296
   overview, 293

View controllers (VC), 51–74
   accelerometer and, 429
   associating to views, 142
   button action, 51–55
   connecting to view, 498f
   creating new, 64–68
   creating programatically, 141
   editing in IB, 73–74
   implementing, 62–64
   MovieEditorViewController, 72
   navigation and, 109–110
   outlets and actions for, 58–60
   overview of, 51
   popping, 115
   properties and customizing, 110–115
   tabs for, 117, 121–125
   as text field's delegate, 42
   tying two together, 115
   user interface, building, 68–72
View-based Application template, 11, 12, 236
viewWillAppear:, 63
viewWillDisappear:, 151
Voice chat, 277, 290–292
Volume level meter, 343, 345

W

Web browsers, 235–240, 241f
Web services, 255f, 257f, 263f, 253–264
Web sharing, 247f
"Where's the Wiimote Using Kalman Filtering to Extract Accelerometer Data" (Rasco), 439
"Wow factor", 533
Writing, to files, 158–161

X

Xcode
   adding MediaPlayer, 294
   as app bundle, 13
   application definition statement, 530
   autocompletion feature, 493
   Build and Go, 15
   Build Results window, 489f
   code navigation, 18
   Debugger Console, 497
   Debugger window, 503f
   documentation for, 10
   Documentation Viewer, 494f
   editor pane, 13
   Hello iPhone project, 12f, 11–13

Interface Builder, 13–18
    user interface, adjusting, 16–18
importance of, 29
installation, 9–11
libXML2 headers and, 474
refactoring in, 133, 134f
Research Assistant, 495, 495f
running apps, 19–21
Shark, 512f, 513f, 514f, 512–516, 517f
syntax highlighting, 488, 493
templates for, 11, 119
    toolbar, 12
    Utility Application template, 165, 166
.xib file types, 16, 496
XML parsing, 255f, 257f, 263f, 253–264

Z

Zero Configuration Networking, The Definitive Guide (O'Reilly), 277
Zombie objects, 506f, 508f, 505–510
Zooming, 382
zPosition, 417

The Pragmatic Bookshelf

Available in paperback and DRM-free PDF, our titles are here to help you stay on top of your game. The following are in print as of December 2009; be sure to check our website at pragprog.com for newer titles.

Title	Year	ISBN	Pages
Advanced Rails Recipes: 84 New Ways to Build Stunning Rails Apps	2008	9780978739225	464
Agile Coaching	2009	9781934356432	250
Agile Retrospectives: Making Good Teams Great	2006	9780977616640	200
Agile Web Development with Rails, Third Edition	2009	9781934356166	784
Augmented Reality: A Practical Guide	2008	9781934356036	328
Behind Closed Doors: Secrets of Great Management	2005	9780976694021	192
Best of Ruby Quiz	2006	9780976694076	304
Core Animation for Mac OS X and the iPhone: Creating Compelling Dynamic User Interfaces	2008	9781934356104	200
Core Data: Apple's API for Persisting Data on Mac OS X	2009	9781934356326	256
Data Crunching: Solve Everyday Problems using Java, Python, and More	2005	9780974514079	208
Debug It! Find, Repair, and Prevent Bugs in Your Code	2009	9781934356289	232
Deploying Rails Applications: A Step-by-Step Guide	2008	9780978739201	280
Design Accessible Web Sites: 36 Keys to Creating Content for All Audiences and Platforms	2007	9781934356029	336
Desktop GIS: Mapping the Planet with Open Source Tools	2008	9781934356067	368
Developing Facebook Platform Applications with Rails	2008	9781934356128	200
Enterprise Integration with Ruby	2006	9780976694069	360
Enterprise Recipes with Ruby and Rails	2008	9781934356234	416
Everyday Scripting with Ruby: for Teams, Testers, and You	2007	9780977616619	320
FXRuby: Create Lean and Mean GUIs with Ruby	2008	9781934356074	240
From Java To Ruby: Things Every Manager Should Know	2006	9780976694090	160
GIS for Web Developers: Adding Where to Your Web Applications	2007	9780974514093	275
Google Maps API, V2: Adding Where to Your Applications	2006	PDF-Only	83
Grails: A Quick-Start Guide	2009	9781934356463	200

Continued on next page

Title	Year	ISBN	Pages
Groovy Recipes: Greasing the Wheels of Java	2008	9780978739294	264
Hello, Android: Introducing Google's Mobile Development Platform	2009	9781934356494	272
Interface Oriented Design	2006	9780976694052	240
Land the Tech Job You Love	2009	9781934356265	280
Learn to Program, 2nd Edition	2009	9781934356364	230
Manage It! Your Guide to Modern Pragmatic Project Management	2007	9780978739249	360
Manage Your Project Portfolio: Increase Your Capacity and Finish More Projects	2009	9781934356296	200
Mastering Dojo: JavaScript and Ajax Tools for Great Web Experiences	2008	9781934356111	568
Modular Java: Creating Flexible Applications with OSGi and Spring	2009	9781934356401	260
No Fluff Just Stuff 2006 Anthology	2006	9780977616664	240
No Fluff Just Stuff 2007 Anthology	2007	9780978739287	320
Pomodoro Technique Illustrated: The Easy Way to Do More in Less Time	2009	9781934356500	144
Practical Programming: An Introduction to Computer Science Using Python	2009	9781934356272	350
Practices of an Agile Developer	2006	9780974514086	208
Pragmatic Project Automation: How to Build, Deploy, and Monitor Java Applications	2004	9780974514031	176
Pragmatic Thinking and Learning: Refactor Your Wetware	2008	9781934356050	288
Pragmatic Unit Testing in C# with NUnit	2007	9780977616671	176
Pragmatic Unit Testing in Java with JUnit	2003	9780974514017	160
Pragmatic Version Control Using Git	2008	9781934356159	200
Pragmatic Version Control using CVS	2003	9780974514000	176
Pragmatic Version Control using Subversion	2006	9780977616657	248
Programming Clojure	2009	9781934356333	304
Programming Cocoa with Ruby: Create Compelling Mac Apps Using RubyCocoa	2009	9781934356197	300
Programming Erlang: Software for a Concurrent World	2007	9781934356005	536
Programming Groovy: Dynamic Productivity for the Java Developer	2008	9781934356098	320
Programming Ruby: The Pragmatic Programmers' Guide, Second Edition	2004	9780974514055	864
Programming Ruby 1.9: The Pragmatic Programmers' Guide	2009	9781934356081	960
Programming Scala: Tackle Multi-Core Complexity on the Java Virtual Machine	2009	9781934356319	250

Continued on next page

Title	Year	ISBN	Pages
Prototype and script.aculo.us: You Never Knew JavaScript Could Do This!	2007	9781934356012	448
Rails Recipes	2006	9780977616602	350
Rails for .NET Developers	2008	9781934356203	300
Rails for Java Developers	2007	9780977616695	336
Rails for PHP Developers	2008	9781934356043	432
Rapid GUI Development with QtRuby	2005	PDF-Only	83
Release It! Design and Deploy Production-Ready Software	2007	9780978739218	368
Scripted GUI Testing with Ruby	2008	9781934356180	192
Ship it! A Practical Guide to Successful Software Projects	2005	9780974514048	224
Stripes ...and Java Web Development Is Fun Again	2008	9781934356210	375
TextMate: Power Editing for the Mac	2007	9780978739232	208
The Definitive ANTLR Reference: Building Domain-Specific Languages	2007	9780978739256	384
The Passionate Programmer: Creating a Remarkable Career in Software Development	2009	9781934356340	200
ThoughtWorks Anthology	2008	9781934356142	240
Ubuntu Kung Fu: Tips, Tricks, Hints, and Hacks	2008	9781934356227	400
iPhone SDK Development	2009	9781934356258	576

More iPhone and Mac books...

Core Animation for OS X/iPhone

Have you seen Apple's Front Row application and Cover Flow effects? Then you've seen Core Animation at work. It's about making applications that give strong visual feedback through movement and morphing, rather than repainting panels. This comprehensive guide will get you up to speed quickly and take you into the depths of this new technology.

Core Animation for Mac OS X and the iPhone: Creating Compelling Dynamic User Interfaces
Bill Dudney
(220 pages) ISBN: 978-1-9343561-0-4. $34.95
http://pragprog.com/titles/bdcora

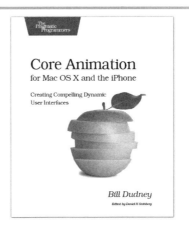

Core Data

Learn the Apple Core Data APIs from the ground up. You can concentrate on designing the model for your application, and use the power of Core Data to do the rest. This book will take you from beginning with Core Data through to expert level configurations that you will not find anywhere else. Learn why you should be using Core Data for your next Cocoa project, and how to use it most effectively.

Core Data: Apple's API for Persisting Data under Mac OS X
Marcus S. Zarra
(200 pages) ISBN: 978-1-93435-632-6. $32.95
http://pragprog.com/titles/mzcd

The Pragmatic Bookshelf

The Pragmatic Bookshelf features books written by developers for developers. The titles continue the well-known Pragmatic Programmer style and continue to garner awards and rave reviews. As development gets more and more difficult, the Pragmatic Programmers will be there with more titles and products to help you stay on top of your game.

Visit Us Online

iPhone SDK Development's Home Page
http://pragprog.com/titles/amiphd
Source code from this book, errata, and other resources. Come give us feedback, too!

Register for Updates
http://pragprog.com/updates
Be notified when updates and new books become available.

Join the Community
http://pragprog.com/community
Read our weblogs, join our online discussions, participate in our mailing list, interact with our wiki, and benefit from the experience of other Pragmatic Programmers.

New and Noteworthy
http://pragprog.com/news
Check out the latest pragmatic developments, new titles and other offerings.

Save on the eBook

Save on the eBook versions of this title. Owning the paper version of this book entitles you to purchase the electronic versions at a terrific discount.

PDFs are great for carrying around on your laptop—they are hyperlinked, have color, and are fully searchable. Most titles are also available for the iPhone and iPod touch, Amazon Kindle, and other popular e-book readers.

Buy now at pragprog.com/coupon.

Contact Us

Online Orders:	www.pragprog.com/catalog
Customer Service:	support@pragprog.com
Non-English Versions:	translations@pragprog.com
Pragmatic Teaching:	academic@pragprog.com
Author Proposals:	proposals@pragprog.com
Contact us:	1-800-699-PROG (+1 919 847 3884)